Palgrave Studies in International Relations

Palgrave Studies in International Relations (the EISA book series), published in association with European International Studies Association, provides scholars with the best theoretically-informed scholarship on the global issues of our time. The series includes cutting-edge monographs and edited collections which bridge schools of thought and cross the boundaries of conventional fields of study. EISA members can access a 50% discount to PSIR, the EISA book series, here http://www.eisa-net.org/sitecore/content/be-bruga/mci-registrations/eisa/login/landing.aspx.

Mai'a K. Davis Cross is the Edward W. Brooke Professor of Political Science at Northeastern University, USA, and Senior Researcher at the ARENA Centre for European Studies, University of Oslo, Norway.

Benjamin de Carvalho is a Senior Research Fellow at the Norwegian Institute of International Affairs (NUPI), Norway.

Shahar Hameiri is Associate Professor of International Politics and Associate Director of the Graduate Centre in Governance and International Affairs, School of Political Science and International Studies, University of Queensland, Australia.

Knud Erik Jørgensen is Professor of International Relations at Aarhus University, Denmark, and at Yaşar University, Izmir, Turkey.

Ole Jacob Sending is the Research Director at the Norwegian Institute of International Affairs (NUPI), Norway.

Ayşe Zarakol is Reader in International Relations at the University of Cambridge and a fellow at Emmanuel College, UK.

More information about this series at
https://link.springer.com/bookseries/14619

Jo Jakobsen

The Geopolitics of U.S. Overseas Troops and Withdrawal

palgrave
macmillan

Jo Jakobsen
Department of Sociology
and Political Science
Norwegian University of Science
and Technology
Trondheim, Norway

Palgrave Studies in International Relations
ISBN 978-3-030-94487-2 ISBN 978-3-030-94488-9 (eBook)
https://doi.org/10.1007/978-3-030-94488-9

Cover credit: guvendemir/Getty Images

This Palgrave Macmillan imprint is published by the registered company Springer Nature Switzerland AG
The registered company address is: Gewerbestrasse 11, 6330 Cham, Switzerland

Preface

I have written a book about the purported tendency of one thing leading to another. In fact, the book itself is a manifestation of this tendency of one thing leading to another, with the end product being almost wholly unintended. It really started years ago. I was co-authoring an entirely unrelated paper, using a measure of U.S. overseas troops as a control variable. In early 2018, I wrote two additional papers, now upgrading the U.S. troop measure to the status of the main independent variable. By then, of course, Donald Trump was president of the United States. That was intriguing, because Trump was a clear-cut *Leaver* (as I call it in the book); he expressed, in his usual boisterous manner, a preference for retrenchment. Many of the troops would thus be coming home.

Except that they wouldn't. There's no way that any major withdrawal of U.S. overseas troops will *actually* happen, I thought to myself at the time. This incited me to start keeping track of developments in various host countries of U.S. troops, though without any further plan. But when time and energy are expended on something, they come to represent a sunk cost. And when costs are sunk, the human mind is inclined to work in such a strange way as to advise sinking even and ever more costs into this something, even if it would likely be wiser just to cut one's losses and find another something to do—a less time-consuming something, or, perhaps preferably, a nothing.

As it turned out, I couldn't leave this project for much the same reasons as the United States cannot leave its empire of bases and troops.

Therefore, I continued to track developments in U.S. troop deployments, withdrawal threats, and retrenchment resistance, as preparation for something that was as yet a bit unclear to me. Then, in spring 2020, I realized that Trump's tenure was about to end, so I shouldn't procrastinate any longer. Hence, I started writing a paper. This was an empirically focused one, to which I added an explanation for why the United States remained where they were. Grounded in structural realism, this explanation made sense to me, but it was incomplete and therefore unsatisfactory; the logic was half-baked, and there were huge missing pieces to the story. By now, however, I had written some 20,000 words, which is totally unsuitable for every respectable journal out there. I decided I had to explore further the theoretical side of things. I spent the rest of 2020 doing just that, reading and researching deeply and widely, and also (I like to believe) thinking deeply and widely. By the end of 2020, it dawned on me that my one-year sabbatical was just around the corner. It was suddenly abundantly clear to me that if *this* project, into which I had sunk considerable costs, wasn't turned into a book, no work of mine would ever appear in any bookstore. Thus I started on the long and frustrating, if also often gratifying, process of writing a book that I had never actually intended to write in the first place. But I'm glad that I did write it, I'm happy to be finished, and I'm quite satisfied with the end result.

The book is an individual project. Still, I have benefited hugely, as I always have done, from having kind and knowledgeable people around me. I would like to thank my editor at Palgrave Macmillan, Anca Pusca, for her firm support of my project from the first time we made contact. The review process has also been of enormous help: the anonymous scholars commenting on my project helped change its direction in a handful of key areas, for which I am extremely grateful. I also benefited greatly from the symposium held in 2019 at University College Dublin and chaired by my friend Krishna Vadlamannati. Back home, at the Norwegian University of Science and Technology (NTNU), I am given the opportunity to devise and work on projects of my own choosing. This I do in a friendly and relaxed environment. Our administration is the best there is. The same goes for the research group to which I belong; the other members of SPECTRE (Studies on Political Economy: Currencies, Trade, Risk & Energy) consistently challenge (and sometimes lambast) and refine my ideas on international security, helping me develop as a scholar. Many people give me invaluable insights from fields of study with which I am only vaguely familiar, which broadens my knowledge and aids my scholarship. Special thanks go to Espen Moe, Indra de Soysa, Jostein Vik, Thomas Halvorsen, Tor G. Jakobsen, Marko Valenta, Zan Strabac, Gunnar Fermann, and Torbjørn Knutsen.

My beloved family—my wife Nina, and my steadily growing kids Marte and Armand—is a patient bunch. And I don't just mean that they are patient with respect to the work I have had to put into this book. They are patient with me in general. And caring. Moreover, Marte helped me weed out mistakes in my references (for which she was benevolently paid the going rate), while Armand helped me keep my head as much on football (or soccer, for any eventual U.S. readers) as on U.S. troops. My kids intermittently refer to me as Professor in Countries That Fight Each Other, a title that surely sounds much better than the rather bland official one.

I want to dedicate this book to two great persons whose help and encouragement, especially in the starting phase of my career, have been absolutely critical. The first is Ola Listhaug—Professor in How Voters Choose Their Leaders –, who is a formidable and very kind man, and the one who changed the direction of my professional life. The other is Jonathon W. Moses, who is Professor in Countries That Try to Become Reasonably Rich. He used to be my supervisor. He is now simply my very good friend.

Trondheim, Norway
November 2021

Jo Jakobsen

About This Book

Why is it so difficult for a great power or a hegemon to retrench its overseas military power? More specifically, why are U.S. military bases and troops still largely where they were five years ago, twenty years ago, or even seventy years ago? Through developing a theory of great-power persistence, this book offers an explanation. Closely aligned with neoclassical realism, the theory adopts a multi-level approach. It argues that the murkiness of the anarchic international system combines with specific psychological inclinations of individuals to produce "better-safe-than-sorry" policies. In the United States, decisions on troop deployments are powerfully influenced by the broader foreign-policy community. Its members tend to be risk-averse and sensitive to the possibility that even minor troop withdrawals might set off harmful geopolitical chain reactions. Preferring the status quo over any uncertain alternative, they want their country to continue to maximize its influence and project its military force abroad in order to steady wobbling geopolitical "dominoes." The theory is put to the empirical test through an elaborate analysis of U.S. overseas troop deployments, withdrawal attempts, and retrenchment resistance during the presidency of Donald Trump (and the first months of the Biden administration), which represents an ideal test case for these mechanisms. Even if U.S. voters elected a retrenchment advocate as president, and despite that the United States is a gradually declining power, the period saw very little change in U.S. overseas troop deployment. The book concludes that, barring any dramatic, unforeseeable international event, the vast network of overseas U.S. military bases and troops is likely to persist for a long time to come.

Contents

About the Author

Jo Jakobsen is a professor at the Department of Sociology and Political Science, Norwegian University of Science and Technology (NTNU), Trondheim, Norway. His doctoral dissertation earned him the 2008 Prize for Young Excellent Researchers in the Humanities and Social Sciences from the Royal Norwegian Society of Sciences and Letters. He has since published widely in the fields of international security, geopolitical risk, and international political economy. He received the Bernard Brodie Prize for the best article published in *Contemporary Security Policy* in 2019.

Abbreviations

AFRICOM	United States Africa Command
CENTCOM	United States Central Command
DMDC	Defense Manpower Data Center
EDI	European Deterrence Initiative
EUCOM	U.S. European Command
GDP	Gross Domestic Product
GDPR	Global Defense Posture Review
ISIS	Islamic State of Iraq and Syria
KFOR	Kosovo Force
MENA	Middle East and North Africa
MOB	Main Operation Base
NATO	North Atlantic Treaty Organization
NSS	U.S. National Security Strategy
SDF	Syrian Democratic Forces
SIPRI	Stockholm International Peace Research Institute
U.K.	United Kingdom
U.S.	United States
YPG	Kurdish People's Protection Units

List of Figures

CHAPTER 1

Introduction

*

The November 2016 election of Donald Trump as the 45th President of the United States threatened to change the perceptions and the realities of America's role in the world, and it augured a contraction of the country's vast network of overseas military bases and forward-deployed troops. Trump's stated preference for retrenchment—understood here as the "irrevocable removal of a state's capabilities regarding a particular region" or host country (Yoder, 2019, p. 131)—was a recurrent theme during the election campaign. He lambasted U.S. allies that were purportedly freeloading on their protector, and he repeatedly threatened to withdraw U.S. troops from numerous host countries.

The Stayers Trump the Leavers

These threats constitute the point of departure of this book, though they are not its main focus. The book instead seeks to answer one very broad, enduring, perhaps enduringly puzzling, and fundamentally important question: *Why is it so difficult for a great power or a hegemon to retrench?* More specifically, why is *the United States*, with respect to its overseas military bases and troops, still largely where it was five years ago, 20 years ago, or even 70 years ago? For if I now foreshadow the

J. Jakobsen, *The Geopolitics of U.S. Overseas Troops and Withdrawal*, Palgrave Studies in International Relations,
https://doi.org/10.1007/978-3-030-94488-9_1

empirical contribution and conclusions of this book, which consist of an analysis of the Trump administration's withdrawal plans and proposals, and the upshot of these, it has to be asserted early on that four years of Donald Trump didn't actually change much. Indeed, the troop withdrawal process in Afghanistan, which Trump started and his predecessor finished (in late August 2021), represented the *only* unambiguous case of a pull-out of U.S. troops in that period (with the partial withdrawal from Iraq serving as a possible second exception). In virtually all of the rest of the world, the network of military bases remained fully intact, and troop numbers either stayed the same or they increased, sometimes vastly so, as they did in Asia–Pacific.

The Great Continuity of U.S. Overseas Bases and Troops

Notwithstanding that the Trump era was not the first instance where a major overhaul of U.S. overseas troop deployments *seemed* to be in the cards, not much ever tend to come of such suggestions. For example, all recent winners of U.S. presidential elections have expressed clear intentions to extricate the United States from the Persian Gulf region, but none of them end up actually doing so (Gause III, 2019, p. 13). Consider also the end of the Cold War and the subsequent collapse of the Soviet Union at the turn of the 1990s. These were epoch-making geopolitical events if there ever were any, and these were events that effectively removed the main rationale for keeping the troops abroad. Yet, they only led to minuscule changes in the geographical pattern of U.S. military bases, even if the number of troops dropped as Russian military power dissipated. New rationales for upholding America's overseas force posture were quickly concocted, as they tend to be when circumstances change. Thus, the military forces of the United States were spread around the world in vast numbers when the Cold War ended; they were spread around the world in vast numbers when the "unipolar era" reached its height in the 1990s; they were spread around the world in vast numbers when the debacles of the Afghan and Iraq Wars had purportedly ended that era, and they are still spread around the world in vast numbers.

When the Cold War commenced for real in 1950, the year of the outbreak of the Korean War, Japan and Germany were by some distance the largest hosts of U.S. overseas (occupation) forces, which they still are (troop numbers are extracted from Kane [2004] and U.S. Department of

Defense [2021]). These two were followed on the list by Guam, a small piece of U.S. territory in the western Pacific Ocean which continues to be a fulcrum of Washington's military-force projection in Asia–Pacific. In Europe, meanwhile, Italy and the United Kingdom hosted thousands of U.S. troops in the years after World War II, which they also do today, and Greece had gained an elevated status in U.S. geostrategic thinking following the 1947 Truman Doctrine, which it has since kept (though now for somewhat different reasons). In the Middle East, Saudi Arabia signed a military-base agreement with Washington in 1946, and it remains a key ally and host of U.S. troops. Similarly, the United States had major military facilities in both Cuba (or, rather, in the leased or occupied territory of Guantánamo Bay) and Greenland in 1950, as it still has. Following the Korean War, moreover, South Korea became the third largest host of U.S. troops, which still is almost two-and-a-half generations hence. Spain as well became a major host in 1954, and the United States continues to have thousands of soldiers stationed on its soil. Of the major hosts of the early 1950s, only then-occupied Austria (which was declared permanently neutral in 1955), Panama (which is still strongly aligned to the United States), and two countries that are long-standing U.S. treaty allies—the Philippines and France—do not any longer host a significant amount of U.S. troops. If we fast forward to the 1990s, Iraq's onslaught on Kuwait, as well as lingering U.S. concerns about Iran, meant that U.S. troops flooded into a few smaller states around the Persian Gulf—namely, Bahrain, Kuwait, and, a few years later, Qatar. All three are still major hosts of U.S. military personnel.

Stayers and Leavers

Among academics, there is little agreement on how much sense it makes to hold on to this network of military bases and to keep an on-the-ground military presence in so many places. One strand of scholars we can subsume under the label of "offshore balancers" or, as I call them herein, the *Leavers* (e.g., Gholz et al., 1997; MacDonald & Parent, 2011; Mearsheimer & Walt, 2016; Posen, 2014; Walt, 2018). They argue that U.S. global military activism is unsustainably costly, largely counterproductive to U.S. national security, and eminently unnecessary especially considering that the United States is geographically shielded by two gigantic oceans. The offshore balancers would want to see a huge share of these troops withdrawn, in particular from Europe and the Middle East.

Instead, they recommend an offshore-balancing strategy under which U.S. military capabilities are primarily to be kept "over the horizon," which would necessitate far fewer bases and troops in other countries.

A second strand of scholars sees things entirely differently (e.g., Brands & Edel, 2019; Brooks & Wohlforth, 2016; Brooks et al., 2012–2013; Brzeziński, 2012; Rovner & Talmadge, 2014). The primacists, or the *Stayers*, claim that relative U.S. geographical seclusion rather hands the United States a very convenient opportunity to project its military power globally. This means that potential threats, of which there are far more than the *Leavers* would ever admit, can be, as they should be, forestalled early and a long way from the U.S. homeland before they turn into unmanageable menaces. This can best be accomplished, say the *Stayers*, by keeping considerable numbers of U.S. troops on the ground in several key countries in practically all regions of the world, though especially in Europe, Asia–Pacific, and around the Persian Gulf.

The *Stayers* of the scholarly community receive massive support from the broader U.S. foreign-policy community. This community is a loose collection of individuals and entities that comprise, *inter alia*, formal institutions of government (such as Congress, intelligence agencies, and the military), consultants, analysts, think tanks, lobbyists, and most of the media in the United States (Friedman & Logan, 2016; Layne, 2017; Porter, 2018; Walt, 2018). This is the establishment whose ideas and preferences help set the agenda of, and give direction to, U.S. grand strategy and foreign policies. For over seven decades, their views have also corresponded closely with the actual reality: The United States has been, as it still is, deeply engaged in global affairs—through its diplomacy, through its plethora of alliances, through deterrence of foes and reassurance of friends, through its leadership in global and regional institutions, through its use of economic tools for geopolitical ends, through its war-fighting, and, vitally, through its far-flung network of military bases and troops abroad.

The Election of a Leaver

This state of affairs represents the status quo, which has changed remarkably little since the Korean War. For many, the thought of one day having to witness a world devoid of U.S. global leadership and overseas bases and troops have bordered on the inconceivable. However, the election of

Donald Trump as U.S. president threatened to alter things quite drastically. His stated preference for retrenchment was a recurrent theme during the election campaign, and it was propounded in a far more boisterous, unforgiven, uncompromising, self-assured, and (admittedly) consistent way than any presidential candidate before him had ever been close to matching. He lambasted U.S. allies that were "not paying their fair share" for "the tremendous security burden" taken on by their protector (*New York Times*, 2016b). Threats to withdraw U.S. commitments toward allies were rife, often quite unambiguously issued, and commonly linked to economic arguments, and they supplemented promises to withdraw troops from countries in which the United States had long been at war.

For example, in a March 2016 interview with *The New York Times* (2016a), Trump affirmed that he would be quite willing to withdraw forces from several countries. The gist of Trump's warnings, as was reflected in another interview with *The New York Times* (2016c) four months later, was that if the United States "cannot be properly reimbursed for the tremendous cost of our military protecting other countries (...) [t]hen yes I would be absolutely prepared to tell those countries, 'Congratulations, you will be defending yourself'." In President Trump's Inaugural Address, on January 20, 2017, he repeated his allegations that America had basically

> subsidized the armies of other countries, while allowing for the very sad depletion of our military. We've defended other nations' borders while refusing to defend our own. And spent trillions and trillions of dollars overseas (...) [making] other countries rich, while the wealth, strength and confidence of our country has dissipated over the horizon. (Politico, 2017)

The tight coupling of pecuniary concerns and base politics—"Of course they should pick up all the expense. Why are we paying for this?" (Ghattas, 2016)—was not something that only came along with the transformation of Trump into a ***politician*** and ***presidential candidate***: Trump the pure ***businessman*** had much earlier expressed his views in similar language. As early as in 1987, during the "Tanker War" subsumed within the larger Iran-Iraq conflict, he had taken out expensive advertisements in major U.S. newspapers arguing that Washington should not continue paying for other states' defense. In an attendant interview with the *Associated Press*, Trump lamented that, for decades, a number of states around the Persian Gulf—as well as Japan—had "been taking advantage of the

United States"; they had massively failed to pay for their own security, making the world laugh "at America's politicians as we protect ships we don't own, carrying oil we don't need, destined for allies who won't help" (Shanahan, 1987). Throwing Japan into the Middle Eastern mix only underscored Trump's view that troop deployments were, more than anything else, a *transaction* not dissimilar to ordinary business deals. Japan was at the time still *the* economically rising power of the world. What sense did it make defending it and stationing all those American troops there when America is "losing hundreds of billions of dollars a year while they laugh at our stupidity"? Trump bemoaned in a 1990 interview with *Playboy Magazine* (Plaskin, 1990).

The Leaver Who Didn't Leave

Some three decades later, but now in a completely different position of authority, President Trump made it into a habit of warning key allies that a failure to renegotiate basing agreements on terms vastly more to the benefit of Washington would lead to the withdrawal of U.S. troops. Japan was again singled out, along with neighboring South Korea (Lee et al., 2018). In Europe, Germany was the main target of Trump's ire. In June 2020, he ordered thousands of U.S. troops to leave this "delinquent" country that "treat us very badly on trade" (Mason & Mohammed, 2020), a withdrawal that was never actually effectuated. The brunt of the withdrawal threats, though, were saved for the Middle East and Afghanistan. U.S. involvement in regional wars—in Iraq, Afghanistan, and, to a lesser extent, Syria—was based on ludicrous decisions, Trump repeatedly said. "We should have never been in the Middle East. To get into the Middle East was the single biggest mistake made in the history of our country" (Myers, 2020). Under his own presidency, Trump promised, the United States would be "ending the era of endless wars," because it was not the job of American soldiers "to solve ancient conflicts in faraway lands that many people have not even heard of" (Holland, 2020).

It soon becomes evident, though, that any plan he may have had to withdraw troops from abroad, and to close overseas bases, had seen scant fulfillment. On this issue, "Trump's actual record doesn't mirror his rhetoric," *The Washington Post* concluded in October 2019 (Tharoor, 2019). A few months later, the same newspaper bitingly stated that Trump was "determined to bring home U.S. military from somewhere," but had thus far "been stymied at virtually any turn" (DeYoung & Ryan, 2020). The last few months of Trump's tenure did not change much either. The exceptions were Afghanistan and Iraq, from each of which a few thousand U.S. troops left, even as a few thousand remained. He did also order a pullout from Somalia near the end of his term. However, Trump was actually the president responsible for putting them there in the first place (Gaffey, 2017), and the troops were not withdrawn from the region but instead relocated to bases in neighboring states (Youssef & Phillips, 2020). Moreover, there were more U.S. troops in Syria when Trump left office than there were when he entered it. Even more tellingly, virtually all non-invasion U.S. troops—that is, those that were stationed at bases in host states that had not been ravaged by recent or ongoing war—remained where they were; indeed, in some regions, such as Asia–Pacific, the numbers were now much larger.

Why did the bases and troops remain in place? For all of his and his administration's seeming whimsicalness, Trump talked like a *Leaver* and he acted like a *Leaver*. Evidence clearly suggests that he really sought to withdraw troops from Afghanistan, Iraq, Syria, Somalia, and Germany, and he also made what was widely perceived, including by the hosts themselves, to be credible threats to leave other countries. Numerous U.S. allies—in the North Atlantic Treaty Organization (NATO), in East Asia, and around the Persian Gulf—were certainly not taking lightly the repeated signals that the U.S. President considered abrogating long-standing military and security agreements. Public opinion in the United States could scarcely have been the cause of base-policy continuation either: voters are notoriously and enduringly ambivalent on the issue of U.S. overseas troops (Kagan, 2012, p. 14; Macintyre, 2019; Pew Research Center, 2016), and Trump had, in the presidential election, benefited greatly from that half of voters who preferred retrenchment.

More plausible candidates for explaining the geostrategic persistence of the United States are institutional inertia and the sheer physical realities of military bases and sunk costs. In combination with the preferences of domestic actors in the United States—viz., the predilections of the broader foreign-policy community—they represent obvious barriers against change. A recent study by Andrew Stravers (2021), for example, attributes variation in U.S. overseas troop deployments in part to the political make-up of Congress. As we shall see later, however, the impact of domestic actors on U.S. overseas military-base and troop policies and patterns can only rarely be disentangled from *system-level* properties and developments. These include, among other things, the impact a troop withdrawal *might* have on global and regional balances of power and on U.S. reputation and credibility. It is the fear that such broader detrimental effects will result even from relatively minor troop withdrawals that in large part guides and informs the foreign policy establishment's arguments in favor of continuity and the preservation of the status quo.

The Literature on U.S. Overseas Bases and Troops

Neither can the specialized literature on U.S. foreign military bases and troops offer much help regarding the causes of continuity and change. Stravers's study notwithstanding, relatively scant research has hitherto been conducted on the *drivers* of the base network, whereas a growing number of journal articles have investigated (a variety of) its *effects*. Some broader, book-length works on U.S. base politics do exist, of course. These include the two studies by Robert Harkavy (1982, 1989) from the 1980s. Another excellent monograph is Kent Calder's *Embattled Garrisons*, an exposé of the base politics of the United States wherein the author also explores the many strategic advantages, and the costs, that accrue to the United States from these bases (Calder, 2007). Furthermore, in an edited volume by Carnes Lord and Andrew Erickson, several country and area experts contribute with case analyses that cover all the major hosts in Asia (Lord & Erickson, 2014). In a recent book, Sebastian Schmidt (2020) dwells deeply into the underexplored topic of the origins of U.S. military-base agreements with sovereign states during and immediately after the World War II. Moreover, quite a few studies center primarily on how U.S. bases negatively affect host countries (and sometimes also the United States itself), and they often also explore resulting host-society opposition to the stationing of foreign troops. These are, for example, the main themes of *The Bases of Empire* (Lutz, 2009), of *Base Nation* (Vine, 2015), and of numerous journal articles (e.g., Cooley & Nexon, 2013; Vine, 2019). Others offer case analyses of specific countries or territories, such as Spain (Cooley & Hopkin, 2010), South Korea (Avey et al., 2018), and Diego Garcia (Erickson et al., 2010).

Recent years have seen quite a rapid expansion of the empirical literature that analyzes various effects of U.S. troop deployments quantitatively. Most of these studies draw on data from the Pentagon's U.S. Defense Manpower Data Center (U.S. Department of Defense, 2021), which were originally and very helpfully organized by Tim Kane (2004, 2006; see also Allen et al., 2021). First out were Glen Biglaiser and Karl DeRouen, who found that U.S. foreign direct investment and trade tend to "follow the flag" (Biglaiser & DeRouen, 2007, 2009). Others have investigated the impact of U.S. military presence on various economic or financial outcomes, such as economic growth and development (Heo & Ye, 2019; Jones & Kane, 2012) and the financial stability of host countries (Aklin & Kern, 2019). Still others have explored the relationship between U.S. troops and various societal and institutional outcomes, such as number of criminal offenses (Allen & Flynn, 2013), institutional quality (Braithwaite & Kucik, 2018), human rights protection (Bell et al., 2017), level of democracy (Stravers & El Kurd, 2020), and transnational terrorism (Braithwaite, 2015). In a recent article published in *The American Political Science Review*, Michael Allen et al. (2020) draw on public opinion data covering 14 countries. They find that attitudes toward the U.S. military presence are shaped in a positive way both by the level of interaction between the local population and U.S. military personnel and by the economic benefits that the host country stands to gain from this presence.

The quantitative part of the literature also contains contributions centering on military-security outcomes. Some such are published by the RAND Corporation. A recent one, authored by Bryan Frederick et al. (2020), finds that deterrence of adversaries is strengthened, and the likelihood of conflict is lowered, by a U.S. military presence, so long as that presence consists of heavy ground forces (as opposed to air and naval forces). These findings in large part corroborate those of an earlier RAND study, which concluded that U.S. forward-deployed troops decrease the likelihood of interstate war even if it increases the risk of low-intensity militarized conflict (short of a shooting war) (O'Mahoney et al., 2018). Others use some measure of the defense efforts of the host as their dependent variable. Carla Martinez Machain and Clifton Morgan find that increased levels of forward-deployed troops decrease the hosts' own troop numbers somewhat, presumably because the U.S. troops work as a "tripwire" that strengthens the credibility of America's security commitments to its allies (Martinez Machain & Morgan, 2013). Additional studies pursue related issues drawing on quantitative data on U.S. troops (Allen et al., 2016, 2017; Jakobsen, 2018; Jakobsen & Jakobsen, 2019).

These contributions notwithstanding, those studies whose main ambition is to account for the geographical patterns of U.S. troops—that is to say, more general analyses of the drivers of continuity and change in U.S.

military-base policies—are much rarer. Nonetheless, a very recent contribution is made by Mark David Nieman et al. (2021). In a cogent article in *The Journal of Politics*, they argue, as its title duly suggests, that troop deployments can in large part be viewed as "an international game of risk." That is to say, great powers in recent history—the United States, the Soviet Union, France, and the United Kingdom—deploy soldiers abroad to expand their respective spheres of influence in vital regions. But they also hope that by doing so they will stop the other great powers from deploying additional troops of their own in that same region. Viewed like this, troop deployments are in an important way defensively motivated, akin to the infamous security dilemma. The authors hold that this often sets off an action-reaction process whereby troops from several great powers tend to cluster both geographically and temporally. What they contend is that what really drives such expansion is the *risk aversion* of the great powers. Another way of putting it is that the great powers fear that a failure to expand by deploying troops will, or *might*, set in motion adverse geopolitical chain reactions that will eventually hamper their own power, influence, and security.

The Theory, the Arguments, and the Chapters

The latter sentence really captures much of the essence of my own arguments, which also form the core of my theory of great-power persistence. This theory is an explanation of why the big states of the world are so reluctant to retrench their overseas military power—of why they prefer staying over leaving even when the costs associated with staying seemingly pile up. It is an explanation of why they tend to be so very sensitive when facing even the mere *possibility* of losing parts, even very small parts, of their accumulated influence.

A Theory of Great-Power Persistence

My arguments are broadly commensurate with neoclassical realism (e.g. Lobell et al., 2009; Ripsman et al., 2016; Rose, 1998). They center, in the first instance, on a system-level understanding of state behavior that highlights the way the negative aspects of systemic impulses tend to override the positive ones when foreign-policy opinions are formed, which decreases the likelihood of retrenchment. My theory builds on a small handful of specific assumptions and related implications for expected state objectives and patterns of behavior. Many of these elements it shares with realist theory, in general, and neoclassical realism in particular. The novelty of my theoretical contribution lies in the amalgamation of the "domino" mechanism, which relates closely to the interconnectedness of the world, and individual-level mechanisms, which are really psychological or "human nature"-associated ones.

To start with the *assumptions* of my theory, these can be summarized in the following way:

1. *Anarchy, relative power, and geography*: The international system is anarchic; the foreign-policy behavior of states is, in the first instance, shaped and constrained by relative material power and the enduring properties of geography.
2. *Systemic interconnectedness*: The international system forms a complex, interconnected strategic whole, wherein adverse geopolitical events have the potential to set off chain reactions and cause further dominoes to fall, which greatly amplifies the effects of the initial event.
3. *Filtering of systemic signals*: Systemic signals and impulses are ambiguous and murky, because the world is ambiguous and murky, and these signals and impulses are filtered through the minds of foreign-policy elites, whose perceptions of the constraints and opportunities imposed by the international system ultimately direct a state's foreign-policy behavior.
4. *Negativity bias*: With respect to the external environment, human beings exhibit a negativity bias rooted in human nature; in particular, they tend to be loss averse around some reference point to which they are psychologically committed. The reference point is heavily shaped by perceptions of geography, history, and relative power.

These assumptions, when taken as a whole, lead to four specific types of *expected state objectives and patterns of behavior*:

1. *Reference point bias*: The foreign-policy decisions of any state generally aim either to uphold its position in the hierarchy of states, if a status quo bias applies, or to move it closer to its desired position, if its reference point is not commensurate with the status quo.

2. *Domino sensitivity*: Global and regional major powers are particularly sensitive toward, and will strive to manage and forestall, possible adverse geopolitical chain reactions or falling dominoes, even if these belong in the realm of the improbable; a typical state is a risk averse actor whose key objective is to minimize the likelihood of experiencing very costly outcomes, especially if its reference point is commensurate with the status quo.
3. *Geographic prioritization*: In their foreign policies, states prioritize among different pieces of geography, primarily seeking to exert power over, and to forestall possible adverse geopolitical chain reactions in, those territories that are perceived as being of significant geostrategic importance.
4. *Influence maximization*: Global and regional major powers seek, primarily for eminently defensive purposes, to shape and control events in—viz., to maximize influence over—the external environment to the extent possible under the constraints posed by geographic realities and (perceptions of) relative power.

The Theory's Scope

These assumptions and expected state objectives and patterns of behavior represent the core of my theory of great-power persistence. "Great power" is here to be interpreted somewhat liberally, so as meaningfully to encompass both global and regional great or "major" powers, including those that are only *aspiring* ones. The relevance of the theory, however, generally increases with the power of a state, even if the assumptions of the theory should have a bearing on all states, strong and feeble ones alike. Small and weak states, though, command only very limited material power resources and are not in a position independently to manage and forestall typical adverse geopolitical chain reactions. Thus, their behavior cannot in any really meaningful and interesting way be explained by my theory.

Even so, an overly stringent definition of the term will leave out too many interesting cases from the purview of the theory. The literature is rife with different understandings of various labels placed on powerful states (Kim, 1992, p. 158, fn. 17; Thompson, 1986). Kenneth Waltz (1979, pp. 158–170), for example, derives his identification of great powers from the polarity of the system. In eras of (global) multipolarity—before 1945—there were four or five of these, some losing and some gaining such status along the way. With the end of the World War II came bipolarity, where only the Soviet Union and the United States counted as great powers. After the demise of the Soviet Union, the U.S. stood firmly alone at the apex of the global hierarchy of states, which reduced the number of great powers to one (Waltz, 2000, pp. 5–6). Others, such as George Modelski (1978, p. 216), focus on "global" or "world" powers (in the maritime-supremacy sense), which are "[e]ntities uniquely dominant in the global system," and of which there have only ever been four (Portugal, the Netherlands, Great Britain, and the United States). John Mearsheimer (2001, p. 5) reserves the great-power label for any state that has "sufficient military assets to put up a serious fight in an all-out conventional war against the most powerful state in the world." This closely resembles the great power of the historian Paul Kennedy (1988, p. 697), which is "a state capable of holding its own against any other nation." This, however, is really unknowable until the state in question actually *has* put up a "serious fight" against the most powerful nation (and the weak do sometimes win or tie against the strong, which should not automatically suffice to promote them into the club of great powers).

With the attendant risk of being (fairly reasonably) accused of proposing quite a plastic conception of great powers, I choose instead to borrow the gist of Zbigniew Brzeziński's very useful definition of "active geostrategic players." This means, in terms of scope, that my theory encompasses "states that have the capacity and the national will to exercise power or influence beyond their borders in order to alter (….) the existing geopolitical state of affairs. They have the potential and/or the predisposition to be politically volatile" (Brzeziński, 1997, p. 40). Plastic though it is, the definition is also a conveniently pragmatic one that fits the often-changing realities of the world and, importantly, its many constituent regions.

Of course, on the one hand, the United States is in our day and time the obvious test case for the assumptions and mechanisms I propose herein. On the other hand, the key factor that separates the United States from others is material power. The assumptions and mechanisms, I contend, are really generally working ones that impact several states in an inherently meaningful way. It is, for example, clear—or so I would argue—that a *prospective* regional great power such as Saudi Arabia is quite heavily imbued by worst-case-scenario thinking and a related urge to project its power abroad. These fears and this urge are linked to beliefs in the *possibilities* of adverse geopolitical chain reactions that *might* create a new, significantly worse reality that does not align with Saudi Arabia's reference point (with the Yemen War being one of many manifestations of Riyadh's "counter-domino" behavior). Indeed, even states that do not really aspire to regional major power status (such as Pakistan) tend to behave in quite a similar fashion, though with a more geographically limited scope courtesy of a relative lack of means. These mechanisms are more conspicuous still regarding the geostrategic behavior of Russia and China.

Importantly, both geopolitical chain reactions and negativity biases are general features of, respectively, the system and the individual. So this is really a theory that seeks to explain (some vital parts of) the behavior of global and regional major powers. But it is also clear that the theory fits best and most consistently with *hegemonic* behavior. This means that the United States is by far the most ideal state to study in the post-Cold War period, even if it is not the only one.

The Book's Structure

The next chapter provides the necessary context for the subsequent three chapters, in which I develop in more detail my theory of great-power persistence. Chapter 2 includes three main sections. First, it presents a brief, theoretically informed review of the rise of the United States to great-power status. Its power potential did not immediately translate into a commensurate amount of influence in world affairs, which can in part be explained by its relative geographical secludedness. Over the long term, however, power resources tend to determine influence-seeking, which it also eventually did for the United States. Here, I also briefly review the history of U.S. overseas bases and troops more specifically, emphasizing the two wars—the 1898 Spanish-American War and World War II—that spurred the creation and expansion of a base empire. I also dwell on the transition to the unipolar era, which was generally marked by continuity with respect to global engagement, even if demands were now for smaller, less troop-intensive bases. The second main section accounts for the grand-strategic options of the United States. Even if most members of the broader U.S. foreign-policy community are ardent *Stayers*, the academic debate, as well as public opinion, is polarized between "primacists" (or *Stayers*) and "offshore balancers" (or *Leavers*). Whereas the former advocate for a continued presence of U.S. troops in important overseas regions, the latter advise (full or partial) withdrawal from most regions. The third main section considers various non-systemic, domestic forces that may contribute to upholding the U.S. military-base network. Here, I discuss the leeway and particularities of President Trump and his administration; the preferences and impact of U.S. public opinion and of vested (economic) interests; and the barriers to troop withdrawal posed by institutional inertia and the predilections of the U.S. foreign-policy establishment.

Chapter 3 first introduces neoclassical realism and the concepts of dominoes and adverse (geopolitical) chain reactions, the latter of which is a term that I borrow from Michael Mandelbaum (1988, Ch. 3). Thereafter, the chapter analyzes the main assumptions of structural realism, especially as they were presented by Kenneth Waltz (1979) in his *Theory of International Politics*. These assumptions, in turn, form a starting point for realism's neoclassical variant, and I derive from them implications for military-force projection abroad. I then outline the core arguments of neoclassical realism, especially emphasizing intervening variables at lower levels of analysis. After reviewing systemic conditions that may incentivize retrenchment, I devote the last main section to an exposition of why system-level impulses still *generally* discourage retrenchment. The bulk of lessons from the broader school of international relations realism, I argue, suggests that states have very sound reasons to pursue "better-safe-than-sorry" foreign policies that aim to maximize influence in order to shape and control the external environment as much as possible. If one studies scholarly debates within realism, for example, on such an issue as the role of the U.S. in the world, realism may seem quite fragmented. However, a more thorough scrutiny of arguments and positions instead leads to the conclusion that the unifying elements are markedly at the forefront. This is so particularly regarding risk averseness and pessimism, and with respect to realism's proclivity to prefer that (or at least to *understand why*) states implement (costly) policies that minimize the likelihood—even if that likelihood is tiny—of experiencing very costly outcomes. Relatedly, "domino sensitivity" implicitly saturates political realism, both analytically and normatively. The latter is somewhat of a puzzle since quite a few renowned realist-affiliated scholars criticize domino thinking rather emphatically (e.g., Layne, 2017; Walt, 2018; see also Jervis & Snyder, 1991).

Chapter 4, though it continues with the focus on vital mechanisms at the system-level, also attends to lower levels of analysis, so as better to account for the mechanisms by which system-level signals are "filtered" through the minds of members of the foreign-policy community. In effect, the chapter combines system-level (domino) theory, which rests on the assumption that the world is interconnected and interdependent, with the psychological driving forces that are emphasized by prospect theory (Kahneman & Tversky, 1979). The latter theory revolves around the tendency of human beings to overweight looming losses more than possible gains and to exhibit a reference point bias, which usually implies that the prevailing status quo is preferred over its alternatives. A closer look at both prospect theory and political realism shows that the former corresponds especially well with the "defensive" strand of the latter (Taliaferro, 2004, Ch. 2). The second main section of Chapter 4 argues that states tend to exhibit considerable threat sensitivity. There, I elaborate on the domino theory or metaphor. I argue that, even if it is often roundly criticized, it aligns well with both structural and neoclassical realism. The third main section accounts for how great powers—throughout history, from ancient Greece through the Habsburg and British Empires, and further through the Vietnam War and beyond—have consistently sought to forestall falling dominoes. The fourth main section accounts for different types of dominoes, and stresses the purported importance of reputation and credibility. I conclude by arguing that the behavior of great powers in general indicates that analysts should acknowledge that threat sensitivity and negativity biases are fundamental—and eminently understandable—behavioral traits constituent of actors in the realm of international politics.

Chapter 5 completes the theory part of the book. The chapter, first, analyzes the role that forward-deployed troops is supposed to have, and indeed *do* have, in forestalling adverse geopolitical chain reactions. Specifically, the stationing of overseas troops deters adversaries, reassures friendly and allied countries, and alleviates geopolitical risks to the world and regional economies. These troops, thus, ostensibly serve to steady sundry types of wobbling dominoes. The better part of Chapter 5, however, links the literature on classical geopolitics with the issue of the *location* of overseas military bases and troops. In other words, it confronts the question of which regions and countries and pieces of territory are and should be prioritized by great powers whose resources are not unlimited. Classical geopolitics—in the versions of Alfred Thayer Mahan (1890), Halford Mackinder (1904, 1942 [1919], 1943), and Nicholas Spykman (2017 [1942])—provides very plausible explanations as to why power is projected where it is. Lastly, Chapter 5 identifies and summarizes the implications for the empirical analysis that follow my theory. In particular, the expectation is that the still-unmatched power of the United States, combined with a prevailing status quo that both rests on and consists of a wide and dense network of U.S. bases abroad, help create an environment wherein those voices that warn against the many possible negative consequences from retrenchment usually overwhelm the more sanguine arguments of retrenchment advocates. This ultimately has considerable impact on U.S. foreign policy and geostrategy, which are far more likely to be shaped by *Stayers* than by *Leavers*.

Chapters 6 and 7 represent the book's empirical contribution, where my arguments are put to the test. The period from early 2017 to early 2021 in many ways serves as an ideal test case of the theory of great-power persistence. After the United States had followed a grand strategy of global engagement for some seven decades, U.S. voters elected a self-declared *Leaver* as president of a country whose power, many argued, was in relative decline. Chapter 6 presents, discusses the weaknesses of, and analyzes the official data on active-duty U.S. troops, which the Pentagon publishes on a quarterly basis. Here, I identify the broader patterns that emerge. I first analyze trends from the immediate post-Cold War period until today, before I start focusing exclusively on developments in the Trump era. Both global and regional patterns are exhibited.

Chapter 7, which is the book's longest, goes into more detail, focusing specifically on key cases of proposed withdrawal. In particular, I scrutinize the impact of the rationales and narratives that emerge whenever a troop withdrawal seems to be in the cards. Each and every such case of attempted retrenchment is quickly beset by loud and heavy resistance from the U.S. foreign-policy community, and usually also from host states and U.S. allies more broadly. The chapter analyzes such processes and tensions through five main sections, each of which represents one region. I start with Africa. Thereafter, I analyze developments in the Middle East. I pay special attention to the Trump administration's largely aborted attempts to withdraw troops from Syria, a multifaceted case that amply illustrates many of my main theoretical points. The third main section of Chapter 7 deals with Afghanistan, certainly the most notorious case of attempted withdrawal. After an initial troop surge in 2017, the president later, and on numerous occasions, threatened to close the U.S.'s military bases and pull out his nation's troops. These threats spurred almost unprecedented warnings from the broader foreign-policy community—and from other countries—about the negative geopolitical chain reactions that would likely result, even as the Trump administration in the end succeeded in withdrawing most of the troops. The fourth main section focuses on developments in Europe. I pay due attention to the case of Germany, from which President Trump ordered a withdrawal of U.S. troops that was never actually realized, primarily because it was vehemently opposed by the *Stayers*. Finally, the chapter investigates U.S. troops and withdrawal threats in Asia–Pacific. The main story here is that even if President Trump threatened—sometimes implicitly, a few times more explicitly—to abrogate U.S. defense agreements with, and to pull out U.S. troops from, Japan and South Korea, the upshot of four years of Trump was quite the opposite: U.S. troop numbers *rose* considerably in the period under study. In fact, with respect to Asia–Pacific, even self-avowed *Leavers* tend to be *Stayers*.

Chapter 8, which is the last one, does not entirely leave the empirical side of things behind. Its first main section tracks developments in the first year of the new Biden administration. Overall numbers and developments are analyzed. But the main story is yet again Afghanistan. That is a case, the *only* such case, of complete withdrawal, even if it did not happen on Trump's watch. (The last U.S. soldier was evacuated from the country on August 30, 2021.) Afghanistan is also a case that, quite tragically, illustrates some of the chain reactions that can indeed manifest once a withdrawal process has begun. After the second main section has presented the main conclusions from my study, the final paragraphs are devoted to a discussion of forces and events that *may* make the United States retrench in a meaningful way. My conclusion is that, barring any dramatic, unforeseen, and unforeseeable event of a "black swan" type, the vast and long-standing network of U.S. military bases and forward-deployed troops is likely to persist for a very long time to come.

References

Aklin, M., & Kern, A. (2019). Moral hazard and financial crises: Evidence from American troop deployments. *International Studies Quarterly, 63*(1), 15–29.

Allen, M. A., & Flynn, M. E. (2013). Putting our best boots forward: US military deployments and host-country crime. *Conflict Management and Peace Science, 30*(3), 263–285.

Allen, M. A., Flynn, M. E., & Martinez Machain, C. (2020). Outside the wire: U.S. military deployments and public opinion in host states. *American Political Science Review, 114*(2), 326–341.

Allen, M. A., Flynn, M. E., & Martinez Machain, C. (2021). US global military deployments, 1950–2020. *Conflict Management and Peace Science*. Advance online publication. https://doi.org/10.1177/07388942211030885

Allen, M. A., Flynn, M. E., & VanDusky-Allen, J. (2017). Regions of hierarchy and security: US troop deployments, spatial relations, and defense burdens. *International Interactions, 43*(3), 397–423.

Allen, M. A., VanDusky-Allen, J., & Flynn, M. E. (2016). The localized and spatial effects of US troop deployments on host-state defense spending. *Policy Analysis, 12*(4), 674–694.

Avey, P. C., Markowitz, J. N., & Reardon, R. J. (2018). Do US troop withdrawals cause instability? Evidence from two exogenous shocks on the Korean Peninsula. *Journal of Global Security Studies, 38*(1), 72–92.

Bell, S. R., Clay, K. C., & Martinez Machain, C. (2017). The effect of US troop deployments on human rights. *Journal of Conflict Resolution, 61*(10), 2020–2042.

Biglaiser, G., & DeRouen, K. Jr. (2007). Following the flag: Troop deployment and U.S. foreign direct investment. *International Studies Quarterly, 51*(4), 835–854.

Biglaiser, G., & DeRouen, K. Jr. (2009). The interdependence of U.S. troop deployments and trade in the developing world. *Foreign Policy Analysis, 5*(3), 247–263.

Braithwaite, A. (2015). Transnational terrorism as an unintended consequence of a military footprint. *Security Studies, 24*(2), 349–375.

Braithwaite, A., & Kucik, J. (2018). Does the presence of foreign troops affect stability in the host country? *Foreign Policy Analysis, 14*(4), 536–560.

Brands, H., & Edel, C. (2019). *The lessons of tragedy: Statecraft and world order*. Yale University Press.

Brooks, S. G., & Wohlforth, W. C. (2016). *America abroad: The United States' global role in the 21st century*. Oxford University Press.

Brooks, S. G., Ikenberry, G. J., & Wohlforth, W. C. (2012–2013). Don't come home, America: The case against retrenchment. *International Security, 37*(3), 7–51.

Brzeziński, Z. (1997). *The grand chessboard: American primacy and its geostrategic imperatives*. Basic Books.

Brzeziński, Z. (2012). *Strategic vision: America and the crisis of global power*. Basic Books.

Calder, K. E. (2007). *Embattled garrisons: Comparative base politics and American globalism*. Princeton University Press.

Cooley, A., & Hopkin, J. (2010). Base closings: The rise and decline of the US military bases issue in Spain, 1975–2005. *International Political Science Review, 31*(4), 494–513.

Cooley, A., & Nexon, D. H. (2013). "The Empire will compensate you": The structural dynamics of the U.S. overseas basing network. *Perspectives on Politics, 11*(4), 1034–1050.

DeYoung, K., & Ryan, M. (2020, July 21). Trump is determined to bring home U.S. military forces from somewhere. *Washington Post*. https://www.washingtonpost.com/national-security/trump-is-determined-to-bring-home-us-military-forces-from-somewhere/2020/07/20/c2664fb8-c851-11ea-b037-f9711f89ee46_story.html

Erickson, A. S., Walter, L. C., III., & Mikolay, J. D. (2010). Diego Garcia and the United States' emerging Indian Ocean strategy. *Asian Security, 6*(3), 214–237.

Frederick, B., Watts, S., Lane, M., Doll, A., Rhoades, A. L., & Smith, M. L. (2020). *Understanding the deterrent impact of U.S. overseas forces*. RAND Corporation.

Friedman, B. H., & Logan, J. (2016). Why Washington doesn't debate grand strategy. *Strategic Studies Quarterly, 10*(4), 14–45.

Gaffey, C. (2017, April 19). Why is Trump sending more U.S. troops to Somalia? *Newsweek*. https://www.newsweek.com/us-troops-somalia-donald-trump-al-shabab-586004

Gause, F. G., III. (2019). Should we stay or should we go? The United States and the Middle East. *Survival, 61*(5), 7–24.

Ghattas, K. (2016, May 8). Trump v Clinton on foreign policy. *BBC*. https://www.bbc.com/news/election-us-2016-36232271

Gholz, E., Press, D. G., & Sapolsky, H. M. (1997). Come home, America: The strategy of restraint in the face of temptation. *International Security, 21*(4), 5–48.

Harkavy, R. E. (1982). *Great power competition for overseas bases: The geopolitics of access diplomacy*. Pergamon Press.

Harkavy, R. E. (1989). *Bases abroad: The global foreign military presence*. Oxford University Press.

Heo, U., & Ye, M. (2019). U.S. military deployment and host-nation economic growth. *Armed Forces & Society, 45*(2), 234–267.

Holland, S. (2020, June 13). Trump to West Point grads: 'We are ending the era of endless wars.' *Reuters*. https://www.reuters.com/article/usa-trump-wars/trump-to-west-point-grads-we-are-ending-the-era-of-endless-wars-idINL1N2DQ05U

Jakobsen, J. (2018). Is European NATO really free-riding? Patterns of material and non-material burden-sharing after the Cold War. *European Security, 27*(4), 490–514.

Jakobsen, J, & Jakobsen, T. G. (2019). Tripwires and free-riders: Do forward-deployed U.S. troops reduce the willingness of host-country citizens to fight for their country? *Contemporary Security Policy*, *40*(2), 135–164.

Jervis, R., & Snyder, J. (Eds.). (1991). *Dominoes and bandwagons: Strategic beliefs and great power competition in the Eurasian rimland*. Oxford University Press.

Jones, G., & Kane, T. (2012). U.S. troops and foreign economic growth. *Defence and Peace Economics*, *23*(3), 225–249.

Kagan, R. (2012). *The world America made*. Vintage.

Kahneman, D., & Tversky, A. (1979). Prospect theory: An analysis of decision under risk. *Econometrica*, *47*(2), 263–291.

Kane, T. (2004). *Global U.S. troop deployment, 1950–2003*. Heritage Foundation. Retrieved March 8, 2020, from https://www.heritage.org/defense/report/global-us-troop-deployment-1950-2003

Kane, T. (2006). *Global U.S. troop deployment, 1950–2005*. Heritage Foundation. Retrieved March 26, 2021, from https://www.heritage.org/defense/report/global-us-troop-deployment-1950-2005

Kennedy, P. (1988). *The rise and fall of the great powers*. Random House.

Kim, W. (1992). Power transitions and great power war from Westphalia to Waterloo. *World Politics*, *45*(1), 153–172.

Layne, C. (2017). The US foreign policy establishment and grand strategy: How American elites obstruct strategic adjustment. *International Politics*, *54*(3), 260–275.

Lee, C. E., Kube, C. Welker, K., & Ruhle, S. (2018, April 30). Kelly thinks he's saving U.S. from disaster, calls Trump 'idiot,' say White House staffers. *NBC News*. https://www.nbcnews.com/politics/white-house/kelly-thinks-he-s-saving-u-s-disaster-calls-trump-n868961

Lobell, S. E., Ripsman, N. M., & Taliaferro, J. W. (Eds.). (2009). *Neoclassical realism, the state, and foreign policy*. Cambridge University Press.

Lord, C., & Erickson, A. S. (Eds.). (2014). *Rebalancing U.S. forces: Basing and forward presence in the Asia-Pacific*. Naval Institute Press.

Lutz, C. (Ed.). (2009). *The bases of empire: The global struggle against U.S. military posts*. New York University Press.

MacDonald, P. K., & Parent, J. M. (2011). Graceful decline? The surprising success of great power retrenchment. *International Security*, *35*(4), 7–44.

Macintyre, B. (2019, October 18). America's battle to resist the lure of foreign wars. *The Times*. https://www.thetimes.co.uk/article/americas-battle-to-resist-the-lure-of-foreign-wars-jz7pj5xp6

Mackinder, H. J. (1904). The geographical pivot of history. *The Geographical Journal*, *170*(4), 421–444.

Mackinder, H. J. (1942). *Democratic ideals and reality*. National Defense University Press. (Original work published 1919.)

Mackinder, H. J. (1943). The round world and the winning of the peace. *Foreign Affairs, 21*(4), 595–605.

Mahan, A. T. (1890). *The influence of sea power upon history, 1660–1783*. Little, Brown.

Mandelbaum, M. (1988). *The fate of nations: The search for national security in the nineteenth and twentieth centuries*. Cambridge University Press.

Martinez Machain, C., & Morgan, T. C. (2013). The effect of US troop deployment on host states' foreign policy. *Armed Forces & Society, 39*(1), 102–123.

Mason, J., & Mohammed, A. (2020, June 15). Trump says U.S. to pull some troops from Germany over NATO spending feud. *Reuters*. https://www.reuters.com/article/us-usa-germany-military-trump/trump-says-u-s-to-pull-some-troops-from-germany-over-nato-spending-feud-idUSKBN23M2VG

Mearsheimer, J. J. (2001). *The tragedy of great power politics*. W. W. Norton.

Mearsheimer, J. J., & Walt, S. M. (2016). The case for offshore balancing: A superior U.S. grand stategy. *Foreign Affairs, 95*(4), 70–83.

Modelski, G. (1978). The long cycle of global politics and the nation-state. *Comparative Studies in Society and History, 20*(2), 214–235.

Myers, M. (2020, August 6). Pentagon not confirming any plans to halve US troops in Afghanistan by the election. *Military Times*. https://www.militarytimes.com/news/your-military/2020/08/06/pentagon-not-confirming-any-plans-to-halve-us-troops-in-afghanistan-by-the-election/

New York Times. (2016a, March 26). *Transcript: Donald Trump expounds on his foreign policy views*. https://www.nytimes.com/2016/03/27/us/politics/donald-trump-transcript.html

New York Times. (2016b, April 27). *Transcript: Donald Trump's Foreign policy speech*. https://www.nytimes.com/2016/04/28/us/politics/transcript-trump-foreign-policy.html

New York Times. (2016c, July 21). *Transcript: Donald Trump on NATO, Turkey's coup attempt and the world*. https://www.nytimes.com/2016/07/22/us/politics/donald-trump-foreign-policy-interview.html

Nieman, M. D., Martinez Machain, C., Chyzh, O., & Bell, S. R. (2021). An international game of risk: Troop placement and major power competition. *Journal of Politics, 83*(4), 1307–1321.

O'Mahoney, A., Priebe, M., Frederick, B., Kavanagh, J., Lane, M., Johnston, T., Szayna, T. S., Hlavka, J. P., Watts, S., & Povlock, M. (2018). *U.S. presence and the incidence of conflict*. RAND Corporation.

Pew Research Center. (2016, May 5). *Public uncertain, divided over America's place in the world*. Retrieved August 17, 2021, from https://www.people-press.org/2016/05/05/public-uncertain-divided-over-americas-place-in-the-world/

Plaskin, G. (1990, March 1). The Playboy interview with Donald Trump. *Playboy Magazine*. https://www.playboy.com/read/playboy-interview-donald-trump-1990

Politico. (2017, January 20). *Full text: 2017 Donald Trump inauguration speech transcript*. https://www.politico.com/story/2017/01/full-text-donald-trump-inauguration-speech-transcript-233907

Porter, P. (2018). Why America's grand strategy has not changed: Power, habit, and the U.S. foreign policy establishment. *International Security*, *42*(4), 9–46.

Posen, B. R. (2014). *Restraint: A new foundation for U.S. grand strategy.* Cornell University Press.

Ripsman, N. M., Taliaferro, J. W., & Lobell, S. E. (2016). *Neoclassical realist theory of international politics*. Oxford University Press.

Rose, G. (1998). Neoclassical realism and theories of foreign policy. *World Politics, 51*(1), 144–172.

Rovner, J., & Talmadge, C. (2014). Hegemony, force posture, and the provision of public goods: The once and future role of outside powers in securing Persian Gulf oil. *Security Studies, 23*(3), 548–581.

Schmidt, S. (2020). *Armed guests: Territorial sovereignty and foreign military basing*. Oxford University Press.

Shanahan, J. (1987, September 2). Trump: U.S. should stop paying to defend countries that can protect selves. *Associated Press*. https://apnews.com/article/05133dbe63ace98766527ec7d16ede08

Spykman, N. J. (2017). *America's strategy in world politics: The United States and the balance of power.* Routledge. (Original work published 1942.)

Stravers, A. (2021). Pork, parties, and priorities: Partisan politics and overseas military deployments. *Conflict Management and Peace Science, 38*(2), 156–177.

Stravers, A., & El Kurd, D. (2020). Strategic autocracy: American military forces and regime type. *Journal of Global Security Studies, 5*(3), 427–442.

Taliaferro, J. W. (2004). *Balancing risks: Great power intervention in the periphery*. Cornell University Press.

Tharoor, I. (2019, October 11). Trump may be against 'forever wars,' but he's no dove. *Washington Post*. https://www.washingtonpost.com/world/2019/10/11/trump-may-be-against-forever-wars-hes-no-dove/

Thompson, W. R. (1986). Polarity, the long cycle, and global power warfare. *Journal of Conflict Resolution, 30*(4), 587–615.

U.S. Department of Defense. (2021). *Military and civilian personnel by service/agency by state/country. U.S. Defense Manpower Data Center.* Retrieved November 23, 2021, from https://dwp.dmdc.osd.mil/dwp/app/dod-data-reports/workforce-reports

Vine, D. (2015). *Base nation: How U.S. military bases abroad harm America and the world*. Metropolitan Books.

Vine, D. (2019). No bases? Assessing the impact of social movements challenging U.S. foreign military bases. *Current Anthropology, 60*(S19), 158–S172.
Walt, S. M. (2018). *The Hell of good intentions: America's foreign policy elite and the decline of U.S. primacy.* Farrar, Straus and Giroux.
Waltz, K. N. (1979). *Theory of international politics.* McGraw-Hill.
Waltz, K. N. (2000). Structural realism after the Cold War. *International Security, 25*(1), 5–41.
Yoder, B. K. (2019). Retrenchment as a screening mechanism: Power shifts, strategic withdrawal, and credible signals. *American Journal of Political Science, 63*(1), 130–145.
Youssef, N. A., & Phillips, M. M. (2020, November 20). U.S. considers moving forces out of Somalia. *Wall Street Journal.* https://www.wsj.com/articles/u-s-considers-moving-forces-out-of-somalia-11605911141

CHAPTER 2

Stayers, Leavers, and U.S. Overseas Troops

*

Let us begin this second chapter by stating the obvious: The United States is still immensely powerful in terms of its command of resources, and it still has an extraordinary amount of global influence. Yet, however one judges the efficacy of various types of power resources—economic strength, military capabilities, cultural and ideological attractiveness—what vitally underpins this influence is the United States' *projection of military power overseas*. In particular, to use the renowned expression of Barry Posen (2003, p. 21), its "command of the commons"—the way it essentially dominates the air, the oceans, and space—serves as "the key enabler of the hegemonic foreign policy that the United States has pursued since the end of the Cold War." Command of the commons, in turn, critically hinges not only on total defense spending and sophisticated military hardware; Posen (2003, pp. 16–19) argues that the military bases abroad, and the concomitant forward deployment of U.S. troops, constitute *the* key piece of the globe-spanning infrastructure that makes it possible for the United States to shape developments in vital regions.

Military bases cause and represent what George Modelski (1978, pp. 229–232) calls "functional network control"; they enable global hegemony, but (mostly and preferably) without the burden that is associated with direct political control over other peoples' territories.

J. Jakobsen, *The Geopolitics of U.S. Overseas Troops and Withdrawal*, Palgrave Studies in International Relations,
https://doi.org/10.1007/978-3-030-94488-9_2

For the sea powers of history—Athens, Carthage, Venice, and their globally-focused successors—gaining political control over land territory was of far less import than securing, along the main sea lanes of communication, conveniently-located islands and onshore maritime bridgeheads or stepping stones; these were the only pieces of territory really worth fighting for (Lambert, 2018, p. 73). The sometimes violent acquiring of such stepping stones, which are especially found along the rimlands of Eurasia, Africa, and the Americas, was what permitted a number of European maritime states—Portugal, the Dutch Republic, and Great Britain—successively to dominate the world's oceans and world trade. These empires of yesteryear were able to establish global hegemonies of sorts from the late fifteenth century and until the United States picked up the mantle in the middle of the last century. This they could not have done without a vast network of bases. More generally, there is a clear limit to how much a state can shape and control others' behavior without the necessary military foundation (Rovner & Talmadge, 2014). To a considerable extent, then, forward-deployed U.S. forces can be seen as a bellwether of U.S. power and influence.

Power and Hegemony and the U.S. Military-Base Network

The U.S. military-base network was immense throughout the four decades when it fought a multifaceted cold war against the Soviet Union, a true great-power rival. It remained immense after as well. Indeed, the end of the Cold War saw the United States propelled into an unprecedented position as the world's sole superpower. Some, in the early 1990s, predicted that other states would soon rise to great-power status and that multipolarity would thereby be restored (Waltz, 1993). This followed analyses in the preceding decade that painted a rather bleak picture of the United States' prospects of stemming its relative decline (Kennedy, 1988, pp. 557–564, 665–692). But the "accidental hegemony," as Barry Posen (2003, p. 19) called it, or the "unipolar moment," in the famous expression of Charles Krauthammer (1990–1991), would instead prove to be anything but fleeting.

Power and Geography

Krauthammer (1990–1991, p. 24) conceived of "moment" in relative terms—"another generation or so." Others as well, such as William Wohlforth (1999, p. 37), talked about unipolarity as "a deeply embedded material condition of world politics that has the potential to last for many decades." And surely, while the global distribution of material resources is gradually changing, and while China is undoubtedly rising, the United States still predominates militarily, technologically, economically, and financially. It is still, in many if not most respects, "unrivaled" (Beckley, 2018a). Its gross domestic product (GDP), measured at official exchange rates, represents 23.6% of world GDP (World Bank, 2021). It also accounts for nearly 40% of world military spending (SIPRI, 2021a). In addition, the legacy of the Cold War and its aftermath continues to act as a force multiplier for the United States: the international institutions that are so much shaped by Washington live on, as do its alliances, its diplomatic relationships, and, for the most part—which this book will dwell on—its forward military deployments. As Hal Brands (2016, p. 339) states in *Making the Unipolar Moment*, these force multipliers placed the vastly powerful America in a "position to punch *above* its own geopolitical weight."

U.S. power is also helped by some highly fortunate facts of *geography*. The country is shielded on each side of the mainland by two vast oceans, both of which link it to world trade routes. By virtue of "the stopping power of water" (Mearsheimer, 2001, p. 126), they function as natural barriers against external attacks. The United States in many ways bears the characteristics of a gigantic island. This greatly fortuitous geographical reality was highlighted long ago by geopoliticians such as Nicholas Spykman (1938a, p. 43), who judged the United States to be the "most favored state in the world from the point of view of location." Even earlier, the acclaimed naval strategist Admiral Alfred Thayer Mahan (1890, pp. 33–35) pointed out that this quasi-island status significantly lessened U.S. concerns about territorial defense while at the same time generating for it a geographical basis for maritime expansion. In that sense, Great Britain could serve as a blueprint: by the time of Mahan's writing, the British Empire had lasted for nearly three centuries, and it had steadily expanded. Relatively shielded from major security threats against the Home Islands, Great Britain was able to disentangle itself from much of the day-to-day military squabbles on the continent, usually only intervening when a potential hegemon threatened to conquer its neighbors across the Channel. At the same time, this relative security enabled Britain to roam around the world oceans, soon dominating international trade by virtue of its magnificent navy.

Thus, Great Britain was both isolationist and interventionist. It was isolationist insofar as its separation from the tumultuous European continent permitted London to engage militarily only occasionally and judiciously. It was interventionist in that its energies and resources could then be spent *globally* at a level practically unattainable to other states and empires, which usually had more geographically proximate worries to attend to. Great Britain had the opportunity and the choice to go global, and it did so. But paradoxically, for Britain as well as for the United States, these realities of geography and power make feasible both expansion and seclusion. And it is not necessarily a given which one of these alternatives produces the largest amount of security. For the United States, moreover, the lure of isolationism was even greater than for Britain, considering the vast distance that separated the former from Europe.

The Emergence of a Superpower

Nicholas Spykman (1938b, p. 236) tersely notes that "geography does not argue. It simply is." And what it is, and what it was, was a feature that eventually allowed the United States to dominate the Western Hemisphere. The declaration of regional hegemonic *aspirations*, which soon gained doctrinal status, was famously made by President James Monroe in 1823. The actual *attainment* of something quite akin to such hegemony was reached at the turn of the twentieth century; for it was the 1898 Spanish-American War that served as the incubator of the United States' initial overseas base construction, which formed the core of Washington's expansionist project. The emphatic U.S. victory in a conflict with theaters in both the Caribbean and the Western Pacific resulted in its acquisition of Puerto Rico, Guam, and the Philippines. In addition, Hawaii was annexed, and Cuba effectively became a U.S. protectorate. The latter gain soon led to Washington obtaining an indefinite lease of Guantánamo Bay, which, to the chagrin of the post-revolution Cuban government, for over a century has served as a vital U.S. naval base conveniently facing the Caribbean Sea. The United States also acquired the eastern part of Samoa from Great Britain and Germany in 1899. Soon thereafter, in 1903, Washington manipulated domestic developments in Colombia, helping Panama obtain independence, which added to the list of U.S. client states and ensured that the now U.S.-controlled Panama Canal would soon reach its completion. By the early twentieth century, therefore, the first batch of global "stepping stones" had been secured by what was undoubtedly a rapidly emerging great power.

Hence, U.S. *regional* hegemony—the implementation of the Monroe Doctrine—was for all intents and purposes accomplished as the twentieth century commenced. But it did not immediately entail any coherent drive for *global* influence. The United States was, overall, an insular nation, and the insularity aspects came for decades to weigh heavily on Washington's geostrategic thinking. With its involvement in the latter stages of the First World War serving as an outlier event, the U.S. by and large opted not to intervene in overseas affairs. Instead, it took advantage of the other alternative allowed for by its favorable geography; viz., to stand aloof from world affairs under the assumption that the country was secure behind its natural barriers.

It was only with the Japanese attack on Pearl Harbor, in December 1941, and the subsequent declaration of war by Germany, that the scales were tipped in favor of another possibility: that regional hegemony, although bolstering U.S. security, did not in fact bring about *absolute* security in an interconnected and interdependent world. Rather, the relative isolation of the United States created opportunities that no other country had, and which the nation should take advantage of—opportunities to go global, to expand, and to *pre-empt* any severe security threats. Thenceforth, the dominant strand in U.S. geostrategic thinking would advise countering burgeoning threats early on and preferably as far away from the U.S. homeland as possible. The new interventionist inclinations came suddenly. They were preceded, however, by decades of steady increases in U.S. absolute and relative economic power. This implied that the United States was a nation with an enormous potential for *global* influence. That potential was famously not yet realized as the interwar period drew to a close. Most notably, it was not yet translated into comparable amounts of *military* power resources. By the eve of the war, therefore, the United States was only a step away from becoming a *real* great power, one with the military capabilities and global influence to match its tremendous potential. Then Pearl Harbor presented it with the incentive to make the step.

The Second World War, more generally, incited the United States to acquire its second batch of stepping stones. In 1940, through the "Destroyer for Bases Agreement," Washington essentially started inheriting significant bits of the British Empire's extensive web of bases, in part to support a United Kingdom that was severely distracted by its war efforts against Nazi Germany (Pettyjohn, 2012, pp. 46–47). These bases were supplemented by agreements with several Latin American countries, from which the United States leased bases and secured overflight rights (Harkavy, 1982). The value of this "leasehold empire" (Sandars, 2000), and of the many islands acquired in the Pacific War (and also of the bases over which the United States had assumed control in Iceland and Greenland [Lutz, 2009, p. 12]), was clear enough to make the United States determined, once the war had ended, "not to repeat its earlier, almost cavalier, disregard for the importance of external bases" (Harkavy, 1989, p. 2).

And determined the United States proved to be. The occupation and reconstitution of (West) Germany and Japan, of course, necessitated the stationing of huge numbers of American soldiers in both countries. This troop presence was supposed to be of limited duration, but it would actually prove to last for generations. The outbreak of the Korean War, in June 1950, further stimulated the projection of U.S. power. Not only did the war lead to the permanent stationing of U.S. troops in South Korea, which supplemented the troops in Japan and the huge U.S. bases in the Philippines; with the broader superpower conflict now dramatically taking hold, it also spurred Washington into sending massive forces to Europe, the main theater and "prize" in the Cold War (Trachtenberg, 2012, p. 111).

The Geography of U.S. Bases and Troops

These bases were overwhelmingly acquired by Washington through voluntary agreements with an array of sovereign host countries (Schmidt, 2020); they were thus an element of, to use the Cold War historian Geir Lundestad's much-cited phrase, America's "Empire by Invitation" (Lundestad, 1986). Partly because most of these agreements were relatively non-contentious, the vast majority of them lasted until the end of the Cold War. Moreover, the onset of the unipolar era did not fundamentally alter the basic (geographical) pattern of U.S. overseas troops deployment. True enough, total numbers of active-duty troops stationed abroad declined from around 460,000 in 1989 to 300,000 in 1992 (Kane, 2004), as a consequence of the end of superpower tensions and the resulting urge to reap some peace dividend therefrom. Many smaller and some larger military facilities were closed, for a variety of reasons, among other places in Panama, the Philippines, and Spain. Even so, the Cold War base posture was generally upheld: Those states that were the most important hosts of U.S. troops in the 1950s were still the most important hosts well into the post-Cold War era.

A major part of the rationale underpinning this continuity was that of locking in the spoils from the victory obtained in the superpower struggle—that is to say, to secure the new, immensely favorable status quo and international order (Brands, 2016, p. 324). It was now less about the territorial defense of host countries; forward-deployed troops were as much seen as a power-projection tool to be sent to areas of operation some distance away from their bases if and when the need arose (Henry, 2006, p. 17). The bases thus functioned as springboards from which U.S. influence would radiate. Important military facilities therefore remained in Washington's hands. In Europe, numbers of active-duty U.S. troops were above 100,000 up until 2005. In fact, Russian leaders acquiesced to this not only because the Soviet Union had been outcompeted by a stronger adversary over which it now had very limited influence, but also because these troops were seen by Moscow as well as a stabilizing factor on the European continent (Service, 2015, pp. 435, 444). U.S. troops also remained in large numbers in Asia. Tens of thousands of soldiers stayed put in their bases in South Korea, as concerns about North Korea's nuclear-weapons ambitions were growing (Roehrig, 2014, p. 71). Of even greater import, during President Bill Clinton's first administration, Washington decided to bolster its already sturdy alliance with Japan, which thenceforth included a pledge to uphold troop numbers in East Asia for the long term (Christensen, 2016, p. 188).

Added to this was an increasing troop presence in the Middle East. Saudi Arabia had granted the United States basing rights as early as in 1946, and Bahrain (at the time a British protectorate) signed a naval-base agreement with Washington in 1949 (Hiro, 2019, pp. 32–33). Following the 1980 Carter Doctrine, in which U.S. president Jimmy Carter stated that military force could or would be used if developments in the region threatened its vital national interests, construction of military facilities commenced in several states in and around the Persian Gulf (Brands, 2016, p. 239). Yet, U.S. presence in the region was still relatively modest until the 1987 "Tanker War" (which was part of the larger Iran-Iraq War) and, more clearly, the 1991 Gulf War and its aftermath, changed calculations in Washington (Gordon, 1994; Kamrava, 2018, p. 39). Iraq's invasion of neighboring Kuwait in the summer of 1990 was certainly a turning point. The immediate reply by Washington was to deploy a massive amount of military-service personnel to Saudi Arabia to deter the Iraqis from undertaking any further expansionist adventures. The troops were sent on Riyadh's invitation—Saudi leaders urged Washington to establish a permanent base in the kingdom out of fear of both Iraq and Iran (Wearing, 2018, pp. 43–44)—, and they remained for a while in the country even after Iraq had been driven back from Kuwait by early 1991.

Despite the success in the (First) Gulf War, concerns in Washington about the Middle East continued to increase in the 1990s. The policy adopted was "dual containment." Rather than attempting to play unruly Iraq off against its equally unruly Iranian neighbor, and instead of "passing the buck" of containment onto other regional states, as had very much been the playbook of the 1980s, now both Iran and Iraq would be checked by the United States itself (Kamrava, 2018, p. 44). This required troops, many troops, and troops that were stationed in military bases throughout the region. And then, of course, the devastating terrorist attacks on September 11, 2001, spurred U.S. invasions of Afghanistan and, shortly after, Iraq. U.S. troop numbers in the region consequently exploded, vastly increasing the perceived and actual importance of a range of states and territories hosting U.S. forces. These included Qatar, Kuwait, Bahrain (home to the U.S. Fifth Fleet), Jordan, Turkey, Djibouti (in the Horn of Africa), and British Indian Ocean Territory (i.e., the island of Diego Garcia).

The general picture is that U.S. bases and troops are where they were. It is also the case that both the pre- and the post-Cold War network of U.S. overseas bases and facilities has been and is vast. According to the Pentagon's 2018 Base Structure Report (which is the latest such report that has been made available), the total value of the military facilities (i.e., buildings and structures) that the United States owns and leases abroad is around $200 billion (U.S. Department of Defense, 2018). Furthermore, this number surely represents an underestimation as it explicitly excludes some important countries (such as Iraq and Afghanistan, though the United States has now left the latter) and fails to report other, known military facilities (such as in Israel and the Philippines [Herrera & Cicchini,

2013, pp. 141–142]). But the exact numbers are not of critical importance; and neither is a precise tally of U.S. military bases abroad. As one scholar noted, this is a question whose answers range "from 34 to 737, to 850, to well over a thousand, depending on a host of factors" (Davis, 2011, p. 216). The conclusion is in any case that the United States spends enormous resources upholding and sometimes expanding its network of overseas military facilities.

This the country and its leaders have done for two-and-a-half generations. Figure 2.1 shows the total number of forward-deployed U.S. military personnel, classified as permanently assigned, active-duty troops, from 1950 until 2021 (as of September 30). All data are based on information from the Pentagon's U.S. Defense Manpower Data Center (Kane, 2004; U.S. Department of Defense, 2021) (Please see Chapter 6 for more information on these data).

What is clearly discernible from these numbers is, first, that significant peaks appear around large wars (that is to say, the Korean War, the Vietnam War, and the more recent wars in Afghanistan and Iraq). Second, adjusted for these wars, the long-term trend is pointing modestly downward. In part, though, the relatively steady decline can be explained by some specific policy decisions. For example, the 2004

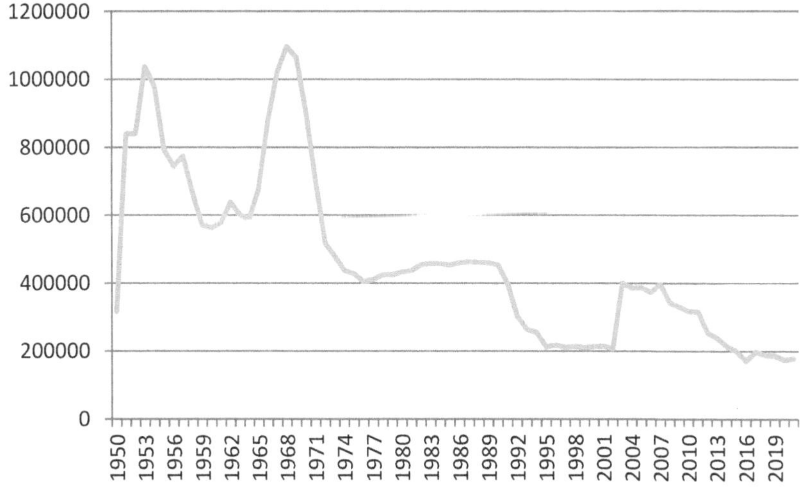

Fig. 2.1 U.S. overseas troops, total, 1950–2021

Global Defense Posture Review (GDPR), and its attendant concept of an Integrated Global Presence and Basing Strategy, aimed at reducing, but not removing, the U.S.'s global military footprint (Szayna et al., 2015, p. 2). In fact, the GDPR called for *increasing* the number of bases or facilities overseas; these were to be smaller, leaner, and more dispersed, allowing for a U.S. presence in areas from which it theretofore had been absent (Cooley, 2005, p. 85). In addition, changes to military doctrines and technology have generally lessened the need for a heavy troop presence on main operation bases (MOBs), increasing instead the demand for, and utility of, "lighter," more flexible, and more dispersed bases (O'Hanlon, 1998–1999, p. 74). These smaller versions of U.S. overseas facilities are colloquially known as "lily-pads." They constitute a less visible form of U.S. presence; they are therefore less contentious and less likely to invite host-society opposition. They can be very suitable nonetheless, functioning as regional access points or springboards from which the U.S. military, typically located at a main operating base nearby, can rapidly move around the region in question if the need arises.

Notwithstanding the expansion of the number of facilities to which the United States have access, the geographical pattern shows a noticeable continuity. It is maybe a bit of an exaggeration to say, with the historian Niall Ferguson (2003, pp. 378–379), that "a map showing the principal US military bases around the world looks remarkably like a map of Royal Navy coaling stations a hundred years ago." However, if one adds to this latter map of the British Empire the U.S. overseas basing status as of 1954, the year after the Korean War ended, or even 1945, the similarities are quite conspicuous.

Power-as-Resources and Power-as-Influence

For it was, to repeat, Pearl Harbor that provided the impetus for the United States to take the history-changing step and go global, a globalization of U.S. military presence that has not since been reversed. That step may have come late, but it did eventually come. And perhaps it would have been inevitable even without Japanese and German imperialist expansion. In general, the command of material resources tends to stimulate attempts to expand abroad, or to maximize influence over other states and over the international system as such (Zakaria, 1998, pp. 3ff.). However, the two conceptions of power—*power-as-resources* and *power-as-influence* (or *outcomes*)—do not march in perfect lockstep. This presents challenges for scholars attempting to gauge the relative power of states. Was the United States a great power in the late nineteenth century?—or should such a label be reserved for the period starting in the middle of the Second World War? That was when U.S. production of combat aircraft, military vessels, and all sorts of war material went into high gear, enabling the United States to shape the direction of the war in both Europe and Asia.

Scholars do not fully agree on this issue. With respect to theoretical definitions, there is a general tendency to view a state's power in relational terms, following Robert Dahl's famous understanding that "A has power over B to the extent that he can get B to do something that B would not otherwise do" (Dahl, 1957, pp. 202–203). Thereby, power is used in a way "that is interchangeable with such terms as 'influence' and 'control'" (Baldwin, 1979, p. 162); it centers on social relations (Barnett & Duvall, 2005), or on psychological ones, whereby, in the words of Hans Morgenthau (2006 [1948], p. 11), it encompasses "anything that establishes and maintains the control of man over man."

Yet, although this is a broadly accepted understanding of power, many international relations scholars point to the challenges that result, particularly with regard to issues of operationalization and measurement. William Wohlforth (1993, p. 4), for example, notes that this confluence of power and influence means that, in studies of the effects of the balance of power, the dependent variable is essentially the same as the independent variable, which precludes meaningful causal analysis. Wohlforth, and many others, employ the fallback position of understanding, and of measuring, power in terms of the command of resources. This is the *power-as-resources* position. It often simplifies the picture by emphasizing economic and military capabilities (Kennedy, 1988), or by referring to economic resources (along with size of population) as representing *potential* or *latent* power, which in turn may or may not be translated into *actual* power in the form of military capabilities (Mearsheimer, 2001, p. 43). Others prefer a wider range of more fine-grained measures when estimating relative power (Beckley, 2018b; Brooks & Wohlforth, 2016a). The basic assumption is in any case the same: a country is powerful and has the ability to shape regional and world events insofar as it controls appropriate resources or capabilities.

Grand Strategies of the United States

This association between power-as-resources and power-as-influence has been at the core of U.S. grand strategy ever since it entered World War II. Granted, arguments resembling the isolationist inclinations of old have never fully disappeared from the national-security discourse. Nevertheless, U.S. grand strategy—the plan or "theory" that explains how the country can best produce security for itself (Walt, 1987, p. 2)—has been remarkably consistent with respect to its core tenet: security and prosperity for the United States are best provided by following a grand strategy of "primacy" or "geopolitical dominance"; that is, a strategy that seeks to preserve unipolarity and the U.S.-shaped international order by outcompeting any rising rivals—which are seen as the biggest potential threat to the United States—economically, militarily, and politically (Posen & Ross, 1996–1997, pp. 32–43).

Pax Americana and the Stayers of the Blob

In other words, the United States has consistently sought to dominate and to shape world affairs, and often also the foreign and domestic policies of countries deemed important to U.S. national interests. Unipolarity, thus, has been translated into a program for global influence, which has both supported and furthered a U.S.-led international order—a U.S. liberal hegemony, or a *Pax Americana*. This state of affairs involves a fairly thick and wide web of generally liberal international rules and institutions that direct relations among states and constrain the behavior of smaller and greater powers alike (often even including the hegemon itself). It also involves the United States taking primary responsibility for the provision of many global and regional "public goods," such as security, free trade, and economic globalization (Ikenberry, 2011, p. 16). All of this is, by the majority of the U.S. foreign-policy establishment, believed to benefit most of the world's states, making key regions prosperous, stable, and secure, thereby also making "the world safer for the United States" (Brooks et al., 2012–2013, p. 11).

This does not mean that the United States has been or is omnipotent. As Christopher Layne (2006, p. 12) tersely summarizes it: "In international politics, the United States does not get all that it wants all of the time. But it gets most of what it wants an awful lot of the time, and it affects other states far more than other states affect it." Neither does it mean that *Pax Americana* or U.S. primacy will last forever, even if the gap in relative power between the United States and others—with respect to both resources and influence—is still substantial. And neither does it mean that there is full agreement on whether the United States should *continue* to follow a grand strategy of primacy. This is so even as one notes the general *absence* of any marked ambivalence on this issue among a clear majority of decision-makers, consultants, think tanks, lobbyists, and the media in the United States (Friedman & Logan, 2016; Layne, 2017; Porter, 2018). In a recent book—*The Hell of Good Intentions*—Stephen Walt (2018) laments starkly the self-serving viewpoints of what is rather derogatorily known as "the Blob." This is the foreign-policy community that is concentrated in the U.S. Capital and which incessantly extols the virtues of liberal hegemony and an activist U.S. foreign policy. This is a foreign policy whose objective is to counter any and all real and imagined threats to U.S. security by maximizing U.S. power and influence over allies as well as over adversaries—in virtually every corner of the world. The foreign-policy elite, Walt (2018, p. 95) bemoans, "is a dysfunctional caste of privileged insiders" who generally ignore alternative viewpoints on foreign affairs.

According to Walt, conformity to prevailing perspectives on U.S. grand strategy is normally a prerequisite for climbing the echelons of the ladder of power and influence in Washington. On the other hand, he also notes that (university) scholars are an exception insofar as they enjoy the job security of tenure and can, and indeed are in some ways tasked with, challenging just this consensus (Walt, 2018, p. 110). The academic debate about U.S. grand strategy is therefore, taken as a whole, richer in scope than is that of (the core of the) Blob. As it is, scholars are distributed in camps ranging from those who strongly advise a continuation of U.S. global presence (e.g., Brands & Edel, 2019; Brooks et al., 2012–2013; Brzeziński, 2012; Rovner & Talmadge, 2014), to those who argue that at least a partial retrenchment would be highly beneficial to U.S. national interests, especially considering that U.S. relative power is in (slow but gradual) decline (e.g., Gholz et al., 1997; MacDonald & Parent, 2011; Mearsheimer, 2011; Posen, 2014). For convenience, we can label the former group *Stayers* and the latter *Leavers*, though with the caveat attached that this neat labeling and separation certainly do not do full justice to the wide variety of nuanced viewpoints existing in both camps.

In their oft-cited *International Security* article from the mid-1990s, a time period during which the purpose of the role of the United States in the new, non-bipolar world was uncertain and much discussed, Barry Posen and Andrew Ross offered a four-category typology of U.S. grand-strategy options (Posen & Ross, 1996–1997). At the one extreme is *neo-isolationism*. This depicts a grand strategy whereunder U.S. troops would not any longer be forward-deployed in large numbers, and where a strict prioritization among the wide variety of national interests would apply. At the other extreme is *primacy*, which is what others variously denote as "global dominance" (Mearsheimer, 2011, p. 18), (liberal) "hegemony" (Walt, 2018), or "deep engagement" (Brooks et al., 2012–2013, pp. 13–14). This grand-strategy option highlights the necessities of overwhelming U.S. power and a wide and deep presence of U.S. troops abroad, to preserve regional and global peace and to counter a broad array of threats to U.S. national security and interests. Sandwiched in-between are two other alternatives—*cooperative security* and *selective engagement*—which are rather closely affiliated with primacy and neo-isolationism, respectively.

The dividing lines between these four grand-strategy options, not least between *pairs* of them, are not entirely clear-cut. For the purposes herein it makes sense to place the *Stayers* under the banner of *primacy* or *hegemony*. As for the other extreme, I opt to alternate between using the labels *Leavers* ("Partial Leavers" certainly sounds too cumbersome!) and "offshore balancers." The latter is akin to promoting a grand strategy of restraint (Gholz et al., 1997; Posen, 2014; Walt, 2019), or of retrenchment (Haynes, 2015), where the latter can be understood "as a policy of retracting grand strategic commitments in response to a decline in relative power" (MacDonald & Parent, 2011, p. 11). The label "offshore balancers" has evidently risen to become something close to a common denominator encompassing those who argue that the United States should significantly scale back its foreign activism, including its troops presence abroad.

The Offshore Balancers

Offshore balancers are highly critical of the grand strategy followed by the foreign-policy establishment in Washington, especially since the end of the Cold War removed the necessity to contain another superpower (Gholz et al., 1997; Layne, 1998; Mearsheimer & Walt, 2016; Posen, 2014; Walt, 2018). They tend to bemoan the U.S. activism of recent years, pointing, among other things, to the disastrous unintended consequences arising from the 2003 invasion of Iraq; the growth of international terrorism spurred by that invasion and by the U.S. presence in the Middle East in general; the proliferation of weapons of mass destruction, particularly the growing nuclear ambitions of states singled out by Washington as adversaries to be defeated; the declining legitimacy of democratic values following the use by the United States of torture and targeted killings and its unwavering support for a selection of authoritarians; the growing assertiveness of Russia in Eastern Europe, purportedly a result of U.S. and Western policies toward Moscow (Mearsheimer, 2014); and the enormous resources that are spent by Washington to uphold the current grand strategy of primacy (Mearsheimer & Walt, 2016, pp. 70–71).

In its stead, offshore balancers propose a grand strategy of restraint, consisting of four main parts. First, they highlight the very fortuitous geographical position enjoyed by the United States. The virtual island status of America makes it more secure than any other major power, for the effect of power diminishes with distance. The United States is also, by a veritable margin, the most powerful state in its own region. This enables it, if it so wishes, to focus its attention on internal developments, on the preservation of domestic institutions, and on the production of more butter and less guns. Instead of going abroad searching for demons, offshore balancers contend, the United States should ensure that it remains "the only great power in the Western hemisphere so that its citizens do not have to worry about being attacked, blockaded, or coerced by a nearby rival with comparable capabilities" (Walt, 2019, p. 13).

Second, the United States should, in a much stricter way, *prioritize* among its many foreign-policy objectives, to avoid overextension (Posen, 2013, p. 117). This is so especially considering the gradual decline of U.S. relative power. Not least, it should acknowledge that the only conceivable existential threat stems from the capabilities and policies of other great powers. U.S. grand strategy should adopt a narrower focus that concentrates primarily on what can reasonably be seen as vital regions. These include only the Americas, the Persian Gulf (though mostly on account of its petroleum resources, which may well lose much of their relevance over time), and Europe and East Asia (due to their vast economies and because they encompass several major powers) (Mearsheimer, 2011, p. 11; Mearsheimer & Walt, 2016, p. 72). In the Americas, the United States will clearly not experience anything resembling a peer competitor for the foreseeable future. In the Persian Gulf and in Europe, a regional hegemon is not on the horizon (Russian expansionism can easily be checked by the rich Western European states, if they are not allowed to free-ride on America). Only in Asia, offshore balancers contend, can one realistically point to a potential major challenger that can hamper U.S. security and national interest; so balancing China is the only *major* task to which the United States may need to devote ample military resources—and then only if other regional players do not have the wherewithal and the will to perform this task themselves.

Relatedly, the third proposition of offshore balancers is that Washington should follow a classic *balance-of-power policy*, loosely modeled on what Great Britain did for centuries vis-à-vis Continental Europe. The vast majority of U.S. military assets can and should be kept as a deterrent at home—"over the horizon"—to be deployed overseas only if developments in other regions make the security environment there unraveling for real. Until that eventuality manifests, it would be far wiser if the United States instead "passed the buck" to regional allies and partners—such as Japan, South Korea, Germany, France, Israel, and Saudi Arabia—letting them bear the costs of balancing any menacing regional power (Mearsheimer & Walt, 2016, p. 71). Such a conscious policy of burden-sharing, or even of divide-and-conquer, would help fulfill the same goals that Washington pursues today, but at a much lower cost.

Fourth, the implementation of such a grand strategy would greatly reduce the need for forward-deployed U.S. troops in these regions (Gholz et al., 1997; Posen, 2013, p. 125). Most offshore balancers, to be sure, would recommend keeping some smaller forces in vital regions, especially around the rimland of Eurasia; and any large-scale withdrawal should happen only gradually. Still, only with respect to Asia could one make a plausible case that the power potential of a regional great power—China—could possibly require a significant permanent U.S. presence, as a deterrent and a "tripwire" (Walt, 2018, p. 269). But even there, in the first instance, the United States should seek to pass the buck to regional states whose incentives to balance China will be much more pressing (Schweller, 2018, p. 26; Walt, 2018, pp. 262–263). Instead, Washington should emphasize a broader portfolio of non-military assets to counteract China ambitions and to support its regional allies—such as diplomacy, intelligence gathering, and geoeconomic tools.

The Primacists

Those who favor continued U.S. primacy disagree with offshore balancers on all four accounts. First, geography does not univocally dictate isolationism. Even if it is true that U.S. seclusion adds considerably to its security, it thereby also presents it with unparalleled latitude for creating and maintaining military forces with a *global* reach. Almost all other major powers throughout history have consistently faced severe security threats at their borders. These threats have especially come in the form of the land forces of other, geographically proximate major powers, thereby vastly increasing requirements for territorial defense (Levy & Thompson, 2005). The United States, on the other hand, have long since established an extended defense perimeter; and it has steadfastly maintained this gigantic buffer zone that has worked to push the threats to its security far away from the Western Hemisphere.

Second, the costs undertaken by upholding this extended defense perimeter are, according to the primacists, more than offset by the benefits that accrue (Brands, 2015, pp. 2–3). The crux of the matter is to forestall the risks and threats *before* they materialize and spread (Brands & Edel, 2019; Mandelbaum, 1988, Chapter 3). This should be done by way of forward presence and advanced action whose scope goes beyond offshore balancers' more narrow focus on aspiring regional hegemons. Potential threats to U.S. security are plentiful, and many of them are as yet unknown. Without deep U.S. engagement, primacists hold, what seem like small problems today can easily metastasize and eventually threaten U.S. security: an ostensibly peaceful nuclear-development program can be turned into a full-blown nuclear-weapons capability; a failing state today can become a haven for transnational terrorists tomorrow; the overturning of a democratic regime can lead to the regional diffusion of autocracy; a diplomatic quarrel between two states can escalate into a shooting war; and an expanding regional major power can use its increased power to expand even further, just as Japan did before (and after) Pearl Harbor. A topic to which I will repeatedly return in later chapters, primacy is in many ways an insurance policy that seeks to nip such developments—the known and the unknown—in their bud. As Stephen Brooks et al. (2012–2013, p. 9) write in their *International Security* article that bears the telling title "Don't Come Home, America": "The United States' globe-girdling grand strategy is the devil we know, and retrenchment advocates effectively identify some of its risks and costs.

A world with a disengaged United States is the devil we don't know, and [it] present[s] much greater risks and costs."

Third, primacists consequently advocate a forward-leaning posture, and they essentially highlight the benefits of power or influence maximization. Geographically secure, the United States should stay deeply embedded in other regions. This, primacists say, increases the credibility of deterrence against adversaries; it helps reassure friends and allies that Washington is genuinely committed to upholding world and regional orders, thereby enhancing stability and alleviating the (perceived) risks of conducting economic exchange; it acts as a restraint on partner countries whose foreign and domestic policies would likely become more reckless without the presence and dependable engagement of the United States (Cha, 2009–2010); and, more generally, far from primarily being a grand strategy geared toward fighting and winning wars, primacy instead helps the United States reach its objectives "without recourse to war" (Huntington, 1993, p. 70). It is an offensive posture, but for eminently defensive purposes (Brooks & Wohlforth, 2016b, pp. 9–10): active engagement abroad, or so is the assertion made, makes America more secure and prosperous. The mechanisms involved are therefore not the same ones as under the notorious security dilemma, where one's offensive moves elicit offensive countermoves by others, decreasing security for all. To the contrary, primacists emphasize, U.S. engagement and U.S. hegemony are actively solicited by most of the world's states, which essentially see the United States as a non-threatening sea power whose global influence is to be embraced for the security and stability it brings (Levy & Thompson, 2010).

Fourth, the implication of all this is that primacists advise a *continuation of U.S. military presence* in important regions and countries. There is no disagreement between the primacists and offshore balancers that the most vital areas to U.S. national security are Europe, the Persian Gulf, and East Asia (along with the Americas, where the U.S. rules supreme even without a heavy troop presence). But the two camps differ sharply on the question of the necessity of keeping large amounts of forward-deployed soldiers. The offshore balancers want most of them withdrawn, in particular from the Gulf and from Europe; the primacists want them to stay because they alleviate a range of risks and threats that offshore balancers are reluctant to acknowledge as key impediments to U.S. security in the first place (Brands, 2015, p. 56). For the primacists, U.S. military bases in Germany, Italy, Spain, and Great Britain, among other European host countries, act as a warning to Russia that its growing assertiveness comes with significant costs; they allow for a rapid reaction to any security crisis in Europe and in nearby regions; and they help shape the foreign-policy preferences of the Europeans so that they align more closely with Washington's perspectives and objectives than they otherwise would. U.S. bases in Kuwait, Qatar, Bahrain, and elsewhere in the Middle East dampen the risk of conflict in an area notoriously known for interstate rivalry and war; they deter Iran's ambitions to develop nuclear weapons and export its revolution; and they greatly help keep *jihadi* terrorism in check. And in Asia, tens of thousands of U.S. troops stationed in South Korea, Japan, and on the Pacific island of Guam ensure that China's expansionist drive is kept within limits; they ease regional security competition; and they increase the expected costs for North Korea of making overly audacious foreign-policy moves.

These are the analytical and normative claims of offshore balancers (the *Leavers*) and primacists (the *Stayers*), respectively. The descriptive *reality* is more unambiguous: the United States does engage deeply with the world, and it has done so for three-quarters of a century. Its global involvement is also multifaceted. It spans the whole gamut of foreign-policy tools and acts: diplomacy; commitment to and often leadership of international institutions; the provision of global public goods; management of a number of defense and security alliances; peace brokering; the use of economic carrots and sticks for geopolitical objectives; the issuing of threats against bad actors and the promises of rewards for those who do good; intelligence partnerships; technological leadership; security and military assistance to partner countries; the involvement in overseas wars; and, not least, the presence of hundreds of thousands of U.S. troops stationed at overseas military bases scattered around the world.

Domestic Forces Counteracting Retrenchment

As is the case for most phenomena in international affairs, this long-lasting continuity—which we can call *hegemonic* or *great-power persistence*—is down to several forces located at more than a single level of analysis, though some are bound to be more critical than others. My own explanation or theory, which I will present in more detail over the next three chapters, revolves greatly, albeit not exclusively, around features and mechanisms at the system level. Closely affiliated with the neoclassical realist school, its main argument is that foreign-policy and geostrategic behavior is primarily formed as a reaction to rather shadowy system-level signals that are filtered through the minds of risk-averse decision-makers. Drawing also on prospect theory, I argue that these decision-makers, like human beings in general, are inclined to overweight looming losses more than possible gains and to prefer the prevailing status quo over its alternatives. The broader foreign-policy community regularly employ worst-case-scenario thinking in matters of national security; they have long been particularly prone to emphasize heavily the *risk* of adverse geopolitical chain reactions being set in motion—in what is crucially an anarchical and interdependent system—if forward-deployed troops are withdrawn from vital regions.

As Sebastian Schmidt (2020, p. ix) rather curtly puts it in a recent book: "For Americans, the idea of having military bases all over the globe was not any more out-of-the-ordinary than drinking coffee in the morning." Retrenchment proposals are scoffed at by the "Blob" as something very unnatural and quite damaging. Virtually any move to withdraw troops from anywhere is almost certain to incite loud warnings by the bulk of the foreign-policy establishment. The members of this broad elite vociferously stress all the various things that *might* go wrong and that *might* harm U.S. national security if troops are pulled out. These voices, and the consistent risk- and change-averse predilections they harbor, pose a formidable barrier against any real change to U.S. grand strategy. In general, great powers rarely retrench their overseas military power voluntarily; more often, they seek to expand; and, more often still, they attempt to uphold the status quo and their prevailing power positions, considering that their relative command of power resources normally changes only very slowly.

My theory, thus, takes as its point of departure system-level properties. This is not to deny that other forces play their part in upholding the U.S. military-base network. Because they do, although by how much is surely debatable, and their impact is bound to vary among cases. This is an issue that will receive its due treatment in the sections below, as well as in the empirical analysis. One thing we *can* quite confidently assert here, though, is that the United States is only rarely forced by host societies or governments to dismantle or to leave bases. That is to say, the "push" factors have, so far at least, carried relatively limited weight in the grand scheme of things.

Trump: The Firebrand Upholder of Continuity?

The period from early 2017 to early 2021 in many ways represents an ideal test case for the issues and mechanisms I propose. For the first time since U.S. leaders embarked on a grand strategy of global engagement about two-and-a-half generations ago, U.S. voters elected a self-avowed, clear-cut *Leaver* as president. This, furthermore, coincided with widespread perceptions that the United States was a power in relative decline. With only two possible exceptions—the end of the Second World War and the end of the Cold War—the last three-quarters of a century have not witnessed any other time period where the issue of retrenchment has been nearly as relevant and high up on the policy agenda as it was in the era of Trump.

Donald Trump was inaugurated as the 45th President of the United States on January 20, 2017. As accounted for in Chapter 1, his application for the job contained a glaring and unprecedented dose of retrenchment promises. For him, the U.S. military-base network represented a brand of morning coffee that was exorbitantly costly and all but superfluous, if not even downright harmful. It was coffee at the price of the most expensive champagne, and it contained lethal poison to boot. In more than one sense, therefore, Donald Trump actually resembled an offshore balancer. His was *seemingly* a foreign policy of change and of restraint; it was unconventionally conveyed by the standards of the Blob, but it was still one that was, on the face of it, not without promise for the *Leavers*. According to Stephen Walt (2018, p. 11), a distinguished offshore balancer, Trump's core message was that, under his tutelage, America would follow a vastly different grand strategy. But America didn't get a different grand strategy, was Walt's *ex-post* verdict. What they instead got was more of the same from a president essentially reined in by the Blob. U.S. strategy was still misguided, but now it also had as "captain of the ship of state (…) an ill-informed and incompetent skipper lacking accurate charts, an able crew, or a clear destination [whose] handling of defense policy was simply 'business as usual,' with a bit more money and a few more bombs" (Walt, 2018, pp. 219, 233).

Several other scholars also denied that there was much or any systematic thought behind the new president's rhetoric and policies. Emphasizing more the *lack* of any overarching grand strategy, they pointed out the erratic and unpredictable nature of Trump's leadership and initial foreign policies. "Uncertainty, unpredictability and confusion" would mark this presidency, early prognoses went (Huxley & Schreer, 2017, p. 82). To the extent that there was some direction in Trump's foreign policies, that direction did not constitute any *Weltanschauung*; for the president, foreign policy was purely a status-driven endeavor where short-term thinking, the craving for respect, and extreme symbolism substituted for policies of any substance (Wolf, 2017).

Of interest here is also the observation, as claimed by some analysts, that Trump's initial months in office had resulted in little change in a foreign policy that was purportedly "remarkably unremarkable" (Abrams, 2017, p. 10). In an early study of nineteen foreign-policy cases from the first year of the Trump presidency, Paul MacDonald (2018) found that about half of these represented clear continuity from policies pursued by the preceding administration. The pre-election rhetoric had seemingly given way to a more moderate way of conducting foreign affairs. The new administration, some held, had engaged in a process of "emergent learning" (Popescu, 2018). Faced with the realities of the job as president, Trump did not withdraw any troops from Afghanistan, but instead sent more; attacks against the Islamic State in Syria and Iraq continued; no security alliance was abrogated; NATO's eastern flank was soon bolstered as an additional deterrence against Russia; and so on.

What also soon surfaced to a greater degree than previously was the ***militarism*** of the president. This was perhaps an innate characteristic of Donald Trump. In a 1990 interview with *Playboy Magazine*, he expressed a strong belief "in extreme military strength," asserting that a future President Trump(!) would "have a huge military arsenal, perfect it, understand it" (Plaskin, 1990). The president's December 2017 National Security Strategy (NSS), whose purpose is to communicate the direction of the government's national-security vision, also focused on the need to "retain military overmatch" (The White House, 2017, p. 28). In fact, in that and in most other respects, it displayed considerable continuity with previous such documents. It is always the case that an NSS contains ample words to the effect that the United States must ensure that "our military power is second to none"; adversaries are to be quite certain "that we can and will defeat them – not just punish them if they attack the United States," so "that they cannot accomplish objectives through the use of force or other forms of aggression" (The White House, 2017, pp. 26, 28). This signals that primacy, in the sense of upholding a level of military capabilities "second to none," is an ingrained feature in U.S. national-security thinking. The "pronounced militarist bias" of Trump and his administration, then, suggested that foreign- and security-policy orthodoxy was not necessarily under serious threat after all (Harris, 2018, p. 612). Primacy or hegemony was to be upheld, it looked like, even if it would not any longer amount to a *liberal* hegemony; rather, in the words of Barry Posen (2018, p. 25), it would be "illiberal hegemony," or "[p]rimacy without a purpose." What could change, however, was the emphasis placed on the deployments of forces *overseas*, information about which the National Security Strategy contained less.

It *is* possible to devise a military posture that allows for overmatch in military capabilities while it simultaneously keeps the number of forward-deployed troops low. But major paradoxes would nonetheless apply (Harris, 2018). For example, what is the *objective* or *purpose* of this overwhelming military force? *How* and *where* is this $700-billion military power to be used? If the United States is inherently secure behind two world oceans in a region with no aspiring great power, what is the point of keeping—indeed increasing the strength of—armed forces whose total budgets surpass the military expenditures of the next ten largest militaries in the world *combined* (SIPRI, 2021a). If a significant part of these forces aren't to be forward-deployed so as more credibly to deter and to contain burgeoning threats far away from America, what are they really there for in the first place? The caveat that here arises with respect, first, to the empirical question of whether or not retrenchment under Trump was something that could plausibly be expected and, second, to the theoretical issue concerning the main driving forces of continuity (and/or change), is that Trump's was an administration that might have been so whimsical and incoherent, and so lacking in grand-strategic direction—but also so bent on militarism—as to render retrenchment, for all intents and purposes, a non-issue.

Trump's Transactional Diplomacy

Alternatively, coherence and direction might have applied to the Trump administration, but these were features masked by the conscious creation of a mismatch between rhetoric and actual objectives. One possible interpretation of the Trump presidency is that withdrawal was not really a serious goal per se; instead, the reasoning underpinning the ample warnings that troop withdrawals were in the cards might have been based purely on pecuniary concerns.

This "transactional" perspective certainly permeated debates, speeches, Twitter messages, and negotiations with host countries (Cha, 2019). In Trump's view, the matter of U.S. soldiers abroad concerned, perhaps more than anything else, agreeing upon formulas for the sharing of basing costs that worked much more in the favor of the United States. References to the costs of stationing troops overseas, and to the need for changes in burden-sharing, abounded both before and during his presidency. In Trump's words, "it's so sad, but we have $7 trillion in the Middle East" (The White House, 2018), and becoming involved there in the first place was the "worst decision ever made" (France24, 2019). About South Korea, the United States purportedly "lose money on trade, and we lose money on the military. (...) [A]t some point (...) I want to bring our soldiers back home" (Corouсher, 2018). With respect to Japan, "they have to pay us or we have to let them protect themselves" (Gerzhoy & Miller, 2016). And, directed more generally at major hosts of U.S. troops: "We defend Japan. We defend Germany. We defend South Korea. We defend Saudi Arabia. We defend countries. They do not pay us what they should be paying us because we are providing a tremendous service and we're losing a fortune" (Hu, 2016).

Still, on balance, the evidence suggests that the withdrawal threats were very real ones indeed, and that the ultimate failure to retrench in any significant way was caused by forces that were only to a limited degree associated with the transactional—or the whimsical—nature of Trump and his administration. For one, the financial concerns of the president were, more often than not, mixed with serious question marks being raised about the actual *purpose* of U.S. commitments abroad. For example, in December 2018, while visiting U.S. troops in Iraq, President Trump lamented that America is "spread out all over the world (...) in countries most people haven't even heard about. Frankly, it's ridiculous. (...) It's time to get our young people out" (Stokols & Wilkinson, 2018).

The *explicit threats to pull out troops* were clearly one of the key factors which separated Trump from previous administrations. Still, accusations that allies and hosts of U.S. troops free-ride on America, and that the share of the defense burden taken on by the United States is disproportionately large, are certainly nothing new. In particular, the relationship between the United States and the rest of the North Atlantic Treaty Organization (NATO) has seen a "perpetual burden-sharing debate" (Ringsmose, 2010, p. 319) at least since the 1960s (Tonelson, 2000). A relevant question then becomes whether the threats of a withdrawal of

U.S. troops primarily constituted a bargaining tactic; that is, whether the threats were credible. The empirical analysis in later chapters will provide some answers. With respect to allies' perceptions, these certainly didn't foreclose the possibility that the threats indeed *were* inherently credible, and that retrenchment was a real objective (Cha, 2019; Malnick, 2018). This argument is supported by those scholars who early on stressed the very real possibility of a "Fortress America" being constructed by the Trump administration (Brands, 2017), involving at least a major *reduction* of U.S. force projection in key regions (Clarke & Ricketts, 2017). One strand of analysts pointed out that Trump had for decades expressed opinions suggesting that he really held a fairly consistent worldview, something of a grand-strategic thinking (Clarke & Ricketts, 2017). Long-held beliefs of his, which were unmistakably carried over into his presidential term, included that allies were taking advantage of America and had been doing so for decades. Trump had also long argued that multilateral institutions, including defense alliances, embedded within a world constructed on liberal internationalist principles did little but work *against* U.S. national interests, so they needed to be upended (Cooley & Nexon, 2020, Chapter 7). Hence, Trump, according to some, was set on a grand strategy not of pure isolationism, but rather *neo-isolationism* (Cha, 2016); or, in what are perhaps more moderate expressions, "principled realism" or "populist sovereignty" (Ettinger, 2020). Walter Russell Mead (2017, pp. 3–4) called it "Jacksonian populist nationalism," an ideological brand rooted in American history and culture that was "only intermittently concerned with foreign policy," insisting instead that "the role of the U.S. government [was] to fulfill the country's destiny by looking after the physical security and economic well-being of the American people in their national home."

However it was labeled, U.S. foreign policy could thenceforth really be expected to place America's "sovereignty" and prosperity first, with a businessman at the wheel who would confront issues in international politics with a view of using his or his country's leverage in order to get a better deal—whatever the issue at hand. This transactional approach would not only be directed against adversaries and competitors; for neither could allies any longer expect economic and security deals, or Washington's benevolence, to be carved in stone. A withdrawal of security guarantees and troops could now credibly be threatened, a possibility that would presumably make allies such as Germany and South Korea bend to

America's will (Hadar, 2017). It was now time for a *unilateralist* U.S. approach to world affairs.

Yet, the proof of the pudding is mainly in the eating. At the end of the day, the issue about withdrawal-threat credibility is an empirical one that will be treated at more length in Chapters 6 and 7. Suffice it for now to point out that, for purposes of testing my arguments, it is of great interest that the wider foreign-policy community (as well as host countries and allies) have *reacted* or *acted* on the belief that Trump and his team were indeed aiming to withdraw troops, which they in most instances certainly have done. This is notably reflected by the narratives formed in the cases of particular relevance. It is admittedly clear that a few of these withdrawal suggestions were quite far from being realized (Qatar is arguably one such case), and some were probably driven in part by bargaining-leverage concerns (Japan likely falls into this category). In the latter case, though, unless Trump's bargaining strategy was solely one of consistent and audacious bluffing, an unsatisfactory conclusion to the bargaining process should result in him leaving the negotiating table, which here implies retrenchment. Or, as Trump himself has argued more than once, and not necessarily unreasonably: "you always have to be prepared to walk" (Bump, 2016).

Quite a few other cases, however—including South Korea, Iraq, Syria, Afghanistan, Somalia, and Germany—are more clear-cut "ideal-type" ones for these purposes. For example, In Syria, a withdrawal was ordered and partly executed, but then almost immediately (if only partially) reversed (DeYoung et al., 2019; Schmitt & Haberman, 2019). The reversal of policy followed, and was no doubt in large part caused by, a plethora of warnings, coming from a variety of angles and actors, about the avalanche of negative geopolitical consequences that *might* ensue from retrenchment. The initial decision *not* to withdraw from Afghanistan in 2017–2019 was, according to Trump (and this seems very plausible), taken only because he actually did listen to and heed such warnings and narratives himself (BBC, 2017; Mehta & Copp, 2017). But as we now know, the Afghan withdrawal was only paused, not abandoned, as a core objective of the president; and by the end of his tenure, it was had progressed significantly. By way of a conclusion, therefore, there is much reason to brand Donald Trump the president that really sought retrenchment—he did and he does belong to the *Leavers*.

Institutional Inertia and Domestic Constraints on a Leaver

Granted, even if one accepts the view, as I do here, that Trump was bent on carrying out a significant withdrawal of troops, this does not in itself negate the possibility that other non-systemic forces can plausibly account for the observed lack of retrenchment. A president is certainly not unconstrained. Coming into the job, he or she needs to adjust to a domestic structure of national security that has remained fairly constant for decades (Dombrowski & Reich, 2018, p. 60). This shapes decision-making in the direction of continuity, even if the amount to which it does so varies among presidents. There is a limit to how much a president can ignore the expertise that is contained in this structure when the many different bits of U.S. foreign policy are to be implemented.

Stephen Walt, for his part, would hesitate to use the word "expertise" at all. He prefers instead to alternate between the rather demeaning label of "the Blob" and the more neutral phrases of "foreign policy community" or "foreign policy professionals," by which he means "*individuals and organizations that actively engage on a regular basis with issues of international affairs*" (Walt, 2018, p. 95, italics in the original). In *The Hell of Good Intentions*, Walt vehemently contends that U.S. foreign policy is on the wrong path. And it is so, he says, because of just such constraints on leadership that stem from a thoroughly established set of domestic voices and interests that work in tandem to uphold continuity in foreign affairs. These voices and interests comprise, inter alia, of formal institutions of government (including, e.g., Congress, the intelligence community, and the U.S. military); membership organizations; think tanks; interest groups and lobbyists; the media; and large parts of the academic sector. Even if this is a rather motley crew, they all share the aim of keeping U.S. foreign policy on the track it has followed for two-and-a-half generations: preserving U.S. primacy, especially in terms of military preponderance; keeping or even expanding U.S. spheres of influence globally; and promoting liberal norms such as democracy and human rights, again globally. This world-spanning agenda certainly does not mesh well with any notion of retrenchment or withdrawal of U.S. overseas troops. The "marketplace of ideas" in foreign affairs, Walt claims, is rigged by the Blob. The arguments used to ensure foreign-policy continuity, he goes on, are largely self-serving, whether consciously so or not: they help a plethora of people in the broader foreign-policy community keep their jobs and enhance their careers. Insofar as his analysis is close to the mark, therefore, the observed continuity of U.S. troop deployments may primarily be caused by domestic-level forces—or even individual-level ones, if one fully embraces the view that what first and foremost drives continuity is the selfish-cynical, career-furthering motivations of this group of professionals.

Real motivations and intentions cannot be fully unveiled, however. What we should, in the first instance, instead focus on are the words and the deeds of this elite. And it is quite clearly the case, as Walt spends ample time on in his book, that these domestic constraints on the president's foreign-policy discretion are also closely linked to *international* events and developments—or, more often, to *potential* or *hypothetical* international events and developments. *The things that might go wrong* if the foreign policy changes are a—or even *the*—selling point of U.S. foreign-policy professionals (the sales aspect Walt deals specifically with in the fourth chapter of his book, entitled "Selling a failing foreign policy"). As another scholar notes, "because international politics is a complicated and unpredictable realm, presidents may be hesitant to depart dramatically from their predecessors. Presidents may be dissatisfied with the status quo, but they may be unwilling to shift policies too quickly or decisively" (MacDonald, 2018, p. 407). The "complicated and unpredictable realm" of foreign affairs is a reality that is unrelentingly grasped by the foreign-policy community, who use it to focus policymakers'—including the president's—attention to the overabundance of threats that are seemingly out there. To return to Walt, the Blob exploit this uncertainty: they inflate threats, big and small alike; they overstate the power of adversaries; and they understate the capabilities of allies (Walt, 2018, Chapter 4). In short, they make the highly unrealistic claim that a retrenching America is an America that will soon witness today's threats multiply and metastasize, to the clear and likely irreversible detriment of the security of the country.

For the most part, I agree with Walt's assessment that the U.S. foreign-policy community is inclined to engage in threat inflation. I would still nuance two of his points. First, with respect to levels of analysis, I argue that the international events themselves—including both those which have materialized and those which are still only in the realm of possibilities—can and should be viewed as the main driving forces of foreign-policy continuity. But importantly, actual and hypothesized events and developments on the international arena are *filtered* through the minds of (mostly but not exclusively) U.S. domestic actors, who interpret these events, or events-to-be, who exaggerate them (as is Walt's reasonable claim), and who utilize them in order to decidedly contribute to shaping U.S. foreign policy. Second, and on this our disagreement is probably clearer, I am inclined to empathize with the Blob to a greater degree than is Walt. I believe that members of the foreign-policy community generally and quite genuinely believe what they say and what they fear. And I also find it quite understandable, and often justifiable, that they tend to inflate threats, that they recommend that their country and its president often act on worst-case-assumption scenarios, and that they express a very heightened sensitivity toward *possible* adverse geopolitical chain reactions and falling dominoes. As the next three chapters will hopefully explicate, we possess ample theoretical (and empirical) knowledge that, when taken as a whole, can help show and explain that such a pessimistic stance, and such an inclination to see dangers were perhaps there aren't many or any, is not necessarily unreasonable—or predominantly self-serving—after all.

Whether or not they are rooted in judgments about system-level events and developments, it is nonetheless undoubtedly so that domestic constraints are important. On the other hand, the president, who is also the Commander-in-Chief, does have considerable leeway in foreign affairs, and usually more so than in most areas of domestic policymaking. In recent years, moreover, political polarization has often left the U.S. Congress paralyzed, effectively increasing the power of the executive. The incoming president, of course, brings his or her own personality into the job. For most new presidents, that does not lead to any radical changes in the country's direction; but for others, those to whom Joseph Nye refers as "unfiltered" presidential choices, structural constraints have less effect on their behavior and agenda. Personalities as different as Abraham Lincoln and Donald Trump may be placed in this category of leaders whose "range of (…) performance is wider" (Nye, 2019, p. 78). Moreover, the president assembles his own team of staffers, which come in addition to all those political appointees that the U.S. bureaucratic structure permits (MacDonald, 2018, p. 405). International relations scholar and former U.S. foreign-policy advisor Victor Cha (2019, p. 173) poses the question of whether anyone could really stop Donald Trump from pulling out U.S. forces from a country such as South Korea. Cha's own answer is simply that no one really can, and "that there is very little standing in his way. As the Commander-in-Chief, the president exercises final and unchallenged say over military matters."

U.S. Public Opinion

The U.S. public represents an additional domestic-level force that may increase the executive's inclination to uphold the foreign military-base network. For example, in a September 2020 poll conducted by the Chicago Council of Global Affairs, a majority of respondents expressed a wish to see the United States *more* rather than *less* involved in global affairs. On that question, and on questions regarding the maintenance and strengthening of alliances, public support for U.S. engagement seems still to be considerable, well "in line with some of the highest readings in [the] survey over the past 46 years" (Daalder & Smeltz, 2020).

Yet, in a poll conducted only a few weeks earlier, three-quarters of U.S. adults said they supported a withdrawal of troops from Afghanistan and Iraq (Kheel, 2020). In fact, and somewhat bewilderingly, other periodic surveys suggest that more Americans than ever agree with the statement that the "U.S. should mind its own business internationally" (Brooks & Wohlforth, 2016b, p. 6; see also Macintyre, 2019). It seems that the U.S. public is generally, in Robert Kagan's words, "torn to the point of schizophrenia" on the issue of Washington's foreign policy and global responsibilities. And it is very likely that it is just this schizophrenia that is reflected in the polls referred to above, as well as in other, earlier surveys of U.S. public opinion (Pew Research Center, 2016). Americans, says Kagan (2012, p. 14),

> are reluctant, then aggressive; asleep at the switch, then quick on the trigger; indifferent, then obsessed, then indifferent again. They act out of a sense of responsibility and then resent and fear the burden of responsibility they have taken on themselves. Their effect on the world, not surprisingly, is often the opposite of what they intend. Americans say they want stability in the international system, but they are often the greatest disrupters of stability. They extol the virtues of international laws and institutions but then violate and ignore them with barely a second thought. They are a revolutionary power but think they are a status quo power. They want to be left alone but can't seem to leave anyone else alone.

This is what Henry Kissinger (2014, p. 279) calls the "congenital ambivalence" of U.S. public opinion, a characteristic of U.S. voters that, insofar as it is a reasonable one, makes it hard to argue that making promises of continued primacy represents any *sure* way of winning elections.

Americans seem to want engagement and non-engagement—often simultaneously! Global engagement does have significant backing in U.S. public opinion. However, so has retrenchment. As presidential candidate, Trump chose to pander to the latter of this duality of sentiments, propounding a message that certainly resonated with a major share of voters. This share is perhaps of particular import when wars are perceived to go badly. In fact, the question of "why are we really there?" is one that is not only asked about the big wars America fights and doesn't win, such as in Vietnam, Afghanistan, and Iraq; it also arises when the United States suffers fatalities, even if only a handful, in smaller conflicts equally symbolic of the country's globe-spanning activities, such as in Somalia (in 1993) and Niger (in 2017). Cases like these perhaps both reflect and help cause the U.S. public's "schizophrenia" with respect to their nation's global engagement.

Donald Trump was elected president in no minor part because his preferences resonated well with the retrenchment sentiments of a large share of U.S. voters who had become disillusioned with the foreign-policy elites in Washington (Sanger, 2019). Parts of Trump's message on retrenchment, it should also be said, did not represent any complete break with his predecessor's policies. For example, Barack Obama consistently refrained from placing substantial numbers of U.S. troops in Syria. This appeared to signal, more broadly, that "the days of sizeable US troop deployments in the region were over, even to deal with a clear security threat" (Phillips, 2016, p. 212). Nevertheless, Trump proceeded to take the retrenchment rhetoric several steps further, even repeatedly threatening to withdraw U.S. commitment to the North Atlantic Treaty's critical Article 5 (Birnbaum & Rucker, 2018; *New York Times*, 2017). This was as radical a proposal and threat as they come, but it did not seem to scare away the majority of voters.

Vested Interests at Home and Abroad

Finally, there also exist a variety of vested interests that want to keep the military-base infrastructure in place, or even to expand it. These interests range from private firms engaged in base construction, maintenance, or services (Herrera & Cicchini, 2013); to local communities in host countries dreading the effects that base closure would have on the economy and on employment rates (Allen et al., 2020; Blankenship & Joyce, 2020); and to what Kent Calder (2007, p. 36) eloquently refers to as "[i]nertial bureaucratic and political forces, operating on a logic akin to Newton's second law on thermodynamics, [which] cause bases to continue in motion, once conflict has initiated or rationalized them, until stopped by some more powerful outside force."

As for economic interests, it has been quite common to supplement base treaties with other types of agreements between the United States and host countries. These include the signing of preferential trade agreements and deals requiring Washington to purchase goods and services from local economic actors (Blankenship & Joyce, 2020). They are all arrangements that create economic relationships, and hence dependencies, between sender and host. Usually the dependencies are asymmetrical, meaning that the host country or communities stand to lose quite heavily if the United States withdraws, whereas the direct economic stakes for actors in the United States are comparatively meager. For example, when President Trump in the summer of 2020 announced his intention to withdraw a large share of the U.S. troops stationed in Germany, this created widespread *economically*-based fears in German *Länder* such as Bavaria and Baden-Württemberg. One town that was set to be particularly affected was Grafenwöhr in Bavaria, where almost half of the population is dependent on the U.S. military base for employment (Connolly, 2020).

With respect to the interests of *U.S.* economic actors, the arms industry is likely the main benefactor of close-knit security partnerships between the United States and other countries. For example, a quick analysis of arms-trade data from the Stockholm International Peace Research Institute (SIPRI, 2021b), combined with data on U.S. forward-deployed troops, reveals that, after Donald Trump assumed the U.S. presidency, and amid threats that the administration was contemplating a withdrawal of a substantial number of troops from several host countries, U.S. arms exports increased markedly to states hosting more than 250 U.S. troops. The period 2016–2019 saw a 43% increase in arms sales to this group, whereas the same period saw a 50% *decrease* in arms exports from the United States to those countries hosting less than 250 U.S. troops.

On the other hand, while it is clear that many private economic actors have huge stakes in the U.S. military-base network, it is quite unlikely that these actors have much real leverage when it comes to shaping Washington's overall defense and security policies. This is unquestionably the case regarding economic actors and local communities in the host countries. For example, there is nothing to indicate that the *economic* concerns of

German communities or businesses in any way helped cause that withdrawal process to fizzle out, as it eventually did. Neither is it a plausible argument that the U.S. government is "captured" to any major extent by domestic private economic interests in the United States. In fact, the separation between the market and the state is integral to the way U.S. institutions work and U.S. ideology is perceived and internalized by government officials (and voters, for that matter). This observation, for example, forms a common thread in Steve Coll's magnificent book about the recent history of U.S. oil major ExxonMobil (and its relationship to U.S. power and foreign policies); the company has been more frustrated than happy with the support it has received from U.S. diplomats, foreign policymakers, and military and intelligence services, and it has little real influence over these entities (though its sway over some of the host countries in which it operates has been considerable) (Coll, 2012). And even though the U.S. arms industry ultimately reaped some significant rewards from the pressure put by the Trump administration on several Gulf Arab states, this whole process of bargaining and withdrawal threats was orchestrated in Washington, as opposed to in any corporate headquarters. Equally importantly, they were orchestrated and effectuated for quite specific objectives associated with U.S. *national* security and economic interests.

*

It is nonetheless undoubtedly so that the United States draws some significant benefits from deploying troops abroad, including economic ones. In general, "[d]efense commitments and overseas military bases, though costly, augment U.S. influence abroad and contribute to upholding *Pax Americana*" (Jakobsen & Jakobsen, 2019, p. 158). Substantial rents accrue to the hegemon of the system, such as "preferential access to, better knowledge of, and superior bargaining power in global transactions and communications (...) and a capacity to 'set the rules' in world affairs" (Modelski, 1978, pp. 227–228). These are benefits parts of which Washington would have to forego if retrenchment were to progress. At a bilateral level, moreover, most host countries reciprocate in both tangible and intangible ways for the security that the U.S. troops provide (Lake, 2009; Morgan & Palmer, 2003; Morrow, 1991). More wealthy allies, for instance, usually help finance a substantial portion of the basing costs. Military bases and troops typically also give the United States considerable leverage over important political and economic decisions made by the host, especially in the area of foreign, security, and defense policies.

However, these benefits accrue to the United States *courtesy of properties and mechanisms belonging to the system level of analysis.* They concern leverage; bargaining power; global, regional, and bilateral influence; regional and world order; and *Pax Americana*. In short, they concern the prevailing *status quo* in world affairs, which is by the bulk of the U.S. foreign-policy establishment seen as absolutely vital to preserve. This is the status quo that is backed up and bolstered by U.S. material resources, a core part of which is constituted by U.S. military power and the plethora of U.S. overseas military bases and troops. It is the *possibility* that the status quo may start to unravel, which *may* set in motion further adverse geopolitical chain reactions, that underpins this foreign-policy establishment's consistent and forceful opposition to the withdrawal of U.S. troops—of virtually *any* U.S. troops from virtually *anywhere*.

References

Abrams, E. (2017). Trump the traditionalist: A surprisingly standard foreign policy. *Foreign Affairs, 96*(4), 10–16.

Allen, M. A., Flynn, M. E., & Martinez Machain, C. (2020). Outside the wire: U.S. military deployments and public opinion in host states. *American Political Science Review, 114*(2), 326–341.

Baldwin, D. A. (1979). Power analysis and world politics: New trends versus old tendencies. *World Politics, 31*(2), 161–194.

Barnett, M., & Duvall, R. (2005). Power in international politics. *International Organization, 59*(1), 39–75.

BBC. (2017, August 22). *Trump rules out Afghan troops withdrawal*. https://www.bbc.com/news/world-us-canada-41008116

Beckley, M. (2018a). *Unrivaled: Why America will remain the world's sole superpower*. Cornell University Press.

Beckley, M. (2018b). The power of nations: Measuring what matters. *International Security, 43*(2), 7–44.

Birnbaum, M., & Rucker, P. (2018, July 12). At NATO, Trump claims allies make new defense spending commitments after he upends summit. *Washington Post*. https://www.washingtonpost.com/world/europe/trump-upends-nato-summit-demanding-immediate-spending-increases-or-he-willdo-his-own-thing/2018/07/12/a3818cc6-7f0a-11e8-a63f-7b5d2aba7ac5_story.html

Blankenship, B., & Joyce, R. M. (2020). Purchasing power: US overseas defense spending and military statecraft. *Journal of Conflict Resolution, 64*(2–3), 545–573.

Brands, H. (2015). *The limits of offshore balancing*. U.S. Army War College Press.

Brands, H. (2016). *Making the unipolar moment: U.S. foreign policy and the rise of the post-Cold War order*. Cornell University Press.

Brands, H. (2017). U.S. grand strategy in an age of nationalism: Fortress America and its alternatives. *Washington Quarterly*, *40*(1), 73–94.

Brands, H., & Edel, C. (2019). *The lessons of tragedy: Statecraft and world order*. Yale University Press.

Brooks, S. G., Ikenberry, G. J., & Wohlforth, W. C. (2012–2013). Don't come home, America: The case against retrenchment. *International Security*, *37*(3), 7–51.

Brooks, S. G., & Wohlforth, W. C. (2016a). The rise and fall of the great powers in the twenty-first century: China's rise and the fate of America's global position. *International Security*, *40*(3), 7–53.

Brooks, S. G., & Wohlforth, W. C. (2016b). *America abroad: The United States' global role in the 21st century*. Oxford University Press.

Brzeziński, Z. (2012). *Strategic vision: America and the crisis of global power*. Basic Books.

Bump, P. (2016, July 26). Donald Trump's explanation of when he'll walk away from military partners, annotated. *Washington Post*. https://www.washingtonpost.com/news/the-fix/wp/2016/07/26/donald-trumps-explanation-of-when-hell-walk-away-from-military-partners-annotated/

Calder, K. E. (2007). *Embattled garrisons: Comparative base politics and American globalism*. Princeton University Press.

Cha, T. (2016). The return of Jacksonianism: The international implications of the Trump phenomenon. *Washington Quarterly*, *39*(4), 83–97.

Cha, V. D. (2009–2010). Powerplay: Origins of the U.S. alliance system in Asia. *International Security*, *34*(3), 158–196.

Cha, V. D. (2019). The unintended consequences of success: U.S. retrenchment from Korea? *Korean Journal of Defense Analysis*, *31*(2), 165–191.

Christensen, T. J. (2016). *The China challenge: Shaping the choices of a rising power*. W. W. Norton.

Clarke, M., & Ricketts, A. (2017). Donald Trump and American foreign policy: The return of the Jacksonian tradition. *Comparative Strategy*, *36*(4), 366–379.

Coll, S. (2012). *Private empire: ExxonMobil and American power*. Penguin.

Connolly, K. (2020, August 2). 'Simply not OK': Removal of US troops worries German communities. *The Guardian*. https://www.theguardian.com/world/2020/aug/02/removal-of-us-troops-from-germany-will-gravely-affect-local-communities

Cooley, A. (2005). Base politics. *Foreign Affairs*, *84*(6), 79–92.

Cooley, A., & Nexon, D. (2020). *Exit from hegemony: The unraveling of the American global order*. Oxford University Press.

Coroucher, S. (2018, March 15). Did Trump just threaten to withdraw U.S. troops from South Korea? *Newsweek*. https://www.newsweek.com/did-trump-just-threaten-withdraw-us-troops-south-korea-845943

Daalder, I., & Smeltz, D. (2020, September 21). Commentary: Most Americans want more global engagement. *Chicago Tribune*. https://www.chicagotribune.com/opinion/commentary/ct-opinion--20200921-oao7elazsbexbau6yxap47smqi-story.html

Dahl, R. A. (1957). The concept of power. *Behavioral Science, 2*(3), 201–215.

Davis, S. (2011). The US military base network and contemporary colonialism: Power projection, resistance and the quest for operational unilateralism. *Political Geography, 30*(4), 215–224.

DeYoung, K., Lamothe, D., Ryan, M., & Birnbaum, M. (2019, October 26). Trump decided to leave troops in Syria after conversations about oil, officials say. *Washington Post*. https://www.washingtonpost.com/world/us-defense-secretary-mark-esper-says-us-will-leave-forces-in-syria-to-defend-oil-fields-from-islamic-state/2019/10/25/fd131f1a-f723-11e9-829d-87b12c2f85dd_story.html

Dombrowski, P., & Reich, S. (2018). Beyond the tweets: President Trump's continuity in military operations. *Strategic Studies Quarterly, 12*(2), 56–81.

Ettinger, A. (2020). Principled realism and populist sovereignty in Trump's foreign policy. *Cambridge Review of International Affairs, 33*(3), 410–431.

Ferguson, N. (2003). *Empire: How Britain made the modern world*. Penguin.

France24. (2019, September 10). *Trump: US going into Middle East was 'worst decision ever'*. https://www.france24.com/en/20191009-trump-us-going-into-middle-east-was-worst-decision-ever

Friedman, B. H., & Logan, J. (2016). Why Washington doesn't debate grand strategy. *Strategic Studies Quarterly, 10*(4), 14–45.

Gerzhoy, G., & Miller, N. (2016, April 6). Donald Trump thinks more countries should have nuclear weapons: Here's what the research says. *Washington Post*. https://www.washingtonpost.com/news/monkey-cage/wp/2016/04/06/should-more-countries-have-nuclear-weapons-donald-trump-thinks-so/

Gholz, E., Press, D. G., & Sapolsky, H. M. (1997). Come home, America: The strategy of restraint in the face of temptation. *International Security, 21*(4), 5–48.

Gordon, M. L. (1994, October 10). Threats in the Gulf: The military buildup; at least 36,000 U.S. troops going to Gulf in response to continued Iraqi buildup. *New York Times*. https://www.nytimes.com/1994/10/10/world/threats-gulf-military-buildup-least-36000-us-troops-going-gulf-response.html

Hadar, L. (2017, January 2). The limits of Trump's transactional foreign policy. *National Interest*. https://nationalinterest.org/feature/the-limits-trumps-transactional-foreign-policy-18898

Harkavy, R. E. (1982). *Great power competition for overseas bases: The geopolitics of access diplomacy*. Pergamon Press.
Harkavy, R. E. (1989). *Bases abroad: The global foreign military presence*. Oxford University Press.
Harris, P. (2018). Why Trump won't retrench: The militarist redoubt in American foreign policy. *Political Science Quarterly, 133*(4), 611–641.
Haynes, K. (2015). Decline and devolution: The sources of strategic military retrenchment. *International Studies Quarterly, 59*(3), 490–502.
Henry, R. (2006). Transforming the U.S. global defense posture. *Naval War College Review, 59*(2), 12–28.
Herrera, R., & Cicchini, J. (2013). Some notes about the U.S. military bases and personnel abroad. *Journal of Innovation Economics Management, 2*(12), 127–149.
Hiro, D. (2019). *Cold War in the Islamic world: Saudi Arabia, Iran and the struggle for supremacy*. Oxford University Press.
Hu, E. (2016, November 10). Japan and South Korea rattled by Trump's talk of closing U.S. bases. *NPR*. https://www.npr.org/sections/parallels/2016/11/10/501531166/japan-and-south-korea-rattled-by-trumps-talk-of-closing-u-s-bases?t=1584560113023
Huntington, S. P. (1993). Why international primacy matters. *International Security, 17*(4), 68–83.
Huxley, T., & Schreer, B. (2017). Trump's missing Asia strategy. *Survival, 59*(3), 81–89.
Ikenberry, G. J. (2011). *Liberal Leviathan: The origins, crisis, and transformation of the American world order*. Princeton University Press.
Jakobsen, J., & Jakobsen, T. G. (2019). Tripwires and free-riders: Do forward-deployed U.S. troops reduce the willingness of host-country citizens to fight for their country? *Contemporary Security Policy, 40*(2), 135–164.
Kagan, R. (2012). *The world America made*. Vintage.
Kamrava, M. (2018). *Troubled waters: Insecurity in the Persian Gulf*. Cornell University Press.
Kane, T. (2004). *Global U.S. troop deployment, 1950–2003*. Heritage Foundation. Retrieved March 8, 2020, from https://www.heritage.org/defense/report/global-us-troop-deployment-1950-2003
Kennedy, P. (1988). *The rise and fall of the great powers*. Random House.
Kheel, R. (2020, August 6). Poll: About three quarters support bringing troops home from Iraq, Afghanistan. *The Hill*. https://thehill.com/policy/defense/510851-poll-about-three-quarters-support-bringing-troops-home-from-iraq-afghanistan
Kissinger, H. (2014). *World order*. Penguin.
Krauthammer, C. (1990–1991). The unipolar moment. *Foreign Affairs, 70*(1), 23–33.

Lake, D. A. (2009). *Hierarchy in international relations*. Cornell University Press.
Lambert, A. (2018). *Seapower states: Maritime culture, continental empires and the conflict that made the modern world*. Yale University Press.
Layne, C. (1998). Rethinking American grand strategy: Hegemony or balance of power in the twenty-first century? *World Policy Journal, 15*(2), 8–28.
Layne, C. (2006). *The peace of illusions: American grand strategy from 1940 to the present*. Cornell University Press.
Layne, C. (2017). The US foreign policy establishment and grand strategy: How American elites obstruct strategic adjustment. *International Politics, 54*(3), 260–275.
Levy, J. S., & Thompson, W. R. (2005). Hegemonic threats and great-power balancing in Europe, 1495–1999. *Security Studies, 14*(1), 1–33.
Levy, J. S., & Thompson, W. R. (2010). Balancing on land and at sea: Do states ally against the leading global power? *International Security, 35*(1), 7–43.
Lundestad, G. (1986). Empire by invitation? The United States and Western Europe, 1945–1952. *Journal of Peace Research, 23*(3), 263–277.
Lutz, C. (2009). Introduction: Bases, empire and global response. In C. Lutz (Ed.), *The bases of empire: The global struggle against U.S. military posts* (pp. 1–44). New York University Press.
MacDonald, P. K. (2018). America first? Explaining continuity and change in Trump's foreign policy. *Political Science Quarterly, 133*(3), 401–435.
MacDonald, P. K., & Parent, J. M. (2011). Graceful decline? The surprising success of great power retrenchment. *International Security, 35*(4), 7–44.
Macintyre, B. (2019, October 18). America's battle to resist the lure of foreign wars. *The Times*. https://www.thetimes.co.uk/article/americas-battle-to-resist-the-lure-of-foreign-wars-jz7pj5xp6
Mahan, A. T. (1890). *The influence of sea power upon history, 1660–1783*. Little, Brown.
Malnick, E. (2018, July 8). Nato fears Donald Trump may pull troops out of Europe if countries do not increase defence spending. *The Telegraph*. https://www.telegraph.co.uk/news/2018/07/07/nato-summit-fears-trump-may-pull-troops-europe/
Mandelbaum, M. (1988). *The fate of nations: The search for national security in the nineteenth and twentieth centuries*. Cambridge University Press.
Mead, W. R. (2017). The Jacksonian revolt: American populism and the liberal order. *Foreign Affairs, 96*(2), 2–7.
Mearsheimer, J. J. (2001). *The tragedy of great power politics*. W. W. Norton.
Mearsheimer, J. J. (2011, January/February). Imperial by design. *National Interest, 111*, 16–34.
Mearsheimer, J. J. (2014). Why the Ukraine crisis is the West's fault: The liberal delusions that provoked Putin. *Foreign Affairs, 93*(5), 77–89.

Mearsheimer, J. J., & Walt, S. M. (2016). The case for offshore balancing: A superior U.S. grand strategy. *Foreign Affairs*, *95*(4), 70–83.

Mehta, A., & Copp, T. (2017, August 21). Trump Afghanistan strategy calls for more troops, regional pressure. *Military Times*. https://www.militarytimes.com/news/2017/08/22/trump-afghanistan-strategy-calls-for-more-troops-regional-pressure/

Modelski, G. (1978). The long cycle of global politics and the nation-state. *Comparative Studies in Society and History*, *20*(2), 214–235.

Morgan, C. T., & Palmer, G. (2003). To protect and to serve: Alliances and foreign policy portfolios. *Journal of Conflict Resolution*, *47*(2), 180–203.

Morgenthau, H. J. (2006). *Politics among nations: The struggle for power and peace* (7th ed., revised by K. W. Thompson & D. Clinton). McGraw-Hill (Original work published 1948).

Morrow, J. D. (1991). Alliances and asymmetry: An alternative to the capability aggregation model of alliances. *American Journal of Political Science*, *35*(4), 904–933.

New York Times. (2017, May 25). In Brussels, Trump scolds allies on cost-sharing, and stays vague on Article 5. https://www.nytimes.com/2017/05/25/world/europe/trump-brussels-nato.html

Nye, J. S., Jr. (2019). The rise and fall of American hegemony from Wilson to Trump. *International Affairs*, *95*(1), 63–80.

O'Hanlon, M. (1998–1999). Can high technology bring US troops home? *Foreign Policy*, *113*(Winter), 72–86.

Pettyjohn, S. L. (2012). *U.S. global defense posture, 1783–2011*. RAND Corporation.

Pew Research Center. (2016, May 5). *Public uncertain, divided over America's place in the world*. Retrieved January 29, 2021, from https://www.people-press.org/2016/05/05/public-uncertain-divided-over-americas-place-in-the-world/

Phillips, C. (2016). *The battle for Syria: International rivalry in the new Middle East*. Yale University Press.

Plaskin, G. (1990, March 1). The Playboy interview with Donald Trump. *Playboy Magazine*. https://www.playboy.com/read/playboy-interview-donald-trump-1990

Popescu, I. (2018). Conservative internationalism and the Trump administration? *Orbis*, *62*(1), 91–104.

Porter, P. (2018). Why America's grand strategy has not changed: Power, habit, and the U.S. foreign policy establishment. *International Security*, *42*(4), 9–46.

Posen, B. R. (2003). Command of the commons: The military foundation of U.S. hegemony. *International Security*, *28*(1), 5–46.

Posen, B. R. (2013). Pull back: The case for a less activist foreign policy. *Foreign Affairs*, *92*(1), 116–128.

Posen, B. R. (2014). *Restraint: A new foundation for U.S. grand strategy*. Cornell University Press.
Posen, B. R. (2018). The rise of illiberal hegemony: Trump's surprising grand strategy. *Foreign Affairs, 97*(2), 20–27.
Posen, B. R., & Ross, A. L. (1996–1997). Competing visions for U.S. grand strategy. *International Security*, *21*(3), 5–53.
Ringsmose, J. (2010). NATO burden-sharing redux: Continuity and change after the Cold War. *Contemporary Security Policy, 31*(2), 319–338.
Roehrig, T. (2014). South Korea: An alliance in transition. In C. Lord & A. S. Erickson (Eds.), *Rebalancing U.S. forces: Basing and forward presence in the Asia-Pacific* (pp. 67–88). Naval Institute Press.
Rovner, J., & Talmadge, C. (2014). Hegemony, force posture, and the provision of public goods: The once and future role of outside powers in securing Persian Gulf oil. *Security Studies, 23*(3), 548–581.
Sandars, C. T. (2000). *America's overseas garrisons: The leasehold empire*. Oxford University Press.
Sanger, D. (2019, October 7). When 'get out' is a president's national security strategy. *New York Times*. https://www.nytimes.com/2019/10/07/world/middleeast/trump-syria-turkey.html
Schmidt, S. (2020). *Armed guests: Territorial sovereignty and foreign military basing*. Oxford University Press.
Schmitt, E., & Haberman, M. (2019, October 20). Trump said to favor leaving a few hundred troops in eastern Syria. *New York Times*. https://www.nytimes.com/2019/10/20/world/middleeast/trump-troops-syria-turkey.html
Schweller, R. (2018). Opposite but compatible nationalisms: A neoclassical realist approach to the future of US–China relations. *Chinese Journal of International Politics, 11*(1), 23–48.
Service, R. (2015). *The end of the Cold War, 1985–1991*. Macmillan.
SIPRI. (2021a). SIPRI military expenditure database. *Stockholm International Peace Research Institute*. Retrieved September 12, 2021, from https://www.sipri.org/databases/milex
SIPRI. (2021b). SIPRI arms transfers database. *Stockholm International Peace Research Institute*. Retrieved March 19, 2021, from https://www.sipri.org/databases/armstransfers
Spykman, N. J. (1938a). Geography and foreign policy, I. *American Political Science Review, 32*(1), 28–50.
Spykman, N. J. (1938b). Geography and foreign policy, II. *American Political Science Review, 32*(2), 213–236.
Stokols, E., & Wilkinson, T. (2018, December 26). Trump goes to Iraq on his first visit to U.S. troops in a combat zone. *Los Angeles Times*. https://www.latimes.com/nation/politics/la-na-pol-trump-iraq-visit-20181226-story.html

Szayna, T. S., Dreyer, P., Eaton, D., & Saum-Manning, L. (2015). *Army global basing posture: An analytic framework for maximizing responsiveness and effectiveness*. RAND Corporation.

The White House. (2017, December). *National security strategy of the United States of America*. Retrieved April 23, 2020, from https://trumpwhitehouse.archives.gov/wp-content/uploads/2017/12/NSS-Final-12-18-2017-0905.pdf

The White House. (2018, June 21). *Remarks by President Trump at protecting American workers roundtable*. Retrieved January 22, 2021, from https://www.whitehouse.gov/briefings-statements/remarks-president-trump-protecting-american-workers-roundtable/

Tonelson, A. (2000). NATO burden-sharing: Promises, promises. *Journal of Strategic Studies, 23*(3), 29–58.

Trachtenberg, M. (2012). *The Cold War and after: History, theory, and the logic of international politics*. Princeton University Press.

U.S. Department of Defense. (2018). *Base structure report—Fiscal year 2018 baseline*. Retrieved March 15, 2021, from https://www.acq.osd.mil/eie/Downloads/BSI/Base%20Structure%20Report%20FY18.pdf

U.S. Department of Defense. (2021). Military and civilian personnel by service/agency by state/country. *U.S. Defense Manpower Data Center*. Retrieved November 23, 2021, from https://dwp.dmdc.osd.mil/dwp/app/dod-data-reports/workforce-reports

Walt, S. M. (1987). *The origins of alliances*. Cornell University Press.

Walt, S. M. (2018). *The hell of good intentions: America's foreign policy elite and the decline of U.S. primacy*. Farrar, Straus and Giroux.

Walt, S. M. (2019). The end of hubris and the new age of American restraint. *Foreign Affairs, 98*(3), 26–35.

Waltz, K. N. (1993). The emerging structure of international politics. *International Security, 18*(2), 44–79.

Wearing, D. (2018). *AngloArabia: Why Gulf wealth matters to Britain*. Polity Press.

Wohlforth, W. C. (1993). *The elusive balance: Power and perceptions during the Cold War*. Cornell University Press.

Wohlforth, W. C. (1999). The stability of a unipolar world. *International Security, 24*(1), 5–41.

Wolf, R. (2017). Donald Trump's status-driven foreign policy. *Survival, 59*(5), 99–116

World Bank. (2021). *World Bank open data: GDP* (constant 2015 US$). Retrieved November 11, 2021, from https://data.worldbank.org/indicator/NY.GDP.MKTP.KD

Zakaria, F. (1998). *From wealth to power: The unusual origins of America's world role*. Princeton University Press.

CHAPTER 3

Political Realism and Structural Constraints on Retrenchment

*

My theory, whose key tenets were presented in Chapter 1, holds that the murkiness of the rivalry- and conflict-ridden anarchic international system, which in and by itself works to incentivize policymakers to overweight negative outcomes, combines with specific psychological inclinations of individuals to produce "better-safe-than-sorry" policies. Specifically, my theory of great-power persistence argues that states are geared toward avoiding *potentially* catastrophic outcomes; they are particularly sensitive to the possibility that negative international events can set off harmful chain reactions, or tip over a row of "dominoes." The foreign policies of great powers, and certainly the United States, typically exhibit a preference for the status quo and a wish to exert influence in key geographic regions, with the aim of controlling and shaping events as much as possible. This shaping and controlling and exertion of influence tend, in the long term, to be roughly commensurate with a state's power resources. Barring the onset of a much more rapid decline in the United States' relative power position than we have hitherto witnessed, there is little reason to expect it to retrench in any major way anytime soon.

J. Jakobsen, *The Geopolitics of U.S. Overseas Troops and Withdrawal*, Palgrave Studies in International Relations,
https://doi.org/10.1007/978-3-030-94488-9_3

Neoclassical Realism and System-Level Dominoes

In the last main section of Chapter 2, I analyzed various domestic-level forces that may contribute to great-power persistence. Still, the main explanation, and the primary driving forces or root causes of the general status-quo bias in U.S. external affairs and overseas basing pattern, is found at the system level.

Pre-Empting Adverse Geopolitical Chain Reactions

For it is basically the anarchic structure of international politics that allows for the *possibility* of detrimental events to take place in vital corners of the world. Such events, moreover, are often potential instigators of *adverse geopolitical chain reactions*—or of *toppling dominoes*. The leaders and the foreign-policy establishments of powerful countries are apt to try and forestall these chain reactions and to steady the dominoes before they start falling. This is so even if their fears may often be overstated. Powerful states are inclined to expand pre-emptively, if their command of resources grants them the opportunity; and, if they already have expanded, they are inclined to refrain from withdrawing lest they open up a void that in turn creates the *potential* for such adverse chain reactions to occur.

But system-level drivers also combine with the *psychological* inclinations and emotions of humans. Says Michael Mandelbaum (1988, p. 139):

> The policy of preemption, of forestalling challenges, depends on an act of imagination by the strong. The powerful send their forces abroad not so much because of what is occurring at the moment as in anticipation of the possible consequences. They act because of what *could* happen. What they seek to avoid, however, is not purely the product of their imaginations. The dangers are not fantasies. Dominoes can topple; chain reactions do occur.

Chain reactions that *might* occur are essentially products of the imagination, but they are nonetheless staple elements of policy debates about U.S. overseas military bases, which the empirical analysis will demonstrate later. This gives the arguments somewhat of a nebulous character; they draw heavily on counterfactuals in that "they rely on judgments about the bad things that might happen if the U.S. stopped doing what it does," as Hal Brands (2018) puts it in a comment about U.S. military bases in South Korea. This likely arises in part from what scholars have identified as a "negativity bias" often taking precedence in states' foreign and security policies, which tend to rest on the maxim that "bad is stronger than good" (Johnson & Tierney, 2018–2019, p. 96)—and that hypothetical bads are stronger than hypothetical goods. Threats are inflated, dangers are exaggerated, and fear becomes a key driver of major portions of states' behavior. This parallels the security-dilemma mechanism, which also rests on the security concerns of states and individuals, and on their proclivity "to err on the side of caution in their security policies" (Jakobsen & Halvorsen, 2018, p. 158).

Viewed in this way, expansion, intervention, and troop deployments abroad become international politics' version of personal insurance policies (Mandelbaum, 1988, p. 139). Following this "better-safe-than-sorry" thinking, states are thus liable to try as best as they can to *control* and to *shape* their external environment (Gilpin, 1981, pp. 94–95); they are, as some prominent realist scholars contend, prone to maximize power or influence (Mearsheimer, 2001, Ch. 2; Zakaria, 1998, pp. 3ff.).

The Multi-Level Approach of Neoclassical Realism

These are not novel ideas, of course. In this and in the next chapter, I lean in part on Michael Mandelbaum's more general analysis of the driving forces behind U.S. Cold War geostrategy, which he spelled out in the long third chapter of his excellent 1988 book *The Fate of Nations* (Mandelbaum, 1988, Ch. 3). I also align with a strand of realist international relations theory to which Mandelbaum's book belongs, and which Gideon Rose (1998) over two decades ago labeled *neoclassical realism.* His 1998 article "Neoclassical Realism and Theories of Foreign Policy," published in *World Politics*, is a review of five books from the 1990s that shared some important assumptions about the driving forces and dynamics of international politics and the foreign policies of great powers. All of them operated within the tradition of political realism; that is to say, they emphasized the primary causal role of power and the balance of power. However, these five works contained nuances that, in Rose's view, substantially enhanced the explanatory power of the structural version of realism—neorealism—, which had gained a position of academic authority with the 1979 publication of *Theory of International Politics* by Kenneth Waltz (1979).

Following Rose's essay, a growing number of studies have proclaimed to be part of this flourishing strand of political realism (e.g., Kitchen,

2010; Lobell et al., 2009; Meibauer, 2020; Rathbun, 2008; Ripsman et al., 2016; Schweller, 2018). While its proponents tend to point out that there is no *one* coherent neoclassical realist theory, but rather several such *theories* (Taliaferro et al., 2009, p. 10), these works do share some important commonalities. Broadly speaking, they all help bridge the divide between Waltzian structural realism, with its heavy emphasis on system-level explanations, and the less theoretical stringent but empirically richer classical realist tradition of, among others, Hans Morgenthau, E. H. Carr, Reinhold Niebuhr, and Henry Kissinger.

The main issue that other schools take with structural realism is its parsimonious approach, which sees leaders' scope for agency severely circumscribed by constraints primarily located at the system level. Many international relations scholars, including classical and neoclassical realists, claim that this represents an overly simplified "hyperrationalism" that ends up explaining too little (Gray, 1999, p. 164; Kirshner, 2015, p. 177; Rathbun, 2008). Realist theory, they say, become more "realistic" insofar as it also explicitly integrates into its framework "structural modifiers" that can better account for a plethora of behavioral anomalies, such as all those instances where leaders fail to interpret correctly the systemic constraints or pressures under which their decisions are taken (Meibauer, 2020, p. 23; Ripsman et al., 2016, pp. 8–12). In short, neoclassical realists give a more prominent place to *intervening* variables in studies that draw on more than one level of analysis. This is at the core of Gideon Rose's justification for identifying an additional school within the realist tradition. Neoclassical realism, he writes,

> explicitly incorporates both external and internal variables, updating and systematizing certain insights drawn from classical realist thought. Its adherents argue that the scope and ambition of a country's foreign policy is driven first and foremost by its place in the international system and specifically by its relative material power capabilities. This is why they are realist. They argue further, however, that the impact of such power capabilities on foreign policy is indirect and complex, because systemic pressures must be translated through intervening variables at the unit level. This is why they are neoclassical. (Rose, 1998, p. 146)

Some see neoclassical realism as a school opting for a "mid-level" between the systemic and domestic levels (Foulon, 2015; Onea, 2021, p. 276). Most would also bring the individual level into analyses of foreign-policy decisions. This makes for a three-level approach (He, 2017, p. 137; Marsh, 2012, p. 489). Even so, the third image—the system level—does take precedence in neoclassical realist analyses; this makes them, in the first place, outside-in explanations of foreign policy. If we, with Tudor Onea (2021, p. 277), imagine the world as one constituted of two chess-boards—one domestic and one international—neoclassical realism "holds that the first move is made with the international board in mind." Those who ignore the "basic insight" that anarchy and relative power forcefully constrain decision-makers, Gideon Rose (1998, p. 165) asserts, "will often waste their time looking at variables that are actually epiphenomenal." That is to say, neoclassical realism by and large adopts a *Waltzian* starting point. The assumptions and arguments laid out in Kenneth Waltz's *Theory of International Politics*, and in later structural realist works by, among others, Robert Gilpin (1981) and John Mearsheimer (2001), constitute a natural first cut when the events or developments under study represent the "big" issues in international politics.

Systemic Constraints and the Balance of Power

In Waltz's conception of structure, two things are especially noteworthy about the international system (Waltz, 1979, Ch. 6). First, its ordering principle is *anarchy*, which simply entails here that there is no higher authority above the main units that exist in the system—the states. That international relations "continue to be a recurring struggle for wealth and power among independent actors in a state of anarchy," as Robert Gilpin (1981, p. 7) asserts, effectively produces a self-help system consisting of states that are autonomous, functionally undifferentiated actors each of which must always be prepared to fend for itself (Waltz, 1979, pp. 111, 118). The second defining principle of the structure of international politics is the *distribution of capabilities* across the units inhabiting the international system. Capabilities, or power(-as-resources), vary significantly between states. Variation in capabilities, furthermore, yields variation in the types and magnitude of the structural constraints and incentives that states face, which cause variation in how states behave (or should behave, at any rate).

These two simple assumptions—anarchy and the importance of the distribution of power as a shaper of state behavior—result in the famous image of international politics as a "pool table," consisting of balls whose inherent properties (shape, the material they are made of, and so on) are essentially the same. However, the sizes of the balls vary greatly. And, on the proverbial pool table, the smallest balls, and indeed most of the medium-sized balls as well, tend to congregate around the corners where they usually stay put. The largest balls, for their part, can be found roaming about the whole table, often interfering with the smaller balls, sometimes knocking them over, dominating the game and determining how it is being played out. In simple terms, structural constraints at the system level allow powerful states to "do more" than less powerful ones. Interests expand as relative power increases. Still, even the most powerful states cannot or should not get too excited about the power-projection opportunities that arise simply from having great capabilities. There always exist, to a larger or lesser degree, countervailing forces that serve to check and to balance large, expanding, and threatening power.

Waltz's book centered on multipolar and bipolar systems, but it did not discuss patterns of state behavior in a unipolar world, which was the world that followed the collapse of the Soviet Union. However, structural realism more broadly informs us that, in such a situation, a grand strategy advocating major retrenchment would be quite unlikely. Power shapes interests: the more power, the more extensive the interests, and the higher the temptation to increase one's influence in key areas around the globe. And the United States did expand its activities and influence following the end of the Cold War. This was so despite the fact that the level of its activities and influence was already unparalleled at the moment the Soviet Union collapsed. The latter fact worked somewhat to limit the marginal utility of expanding even further; instead, it more clearly enhanced temptations to *uphold the status quo*. Still, the areas left behind by the Soviet Empire represented one obvious opportunity for expansion, though ultimately with the view of bolstering the existing, predominantly U.S.-shaped world order. Not surprisingly, most of Eastern and Central Europe—including the Baltic states bordering Russia—soon became firmly entrenched within the U.S. sphere of influence. Apart from that, arguably important, region, U.S. exploitation of, and expansion during, the "unipolar moment" first and foremost consisted of less overt attempts to fortify and institutionalize the world order of which the United States itself had been the key shaper. The list of examples is long. It included enlarging and further strengthening NATO; bolstering key security alliances around the world; expanding the U.S. military-base network in the Middle East; enhancing the institutions underpinning economic globalization and capitalism; ostracizing (and sometimes attacking) recalcitrant regimes; and, more generally, entrenching its ideological hegemony and agenda-setting powers in the world.

One of the biggest debates among international relations scholars in the post-Cold War era centered on how other major powers would respond to U.S. unipolarity and hegemony (e.g., Brooks & Wohlforth, 2005; Levy & Thompson, 2010; Lieber & Alexander, 2005; Pape, 2005a). Neorealists expected to witness countervailing forces amassing quite rapidly in an attempt to balance the United States (Layne, 1993; Mearsheimer, 1990). In the early 1990s, Waltz (1993) himself anticipated that balancing would soon occur on a major scale. Although depending on what the adverb "soon" really entails, he was probably not entirely in the right; the overall evidence pointed instead to the general *absence* of balancing throughout the first 15–20 years of that period: unipolarity and American hegemony had not yet been seriously challenged.

Waltz and structural realism should nonetheless be exonerated. They *might* have got the timing slightly wrong. However, balancing behavior has now certainly returned in considerable force, spearheaded by Russia's newfound assertiveness and China's extraordinarily rapid rise and growing ambitions. Washington, for its part, has hitherto not indicated much if any willingness to cede ground with respect to its position as the dominant power in neither Europe nor the Asia–Pacific (or around the Persian Gulf). As for Asia–Pacific, what can in some ways be viewed as a case of *dual balancing* between China and the United States—each state endeavors to balance the other in order to preserve or improve its own position—has produced a rivalry that has seen a wide range of flashpoints come to the fore in the last few years. For Kenneth Waltz, and for many others, it is by no means surprising to see a rising major power colliding with an established hegemon. There exists a long list of studies arguing that great-power rivalry and intense balancing are expected outcomes under circumstances where the anarchic structure of the international system mixes with uneven economic and military growth rates among the most powerful states (e.g., Allison, 2017; Gilpin, 1981; Kennedy, 1988; Modelski, 1978; Modelski & Thompson, 1996, Organski & Kugler, 1980; Thompson, 1983). However, the expectations of structural theory do not, of course, imply inevitability. This is especially so since structural pressures and incentives under such circumstances can pull states in more than one direction.

As for the existing hegemon, it will, on the one hand, be loath to give up its hard-won position. The branch of structural realism that is known as offensive realism, whose most notable proponent is John Mearsheimer, holds that states are predisposed to maximizing security by maximizing power and influence. This is, in most instances, hard to square with any retrenchment objectives. Defensive realists, such as Kenneth Waltz (1979, p. 126) and Randall Schweller (1996), argue instead that states, in order to maximize security, usually act not as power maximizers but rather as power "satisficers." For Waltz, a basic objective of any state is to *maintain* its position in the system, which does not sit well with retrenchment either. Thus, for offensive and defensive realists alike, the rise of a challenger (say, China) to the existing leader (say, the United States) should suffice to make the latter quite reluctant to withdraw. This is especially so since the challenger can be expected to move into any vacuum left by the leader's withdrawal, which will only work to increase even further the relative power of the challenger, and to degrade that of the leader.

On the other hand, both defensive and offensive realists view states as rational actors that are sensitive to costs and risks. Differences between these two branches of structural realism should, for this and for other reasons, not be exaggerated (Snyder, 2002). It is true that Mearsheimer's power-maximization states are fundamentally revisionist. Yet, they are also cost–benefit-weighing strategic actors that will rationally choose less expansionist policies—and even retrenchment—if the expected costs and risks of further expansion, or of attempting to uphold the status quo, outweigh the expected gains (Mearsheimer, 2001, p. 37). Being inclined to maximize power does not automatically imply an inclination to expand, and neither is it wholly incompatible with retrenchment. This is particularly the case when a leading power faces off against an able competitor that is both unwilling to cede *its* position and eager to roll back the previous gains of the leader—sometimes even at the risk of war.

Opening Up the Black Box of the State

However, neither observers nor decision-makers can realistically make any precise estimates of expected costs and benefits in what is a realm of uncertainty more than of calculable risk. The direction in which states are moved by structural pressures is not always straightforward, and neither are the calculations and reasonings of states and their leaders. Neoclassical realism takes the assumptions and arguments of structural realism as its point of departure: state leaders must attempt to respond to system-level constraints, opportunities, incentives, and risks lest their country's relative power and national security suffer in the long run. More often than not, they duly do so. But they still do so imperfectly, say neoclassical realists. They make mistakes; they miscalculate the power of their own state and that of their adversaries; they misinterpret others' intentions; their views are shaped or misshaped by the strategic culture of their country; and they are beholden to their constituency, to interest groups, and to domestic elites. The fuzziness of systemic constraints combines, among other things, with the natural inadequacy and idiosyncrasies of human beings, which characterize even the greatest leaders of the greatest powers. This is a multi-level mix that quite often produces outcomes that are anomalous and seemingly irrational—when judged by the strictures of structural realism—but which may nonetheless be the upshot of reasonably thoughtful processes at the domestic and individual levels. Neoclassical realism, thus,

> posits an imperfect "transmission belt" between systemic incentives and constraints, on the one hand, and the actual diplomatic, military, and foreign economic policies states select, on the other. Over the long term, international political outcomes generally mirror the actual distribution of power among states. In the shorter term, however, the policies states pursue are rarely objectively efficient or predictable based upon a purely systemic analysis. (Taliaferro et al., 2009, p. 4)

In other words, variables at the system level, such as relative power, are *translated* or *filtered* through variables at lower levels; the analyst should therefore look beyond the international structure in order to offer comprehensive explanations of foreign-policy behavior (Rose, 1998, p. 165). Depending on what is permitted by the domestic political system, writes Nicholas Kitchen (2010, p. 133), a state leader has at least some degree of foreign-policy discretion. He or she will tend to make decisions based on estimations "of relative power and other states' intentions, but always subject to domestic constraints." The magnitude and contents of these constraints will vary among countries, but it is typically the case that the executive's thinking and decisions are influenced, inter alia, "by the cumulative actions of actors as diverse as policy makers, lobbyists, citizens' groups and businesses, and by the process of bargaining with them."

In his 1998 essay, Gideon Rose (1998, p. 168) observed that the "transmission belt" between systemic constraints and foreign-policy decisions was inadequately developed theoretically. Neither was there much agreement on which intervening variables deserved a place of prominence in neoclassical realist analyses. This is to a large degree still the case, even if recent scholarship has endeavored to develop more systematic and presumably more generalizable models or theories of the "structural modifiers" (Ripsman et al., 2016, Ch. 3). Yet, however one judges the progress that neoclassical realist theory has made in terms of theoretical development and stringency—and maybe the sought-after generalizability is unachievable in any case—its proponents have surely contributed to opening up "neorealism's 'black box' view of the state" (Marsh, 2012, p. 489). This has produced an expedient bridge between classical and structural realism, even if it has sacrificed some of the latter's parsimony and *purported* generalizability on the altar of the former's allowances for context and empirical narrative. The word "purported" really represents the crux of the matter here. The sheer reason why neoclassical realists emphasize intervening variables located at secondary levels is exactly because systemic incentives and constraints, while they may be generalizable in and by themselves, do not yield any clear-cut, predictable pattern with respect to actual decision-making. Fully unitary, rational states, if these were also fully informed, would correctly interpret the implications from systemic characteristics and changes, and an eminently predictable pattern of behavior would ensue. But such states do not inhabit the empirical world.

Domestic Guardians of Primacy

This is presumably why Stephen Walt (2018), a structural realist, stresses the major role of the wider foreign-policy community—the Blob—in the formulation and execution of U.S. grand strategy. The widespread agreement within this group on the benefits arising from a U.S. policy of primacy stems in no minor way from an interpretation of what systemic constraints and opportunities allow. But Walt vehemently argues that this is actually a case of consistent *mis*interpretation: the Blob distorts the messages relayed by the system; they continuously exaggerate the evilness of intentions and the power of states and terrorist groups alike; and, more generally, they tend to inflate any threats originating from overseas. The Blob thereby keeps U.S. foreign policy on a primacist path that is detrimental to U.S. national security in the long run. Elsewhere, Walt makes similar claims about the causal chain in foreign-policy decisions. The central message of *The Israel Lobby and U.S. Foreign Policy*, co-written with John Mearsheimer, is that U.S. (and in fact also Israeli) policymakers, egged on by domestic interest groups, fail to grasp what the systemic factors really imply and prescribe (Mearsheimer & Walt, 2007). Hence, Washington's policies remain fundamentally pro-Israel, but for the wrong reasons and with outcomes that ultimately harm both the United States and Israel.

Gustav Meibauer (2020) refers to this group of professionals, whose views forcefully impact final decisions on foreign affairs, the "carriers" of ideas. The ideas of the foreign-policy elite, he writes, are externalized into policy discussions where they are "wielded instrumentally and strategically in an interactive process to communicate, exchange, and convince by their use in language and text" (Meibauer, 2020, p. 30). Policy deliberations on a given topic involve a competition between different voices whose interpretations of systemic incentives often vary considerably. Interpretations vary because of widespread "contradiction and confusion over how to correctly deduce systemic stimuli" (Meibauer, 2020, p. 32).

Meibauer stresses the divergence of ideas more than does Walt, who argues that these coalesce around a perception of international politics as an inherently dangerous sphere but one whose dangers would multiply if the United States failed to pursue an activist foreign policy. Similarly, Christopher Layne (2017, pp. 269–270) writes:

> Underlying US foreign establishment discourse is the notion of danger – bolstered by the generous employment of threat inflation – which is used to define American national identity by delineating the US from external sources of menace (the "Other"). (...) The foreign policy establishment's discourse emphasizes that the USA must hold the line against external danger, and never back down when challenged. This is why claims about the importance of American "credibility" and "resolve" – and metaphors about falling dominoes – feature so prominently in foreign policy establishment discourse.

As noted earlier, Walt identifies a wide range of professionals making up the Blob, whose common denominator is that their livelihoods are based on thinking, talking, or writing about U.S. foreign policy. Talking and writing (and often also thinking!) about U.S. foreign policy are certainly also the tasks of those parts of the media which report on world affairs. There exists a large body of literature suggesting that the media has a considerable impact on public perceptions of international affairs and foreign policy and, through that, on the foreign-policy decisions themselves (for an overview, see Baum & Potter, 2008). The way issues are *framed* by news agencies matters greatly; in particular, the public reacts much stronger to negative than to positive framing of events (Sheafer & Dvir-Gvirsman, 2010). This was, for instance, evident after the 9/11 terrorist attacks. The frequent use by the media of "emotionally powerful visuals and storylines" fed feelings of threat and anxiety among the U.S. public, increasing its "support for strong political leaders as well as support for punitive policies that tend to escalate conflict" (Gadarian, 2010, pp. 469–470). The media, in other words, might help spawn a more interventionist foreign policy.

In addition, although they are certainly not part of the *U.S.* foreign-policy community, the message propagated by the *host countries* of U.S. troops tend to share similar traits as that of the Blob: it is generally a message that extols an activist U.S. foreign policy and warns against the profusion of detrimental developments that will allegedly follow from retrenchment. Hence, the U.S. Empire was *invited* into Europe, and it was urged to stay virtually for eternity by "a majority of Europeans [who] were and are convinced that the US military presence within their own borders helped keep the peace and develop democracies" (Westad, 2017, p. 2). Europeans, writes Zbigniew Brzeziński (1997, p. 60) in *The Grand Chessboard*, may "deplore American 'hegemony'," but at the same time they "take comfort in being sheltered by it." The bulk of elites and the conservative media in Japan do not similarly signal displeasure with U.S. hegemony, but they do propound the same message with regard to the need to keep American soldiers stationed in their country. According to one scholar, Tokyo has constructed and unrelentingly upheld a strategic narrative that emphasizes the anarchic structure of a conflict-ridden world in which forward-deployed U.S. forces in Japan are fundamentally important as a deterrence against China. Those "who question this deterrence narrative are rendered as naive, even 'reckless'" by the Japanese foreign-policy community, as they fail to acknowledge the existential threat posed by a rising China that is purportedly "waiting for its opportunity to seize Japanese territory" (O'Shea, 2019, pp. 452–453).

System-Level Forces Conducive to Retrenchment

Whether they do so explicitly or only implicitly, Japanese and European leaders point to the *possibility* that what they view as ominous events and developments originating at the system level of analysis may eventually snowball into something very grave if the United States and its forward-deployed soldiers were to leave. In other words, U.S. allies would want the *Stayers* among Washington's elites to win the day—for their own benefit and for the ostensible benefit of the United States itself. As I will return to in the next main section, system-level properties and mechanisms *generally* incentivize staying and not leaving—if we follow the assumptions and implications of the school of political realism as a whole.

However, some of the mechanisms and developments at the system level of analysis may still, under specific conditions, work in the opposite direction, incentivizing retrenchment instead. Some such counterforces appeared to be of greater weight than ever just as the self-avowed *Leaver* Donald Trump announced that he would be running for president. This implies that the appearance of Trump in high politics, and not least his being elected the United States' 45th president, might have been more a *symptom* than a fundamental cause of any eventual alteration to the path that U.S. foreign policy has followed for so long (Layne, 2018). This is what Alexander Cooley and Daniel Nexon (2020, Ch. 7) argue in their recently-published book *Exit from Hegemony?*: Donald Trump's presidency, although surely in and by itself a magnifying force of deeper trends, might primarily have been epiphenomenal as it coincided and coexisted with other forces operating at the system (and domestic) level. These latter forces would then act as permissive causes—as enablers of the president's alleged impulse to scale back U.S. military presence abroad. Two such forces come especially to mind (although none of them should be overstated). The first—and here we lean heavily on lessons learned from the long and dramatic history of the rise and fall of great powers—are *real and perceived changes in relative power* among states. The second has to do with *specific geopolitical developments* within an overall structure that is conditioned by the balance of power. This category encompasses both self-inflicted geopolitical wounds and events and actions affiliated with security-dilemma dynamics.

The Rise and Fall of Great Powers

First, changes in global and regional balances of power could work to curb U.S. foreign-policy activities. It is true that analysts and politicians have been pondering seriously about the possible or likely end of U.S. leadership since the 1970s, without ever actually experiencing the hegemon's *real* downfall. Yet, to invoke once more Cooley and Nexon (2020, Ch. 1), the sub-title of their opening chapter reads: "This time is different." Or it *might* be, at any rate.

The end of the last decade certainly witnessed a more aggressive, militarized posture by an at least partially resurgent Russia—exemplified by its brief 2008 war with Georgia—and a rapidly rising China that also became markedly more assertive in the wake of the global financial crisis (Christensen, 2011; Johnston, 2013). In the Middle East, moreover, a handful of regional major powers began embarking on a far more activist foreign policy. These years also saw a United States with a damaged domestic and international reputation after the failure to complete the Afghanistan and Iraq missions, which fueled doubts about the hegemon's future credibility as a stabilizer in vital regions. The assertiveness of both Russia and China continued to show in the subsequent decade. Moscow's annexation of Crimea and its intervention in Syria's civil war; China's attempts to increase its maritime territory by way of an incremental "gray-zone" strategy (Holmes & Yoshihara, 2017); and Beijing's brazen economic and geopolitical expansion through the Belt and Road Initiative—all of these events and developments served as harbingers of more rivalrous and conflictual interstate relations, and of the very real possibility that the world order was about to change. Cooley and Nexon (2020, p. 82), among others, are adamant that "Beijing and Moscow are already altering the ecology of the international order."

Many political-realist scholars saw these as fully expected developments. The unprecedented power preponderance of the United States following the end of the Cold War couldn't but spur pushback and balancing behavior by other aspiring great powers. It was something lawlike over this mechanism, claimed Kenneth Waltz (1993, p. 77): "Hegemony leads to balance," was his simple assertion. Other prominent realists even refrained from calling it hegemony at all. Henry Kissinger (1994, p. 809), for example, pointed to the growing diffusion of power, which, even if the United States' command of resources was unparalleled, meant that its "ability to employ it to shape the rest of the world has actually decreased." In any case, according to John Mearsheimer (2001, p. 41), even the biggest power out there could not conceivably extend its regional hegemony to encompass other regions as well; the world was simply too large, and the stopping power of water provided too much of an obstacle, for massive power to be projected globally.

Several scholars have written brilliantly and very persuasively about both the causes and the consequences of the rise and fall of great powers (e.g., Gilpin, 1981; Kennedy, 1988; Knutsen, 1999). For example, in his classic *War and Change in World Politics*, Robert Gilpin (1981, Ch. 4) emphasizes that upholding an empire or hegemony generally comes with increasing costs and diminishing benefits, which act as impediments to further expansion. The leading state in the system soon gets caught up in a trilemma revolving around the question of how to create an appropriate balance among security, consumption, and investment needs. Normally, it is the third leg of this tripod that is neglected. Governing and dominating the international system demand ample manpower and other resources: to keep its position, the hegemon "must expend its resources on military forces, the financing of allies, foreign aid, and the costs associated with maintaining the international economy" (Gilpin, 1981, pp. 156–157). This is so even as the hegemon's citizens are certain to demand that their consumption needs be better satisfied. This, then, is not so much a conflict between guns and butter as one between "guns, butter, or productivity" (Gilpin, 1981, p. 166). Productivity stands to lose in this conflict, and what almost inevitably ensues, says Gilpin, is political decline—a decline of relative power vis-à-vis secondary states, the latter of which can climb the hierarchy of power unburdened by the costs of leadership.

This trilemma is also a *leitmotif* running through Paul Kennedy's magisterial *The Rise and Fall of the Great Powers*. He also notes the abundance of examples throughout history of great powers overextending or overstretching themselves; viz., their tendency to choose security, or *perceived* security, almost inescapably discovering "'new frontiers of insecurity' whenever they [want] to draw the line" (Kennedy, 1988, pp. 462–463). In particular, as George Modelski (1978, p. 230) warns, the temptation is to fall into the trap of "territoriality," whereby the hegemon becomes obsessed with defending "fixed positions and other distant frontiers." Over the long run, this acts as a severe drain on its resources. Such a transition from a node-centered strategy toward increasing territoriality, says the naval historian Andrew Lambert (2018,

p. 165), has befallen all sea-based empires throughout history. In other words, overextension abroad tends to combine with economic mismanagement or, at any rate, challenges at home. This places significant strain on the hegemon's previously seemingly unassailable position. As for domestic-economic challenges, these are not at all negligible in today's United States. A relatively sluggish economy and high levels of national debt heighten both conflicts over budget allocations and perceptions that this is an overstretched hegemon, and they might therefore increase incentives to retrench (Mandelbaum, 2019, pp. 94–95). When still a presidential candidate, Donald Trump repeatedly referred to the deficiencies of the U.S. military and economy (e.g., New York Times, 2016). Although he no doubt did so in large part for strategic-electoral reasons, the consistent reiteration of such a message works anyhow to underpin the impression that the United States is indeed a declining power.

Perceptions of Power

Upholding primacy or hegemony and a global engagement at a level similar to what has hitherto been the case requires a balance of power that is still skewed in favor of the United States. Relative power, in turn, rests vitally on both military strength and a sound economic foundation. Yet, the issue is not really as straightforward; estimating relative power is riddled with challenges. Ultimately, "objective" measures of power are filtered through the minds of state leaders and public opinion before they affect foreign policies. Several studies, many of which are associated with neoclassical realist theory, highlight the role of *perceptions* as a key intermediate variable in international politics that conditions the effects of the balance of power on foreign-policy outcomes (for an overview, see Ripsman et al., 2016, Chs. 1–3).

Sometimes, external shocks fundamentally alter perceptions, as when the United States battered Spain in the 1898 war. This was a victory, writes Fareed Zakaria (1998, p. 11), that "crystallized the perception of increasing American power both at home and abroad." Some more recent shocks have had the opposite effect on how U.S. power is gauged. The financial crisis of 2008–2009 is often highlighted as a trigger of shifts in the balance of power (Altman, 2009; Zakaria, 2008). Whether or not this is so, it did at least alter the impressions many held of both U.S. power and the legitimacy of the U.S.-shaped international system, both among rivals (Christensen, 2016, pp. 242–243) and allies (Nye, 2010, p. 143) of America. The crisis, moreover, coincided with a growing sense that the costly U.S. wars in Afghanistan and Iraq were all but lost. However, whether a decline of the United States is overwhelmingly "real" or largely only "perceived" is not necessarily critical with respect to U.S. willingness to continue its grand strategy of primacy. Perceptions and reality feed on each other, potentially hamstringing the superpower and increasing the likelihood that it will turn into a more inward-looking country. In his 2012 book *Strategic Vision: America and the Crisis of Global Power*, Zbigniew Brzeziński (2012, p. 105) made this point, rather presciently anticipating some of the seemingly profound political changes that would take place only a few years later:

> [A] decline in American power would likely undermine the health and good judgment of America's economic and political system. (...) A waning United States would likely be more nationalistic, more defensive about its national identity, more paranoid about its homeland security, and less willing to sacrifice resources for the sake of others' development.

In short, a waning America would be more likely to retrench. This is so even if one can also make the counterclaim that perceptions of U.S. weakness and decreasing credibility may instead "encourage policymakers to hold on to burdensome overseas commitments, despite their high costs in blood and treasure" (MacDonald & Parent, 2011, p. 9). Still, the opposite is possibly the more likely scenario *in the long run*. The authors of the latter quote came to this conclusion after analyzing 18 historical cases of great powers in sudden decline: weakening great powers tend to adjust to the reality of their changing circumstances by scaling back on their foreign engagements; that is, they tend to retrench, even if doing so is an undertaking riddled with difficulties.

Perhaps such a mechanism—*perceptions* of waning power causing retrenchment—is one to which the United States of America is particularly susceptible. Over the span of two-and-a-half generations, the distribution of power has been such as to secure a role for the United States as, at best, a global hegemon and, at worst, a *primus inter pares* among a large group of states whose interrelations have long been cooperative. There are significant socio-psychological costs connected to a situation where the nation faces the possibility of having to coexist with other states of equal or greater stature and power. If it really resembles a "truism," to say it with Walter LaFeber (2008 [1967], p. 25), "that

Americans tend to become political 'isolationists' when they cannot dominate international affairs and 'internationalists' politically when they can," then this might in part be explained by what 75 years of being number one does to the self-image of Americans. It might also in part be the result of the country's fortuitous geographic location to which I have referred several times above: the insularity of the United States reduces its vulnerability even as its relative power declines. Especially "under conditions of low vulnerability," writes Randall Schweller (2018, p. 37), "declining powers [tend] to reduce peripheral commitments and look inward," which yields the unsurprising result "that American citizens, after decades of political consensus, are now questioning their country's grand strategy of deep engagement with the rest of the world." It is within such a context that one may view U.S. voters' willingness, if not necessarily eagerness, to elect as president a figure without any political experience of note—and one who promised to attend far more profoundly to domestic issues and interests at the expense of U.S. involvement in world affairs.

Thus, if one subscribes to the opinion that the United States' share of global power really *is* declining—and this book does so, although it also believes that the rate of relative decline has heretofore been quite slow—the election of Donald Trump could be viewed as more of a symptom than a fundamental driving force of any retrenchment ambitions or behavior.

Self-Inflicted Geopolitical Wounds

Having pointed to the importance of real and perceived changes to the balance of power, a second possible underlying cause encompasses more specific geopolitical developments that may make a less active force projection posture seem warranted. The end of the Cold War is the modern classic such event; it caused a marked reduction in the number of U.S. overseas troops, even if changes in the geographic distribution of bases were relatively muted.

Other geopolitical developments origin even more firmly in Washington's own decisions; they are *self-inflicted geopolitical wounds*. Offshore balancers argue quite emphatically that the U.S. grand strategy itself, and the related forward deployment of military personnel, has actually created at least as many security problems as it has solved, and probably far more (Mearsheimer, 2011; Walt, 2018). The clearest case in point is the Middle

East, especially after the 1991 Gulf War and the 2003 invasion of Iraq. The latter decision spawned a range of unintended consequences: inter alia, it left Iraq a severely destabilized country; it greatly enhanced Iran's regional influence; and it ultimately led to the transmutation of al-Qaeda's Iraqi branch into what proved to be an even more egregiously violent terrorist organization—the Islamic State. One can therefore make the case that such interventions and entanglements in a highly volatile region are self-defeating; among other things, the military presence itself seems to exacerbate existing, or even cause new, terrorism problems (Braithwaite, 2015; Byman, 2019; Pape, 2005b).

On the other hand, increased security challenges, even if they are in large part self-inflicted, can also provide incentives to double down on commitments: As I will return to in Chapter 4, states and their leaders are apt to be driven by some of the same psychological mechanisms as human beings in general, an important one of which is the aversion against accepting losses (Jervis, 2017, Ch. 4; Johnson & Tierney, 2018–2019; Kahneman & Tversky, 1979). Indeed, psychological mechanisms are also markedly in play when military forces are deployed to deter and to counter perceived threats from *state* actors. The primary rationale or justification for the stationing of troops abroad is to increase the perceived costs, and thereby lower the likelihood of, expansion or assertive foreign-policy moves by regional states. In East Asia, this strategy is directed against China and North Korea; in Europe, the target is Russia; in the Middle East, Iran is the state actor that the United States strives to contain. In all three cases, the underlying assumption is that the adversary is a "greedy" state whose ambitions will only grow if its drive to dominate others is not nipped in the bud. This is in essence the *deterrence model* of how to produce security (Glaser, 1992); its prevailing logic is that it is sometimes perfectly rational for non-threatening, status quo-inclined powers, which is how Washington sees itself, to institute containment policies against revisionist states that are bent on changing world and regional orders.

The deterrence model contrasts with the spiral logic of the *security dilemma*. That model emphasizes the self-defeating elements of security-seeking under anarchy (Snyder, 1984). It describes a world where moves to increase one's own security easily spur countermoves by one's adversaries, thereby increasing tensions and decreasing security for all. The upshot of this is that overseas military bases may end up *heightening* the very risks and tensions that they were established and upheld to reduce in the first place. In many instances, expansion and the deployment of troops abroad inevitably spur a response by regional players as they cause "fear that generates counteraction instead of scaring an adversary into submission" (Glaser, 2017, p. 9). Some recent quantitative evidence indicates as much: an analysis by Angela O'Mahoney et al. (2018) concludes that while a U.S. forward troop presence seems to deter high-intensity interstate war, it simultaneously incites interstate tensions and increases the likelihood of conflict at lower levels of intensity.

Many observers argue that recent years have produced visible indications that such mechanisms are in play in the regions deemed to be of vital interest to the United States. John Mearsheimer (2014), for example, is adamant that Russia's 2014 annexation of the Crimean Peninsula and its intervention in Ukraine's civil war were not caused by any offensive-expansionist objectives in Moscow per se; rather, they came about as a primarily defensive response to the *West*'s expansionist ambitions—first and foremost represented by an active NATO-enlargement policy. Furthermore, some scholars see the current international politics of East Asia through the prism of the security dilemma; in this case it is centered on China, on the one hand, and the United States and its allies and overseas bases on the other (Liff & Ikenberry, 2014). As for the Persian Gulf and the broader Middle East region, or so one observer notes, an unusually high number of scholars have, for quite some time, and especially since 9/11, raised their voices (or pens or keyboards, rather) in opposition to the heavy U.S. military presence (Gause III, 2019, pp. 8–10). Arguments as to why the United States should scale back their activities from the Middle East are manifold; yet, the apparently self-defeating element inherent in Washington's policies—the increased risks that follow from the U.S. base posture—is a common thread (Ashford, 2018).

In other words, and to sum up the preceding paragraphs, there exist a small handful of interrelated forces that act to increase the prospects of retrenchment—or the *potential* prospects, to qualify the reasoning a bit. Two such forces—shifting balances of power and specific geopolitical developments, including those that are associated with the more general mechanism of the security dilemma—are primarily situated at the system level. These factors should not necessarily be over-emphasized, though. Balance-of-power changes are normally slow-moving developments; and, as I will demonstrate in Chapter 6, the U.S. commands a share of the world's material resources that is not much smaller than it was during the very height of unipolarity. In addition, security threats that originate in part from U.S. behavior and military presence are really nothing new. In fact, the Cold War saw an abundance of episodes of heightened superpower tension (not to mention the debacle that was the Vietnam War) partly caused by U.S. military-force projection. This was widely and generally accepted in Washington as a price worth paying *in order to minimize the risk of experiencing what was perceived to be the very worst*, if relatively unlikely, *outcome*: Soviet Union and Communist hegemony in key regions of the world—and in the world as such. The United States did not retrench in any meaningful way during the four decades of the Cold War, and neither is it likely to do so in the foreseeable future. Or, put a bit more moderately, the United States is, in the foreseeable future, unlikely to retrench significantly *as a direct result of forces located at the system level*.

To wit, any major changes to the overseas basing pattern of the United States are, in the near term at least, far more likely to come about as a consequence of the decisions made by the president than as a result of forces at other levels of analysis—even if the likelihood of the former is also small. This has particularly been the case in the years spanning the presidency of Donald Trump. The period lasting from January 2017 to January 2021 can therefore partly be seen as a very convenient test case for the independent impact of the individual level on retrenchment, under the powerful constraints imposed by the system level. The next main section will examine theoretically just such constraints, which I contend still have a much more pronounced impact on U.S. geostrategy than the counterforces that I have discussed over the last few pages.

Why Systemic Incentives Discourage Retrenchment

U.S. hegemony and the construction and preservation of U.S. military bases around the world are driven not solely by supply-side factors emanating from the establishment in Washington; they are also, to a great extent, solicited by its many clients (Coker, 2015, p. 46). Ultimately, though, decision-making on these matters is the prerogative of the United States. Even in Washington, however, there are dissenters. Voices calling for restraint and at least partial retrenchment seem to have become more common in recent years, especially as some of the wars in which the United States has been embroiled have largely proven unwinnable (Walldorf & Yeo, 2019). Every now and then—in fact, almost exactly every 20 years, one observer claims (Sestanovich, 2020)—Americans do seem to engage in serious debate about what the *purpose* of it all really is; about whether their country should continue being present in most or all of the corners of the world. Still, there is very little to suggest that the United States ever *does* alter its grand strategy and opts for doing significantly less. The dissenting voices never really carry over into actual policymaking. This prompts the question: Given the existence also of carriers of retrenchment ideas, why do the arguments of the primacists and the *Stayers* always prevail? Why, in such deliberations, do the winners win?

The first and most simplistic answer is that they are more numerous and better resourced than their ideational competitors (Walt, 2018, pp. 112–113). This enables them to establish "discourse dominance"; they are the ones superiorly positioned to frame the foreign-policy issues under consideration and create the boundaries beyond which certain viewpoints—such as (neo-)isolationism—are considered illegitimate (Layne, 2017). But this begs the question: Why are the winners seemingly always affiliated with primacy? And why do they so tenaciously highlight all the bad things that will likely follow from retrenchment, very rarely acknowledging that things *might* as well turn out decently even if the United States scales back its foreign activities?

The Pessimism of Realism

It is here, first of all, that neoclassical realism's penchant for using structural realism as its analytical first cut proves helpful. Indeed, as is well known, political realism more broadly paints a very bleak overall picture of international politics. The international system, realists say, is anarchic; it is an arena for zero-sum thinking; states cannot trust other states; human beings are by nature power-seeking *animus dominandi* (Morgenthau, 2006 [1948]); or they may be moral, but the collection of people in a society imbues that society with an *im*moral "patriotic altruism" that is aimed at dominating other societies (Niebuhr, 2005 [1932], p. 62); states will need constantly to think about their relative power position, as only power can contain power; power should be maximized, which leads to the "tragedy of great power politics" (Mearsheimer, 2001); or, alternatively, only security should be maximized and power only "satisficed," which nonetheless leads states into another tragedy—that of the security dilemma. "The real art of statesmanship," writes Robert Kaplan

(2010, p. 134) in *Monsoon*, is "to think tragically in order to avoid tragedy." In security matters, the signals radiating from the international system, according to political realism, are overwhelmingly negative ones, and they generally work to advise "safety-first" policies whose alternatives carry the risk of severe—sometimes existential—threats against any nation so naive as to ignore the realities of realist reasoning.

The pessimism of such systemic signals is further compounded due to these signals also being partially concealed. "The first lesson the student of international politics must learn and never forget," writes Hans Morgenthau (2006 [1948], p. 22) in his classic *Power Among Nations*, "is that the complexities of international affairs make simple solutions and trustworthy prophecies impossible." For example, much of what is written on international security, and much of actual policymaking in military-security affairs, hinge on estimations of relative power. But power is a feature of international politics that does not easily lend itself to measurement; and *relative* power, or the balance of power, is even more problematic to gauge. What seems in theory to be a perfectly doable exercise is far from doable in the real world. And, notes Henry Kissinger (1994, p. 63) in *Diplomacy*, "[e]ven more complicated is harmonizing one's calculations with those of other states, which is the precondition for the operation of a balance of power."

This is one of the main reasons why scholars reach quite different conclusions regarding whether or not China can soon "catch up" with the United States. Some indicate it has already done so on many of the decisive indicators of power (Allison, 2017, Part 1); by the estimation of others, who emphasize alternative measures, China still has a very long distance to cover (Beckley, 2018; Brooks & Wohlforth, 2016). In any event, it is up to policymakers to settle on conclusions after a filtering process; that is to say, the unclear realities of the distribution of power are further filtered or translated through the minds of decision-makers, whose *perceptions* of the realities of relative power greatly shape their policies. Predicting outcomes only by focusing on the distribution of material resources, therefore, misses the key intervening variable that is whether and how policymakers really *understand* the situation they are in (Rose, 1998, pp. 157–161).

Added to this is the perpetual problem of understanding motivations or *intentions*. The intentions of others—in particular, their *future* ones—cannot confidently be known; other states are hence potential threats, which means that states are prone to fear each other (Mearsheimer, 2001, pp. 48–88). What still applies, therefore, is the *security dilemma*, briefly presented earlier but by necessity revisited here. The security dilemma denotes the self-defeating properties of a situation where "what one does to enhance one's own security causes reactions that, in the end, can make one less secure" (Posen, 1993, p. 28). It was John Herz (1950) who introduced the concept over seventy years ago; and the main elements of the dilemma have since been elaborated by numerous scholars (e.g., Glaser, 1997; Jervis, 1978; Montgomery, 2006; Snyder, 1984). Its roots, however, can really be traced back centuries, as it is closely connected to "Hobbesian" ideas about the inescapable security worries that obtain from living in a world without a sovereign. Hence, the dilemma is structural; it belongs to the system level of analysis and arises independently of unit-level characteristics. Even in a world dominated by status quo-inclined powers, states cannot easily escape the logic that helps produce a highly competitive environment wherein states are induced to worry a great deal about their security. For if they do not do so, and if they fail to implement better-safe-than-sorry policies when doing so would be the prudent course of action, they risk being exploited by states whose intentions might not be as benevolent after all (Jakobsen & Halvorsen, 2018, p. 158).

In that sense, being caught up in a security dilemma is not necessarily something that arises from irrational behavior. As Robert Jervis (1976, p. 62) notes in *Perception and Misperception in International Politics*, the fear-based behavior of states does not result from any "limitations on rationality imposed by human psychology nor in a flaw in human nature, but in a correct appreciation of the consequences of living in a Hobbesian state of nature." Rather, worst-case scenario planning sometimes, or perhaps most times, constitutes prudent insurance against threats to one's security or survival (Glaser, 2004). This fits squarely with the famous Prisoners' Dilemma, an analogous situation where players *rationally* end up at a sub-optimal outcome *so as to be sure* to avoid even worse ones. This can help explain, and it can often justify or at least make us better understand, a host of seemingly anomalous and counterproductive decisions and outcomes in international affairs. The security dilemma does not only manifest as arms races. The quest for power more generally is affiliated with numerous additional acts that can spur countermoves and vicious spirals; these include the stationing of troops abroad (Nieman et al.,

2021) and, more broadly, the exertion of influence over others to alleviate potentially "adverse chain reactions" before these gain momentum (Mandelbaum, 1988, Ch. 3).

Maximizing Influence in a Murky World

The last of the points above, about the exertion of influence, indicates what is really the gist of neoclassical realism's expectations about the foreign-policy behavior of states. Leaders of states, and members of the broader foreign-policy community, do not necessarily endeavor to maximize security per se. Security is anyway an elusive concept in an international-political sphere that is "murky and difficult to read" (Rose, 1998, p. 152). But what decision-makers usually set out to do is to attempt to *shape and control* this murky environment as much as their state's power resources permit. That is to say, they strive to acquire *political influence* over others (Onea, 2012, p. 140). States are thus not first and foremost security maximizers, or maximizers of material resources; they are rather "influence-maximizers," as Fareed Zakaria (1998, pp. 19–20) puts it. In "the anarchic, nonhierarchical international environment," he goes on, "states are driven by the system's competitive imperative: if a state does not attempt to maximize its influence, then another will seize the opportunity in its stead." Similar to the way the conflictual outcomes from interactions among Waltz's security-maximization states come to resemble the equally conflictual outcomes generated by Mearsheimer's power-maximization states, states that seek to maximize influence rarely scale back their international activities voluntarily. Instead, Zakaria (1998, p. 60) asserts, "expansion has a way of begetting further expansion," as it works to enlarge the expanding state's interests and thereby also to increase the perceived *threats* against those interests. States are therefore bound to bump into each other quite often in the international competition for influence.

The scope for expansion, however, is intimately tied to the *material resources* a state possesses. If these do not increase, extending the state's foreign influence much further will prove difficult. There are two alternatives to expansion: a state can either attempt to *uphold the status quo*, or it can *retrench*. As for the former option, both systemic and domestic incentives exist to bolster its attraction. The ambiguity and murkiness that characterize the international system and the signals it sends, coupled with domestic competition over ideas, enhance the lure of doing little or nothing in terms of altering the country's foreign policy or geostrategy (Kitchen, 2010, p. 113). This is especially so if determined leadership is lacking. In such circumstances, a compromise among foreign-policy elites that centers on continuity represents a natural outcome.

Relative Power and the Scope for Retrenchment

A decision to *retrench*, neoclassical realists would say, is far less likely in most instances. For example, Jeffrey Taliaferro et al. (2018) argue that, for a hegemon experiencing relative decline, three conditions will normally have to be met in order for it to acquiesce to a process of "peaceful change" that essentially transfers influence to a rising challenger. Two of these conditions are primarily "second-image" ones, revolving around issues that concern the domestic costs of, and constraints on, resisting the challenger. The third is closely linked to systemic factors. It centers on the question of whether the hegemon *perceives* that change is inevitable and that resistance to such change represents an untenable position. This, in turn, hinges on "the relative distribution of power, the nature of the strategic environment that a hegemon confronts in a particular region, and the degree of systemic clarity regarding threats and opportunities" (Taliaferro et al., 2018, p. 286). In the case of the United States (when facing off against China, Russia, or any other challenger), there is little to suggest that there are widespread perceptions that, *for such reasons*, retrenchment or withdrawal from other regions is *inevitable*. The United States is today by no means confronted with the same brutal power realities as was, for instance, the Soviet Union in the late 1980s. The authors conclude that, in spite of widespread "fears about Washington's retreat from global leadership under President Donald Trump, America still enjoys overwhelming preponderance in economic and military capabilities. That condition will likely continue for the foreseeable future," which implies that there "is no reason to expect the United States to accept peaceful change in the near future" (Taliaferro et al., 2018, pp. 289–290).

In a neoclassical realist world, therefore, conscious and voluntary retrenchment is considerably rarer than is expansion, which is rarer still than conscious (or what is perhaps more often than not less conscious) attempts to uphold the status quo. Indeed, even expansionary drives can often best be judged as status-quo-preserving ventures; offensive moves for eminently defensive purposes are a staple trait of great-power politics. For example, an increase in forward-deployed troops, for balancing purposes, or to fill geopolitical "vacuums" (lest they be filled by one's adversary), are often based on the same better-safe-than-sorry inclination that undergirds policies driven by a wish to preserve things as they are. The increase of the United States' military presence in Central and Eastern Europe in recent years (which actually does not show in the official troop numbers) is by and large a product of the domino-related fear that Russia's new-found assertiveness generates. The same is true for the Asia–Pacific region, where increased activity *is* reflected in official numbers. From Washington's perspective, China's rise needs to be balanced or countered *in order to preserve stability and the prevailing state of things*, which also follows from realist theory. In fact, in those cases where U.S. troop numbers have increased in recent years, this increase is almost always motivated by the objective of preserving the status quo. So this expansion, which is simultaneously real *and* only apparent, is, in terms of the objectives sought, actually less about aggrandizement than it is about keeping things as they are. This is particularly so when troop increases manifest in host countries where the United States already has a physical presence. This is mostly (albeit not always) the case in Eastern and Central Europe, and it is virtually always the case in the Asia–Pacific region. Changes in circumstances—that is to say, an increase in China's and Russia's power and ambitions, along with related security-dilemma mechanisms—spawn U.S. "expansion" to secure its position at its self-declared reference point, which for all intents and purposes equals the status quo.

That upholding the status quo is the primary mission for most states follows from the basic assumption and prediction of the neoclassical realist school. Over the long run, states tend to align their foreign-policy ambitions with the amount of power resources they possess (Rose, 1998, p. 152). As most states do not see their relative power position change much in the short term, and often not much in the longer term either, the basic expectation with respect to foreign-policy strategies and outcomes should usually be one of a continuation of the status quo. This is the empirical tendency or pattern which neoclassical realists predict will appear. Obviously, some countries do not fall neatly in line with such a pattern: rising and declining states are expected, sooner or later, to redefine their policies and adjust their aims and behavior, and state- and individual-level intervening variables can help send a country in an unexpected direction. But, as Zakaria (1998, p. 184) notes, even if several factors obviously help influence the scope of a state's ambitions, "in the first instance" it is determined "by a state's power."

In other words, under the assumption that the United States' command of material resources is roughly at the same level, in relative terms, as it was 5, 10, or 25 years ago—and such an assumption is supported by at least the commonly-employed broad indicators of military budgets and size of the economy—we should not expect to witness any substantial withdrawal of U.S. forward-deployed troops. Individual agency surely also plays a part; but the central issue here is that agency geared toward moving away from the path that is at root shaped by the realities of power faces overwhelming obstacles. These obstacles originate in the system structure, but they are often quite consistently articulated and represented by the broader foreign-policy community.

*

As it is, the foreign-policy community in the United States—the Blob—will be extremely reluctant to acquiesce to the downsizing of U.S. military forces in regions and countries that have for decades been regarded by the very same community as absolutely vital to the national interests of the United States. Retrenchment proposals and decisions by the executive can be expected to be met by stiff resistance by the overwhelming majority of foreign-policy professionals. These will likely set in motion an array of measures aimed at blocking such a development: they will suggest and they will implement legislation to counter it; they will write reports, op-eds, and news articles fretting over the power vacuums that will automatically ensue from any withdrawal anywhere; they will incessantly argue that these vacuums are bound to be filled, perhaps immediately, by the power of Russia, of China, of Iran, of North Korea, or of *jihadists*; they will point to all the threats that are lurking abroad waiting for America to withdraw so they can finally manifest in mayhem; they will insistently inflate these threats to hammer home their point; they will claim that any withdrawal, almost no matter how modest, will harm U.S. credibility enormously, marking the beginning of the end of the American Empire; they will point to the vast and wholly negative economic repercussions of retrenchment, as the transactions of international investors and traders are intimately tied to the confidence they have in U.S. power and global engagement; they will join forces with equally worried state allies of the United States to add to the weight of their message; they will contend that a withdrawal of troops will push many more countries into developing weapons of mass destruction, including nuclear bombs; and they will, when conveying their grave concerns, use doomsday-like rhetorics that really signal that a withdrawal of troops—again, almost irrespective of questions of how many and wherefrom—is not an option that a leader of the United States can or should ever even contemplate. All these things—all these cases of intervention and argumentation and creation of still-hypothetical narratives—are to be expected when the issue of retrenchment is seriously considered by the country's Commander-in-Chief. There is little reason to think that most of the foreign-policy professionals do not believe what they say and what they write. And, it should be added, albeit on my own account, their doomsday predictions, however hypothetical and counterfactual, are not *necessarily* mistaken.

References

Allison, G. (2017). *Destined for war: Can America and China escape Thucydides's trap?* Houghton Mifflin Harcourt.

Altman, R. C. (2009). Globalization in retreat: Further geopolitical consequences of the financial crisis. *Foreign Affairs, 88*(4), 2–7.

Ashford, E. (2018). Unbalanced: Rethinking America's commitment to the Middle East. *Strategic Studies Quarterly, 12*(1), 127–148.

Baum, M. A., & Potter, P. B. K. (2008). The relationships between mass media, public opinion, and foreign policy: Toward a theoretical synthesis. *Annual Review of Political Science, 11*(1), 39–65.

Beckley, M. (2018). The power of nations: Measuring what matters. *International Security, 43*(2), 7–44.

Braithwaite, A. (2015). Transnational terrorism as an unintended consequence of a military footprint. *Security Studies, 24*(2), 349–375.

Brands, H. (2018, June 3). How many US troops are needed in Korea? *Korea Herald*. Retrieved January 11, 2021, from http://www.koreaherald.com/view.php?ud=20180603000063&ACE_SEARCH=1

Brooks, S. G., & Wohlforth, W. C. (2005). Hard times for soft balancing. *International Security, 30*(1), 72–108.

Brooks, S. G., & Wohlforth, W. C. (2016). The rise and fall of the great powers in the twenty-first century: China's rise and the fate of America's global position. *International Security, 40*(3), 7–53.

Brzeziński, Z. (1997). *The grand chessboard: American primacy and its geostrategic imperatives*. Basic Books.

Brzeziński, Z. (2012). *Strategic vision: America and the crisis of global power*. Basic Books.

Byman, D. (2019). *Road warriors: Foreign fighters in the armies of jihad*. Oxford University Press.

Christensen, T. J. (2011). The advantages of an assertive China: Responding to Beijing's abrasive diplomacy. *Foreign Affairs, 90*(2), 54–67.

Christensen, T. J. (2016). *The China challenge: Shaping the choices of a rising power*. W. W. Norton.

Coker, C. (2015). *The improbable war: China, the United States and the continuing logic of great power conflict*. Oxford University Press.

Cooley, A., & Nexon, D. (2020). *Exit from hegemony: The unraveling of the American global order*. Oxford University Press.

Foulon, M. (2015). Neoclassical realism: Challengers and bridging identities. *International Studies Review, 17*(4), 635–661.

Gadarian, S. K. (2010). The politics of threat: How terrorism news shapes foreign policy attitudes. *Journal of Politics, 72*(2), 469–483.

Gause, F. G., III. (2019). Should we stay or should we go? The United States and the Middle East. *Survival, 61*(5), 7–24.

Gilpin, R. (1981). *War and change in world politics*. Cambridge University Press.

Glaser, C. L. (1992). Political consequences of military strategy: Expanding and refining the spiral and deterrence models. *World Politics, 44*(4), 497–538.

Glaser, C. L. (1997). The security dilemma revisited. *World Politics, 50*(1), 171–201.

Glaser, C. L. (2004). When are arms races dangerous? Rational versus suboptimal arming. *International Security, 28*(4), 44–84.

Glaser, J. (2017, July 18). *Withdrawing from overseas bases: Why a forward-deployed military posture is unnecessary, outdated, and dangerous*. Cato Institute Policy Analysis 816. Retrieved February 4, 2021, from https://www.cato.org/sites/cato.org/files/pubs/pdf/pa_816.pdf

Gray, C. S. (1999). Clausewitz rules, OK? The future is the past: With GPS. *Review of International Studies, 25*(5), 161–182.

He, K. (2017). Explaining United States-China relations: Neoclassical realism and the nexus of threat-interest perceptions. *Pacific Review, 30*(2), 133–151.

Herz, J. H. (1950). Idealist internationalism and the security dilemma. *World Politics, 2*(2), 157–180.

Holmes, J. R., & Yoshihara, T. (2017). Deterring China in the "gray zone": Lessons of the South China Sea for U.S. alliances. *Orbis, 61*(3), 322–339.

Jakobsen, J., & Halvorsen, T. (2018). The durability of the security dilemma: An empirical investigation of action-reaction dynamics in states' military spending (1988–2014). *Chinese Journal of International Politics, 11*(2), 153–192.

Jervis, R. (1976). *Perception and misperception in international politics*. Princeton University Press.

Jervis, R. (1978). Cooperation under the security dilemma. *World Politics, 30*(2), 167–214.

Jervis, R. (2017). *How statesmen think: The psychology of international politics*. Princeton University Press.
Johnson, D. D. P., & Tierney, D. (2018–2019). Bad world: The negativity bias in international politics. *International Security, 43*(3), 96–140.
Johnston, A. I. (2013). How new and assertive is China's new assertiveness? *International Security, 37*(4), 7–48.
Kahneman, D., & Tversky, A. (1979). Prospect theory: An analysis of decision under risk. *Econometrica, 47*(2), 263–291.
Kaplan, R. D. (2010). *Monsoon: The Indian Ocean and the future of American power*. Random House.
Kennedy, P. (1988). *The rise and fall of the great powers*. Random House.
Kirshner, J. (2015). The economic sins of modern IR theory and the classical realist alternative. *World Politics, 67*(1), 155–183.
Kissinger, H. (1994). *Diplomacy*. Simon and Schuster.
Kitchen, N. (2010). Systemic pressures and domestic ideas: A neoclassical realist model of grand strategy formation. *Review of International Studies, 36*(1), 117–143.
Knutsen, T. L. (1999). *The rise and fall of world orders*. Manchester University Press.
LaFeber, W. (2008). *America, Russia, and the Cold War, 1945–2006* (10th ed.). McGraw-Hill (Original work published 1967).
Lambert, A. (2018). *Seapower states: Maritime culture, continental empires and the conflict that made the modern world*. Yale University Press.
Layne, C. (1993). The unipolar illusion: Why new powers will rise. *International Security, 17*(4), 5–51.
Layne, C. (2017). The US foreign policy establishment and grand strategy: How American elites obstruct strategic adjustment. *International Politics, 54*(3), 260–275.
Layne, C. (2018). The US-Chinese power shift and the end of the Pax Americana. *International Affairs, 94*(1), 89–111.
Levy, J. S., & Thompson, W. R. (2010). Balancing on land and at sea: Do states ally against the leading global power? *International Security, 35*(1), 7–43.
Lieber, K. A., & Alexander, G. (2005). Waiting for balancing: Why the world is not pushing back. *International Security, 30*(1), 109–139.
Liff, A. P., & Ikenberry, G. J. (2014). Racing toward tragedy? China's rise, military competition in the Asia Pacific, and the security dilemma. *International Security, 39*(2), 52–91.
Lobell, S. E., Ripsman, N. M., & Taliaferro, J. W. (Eds.). (2009). *Neoclassical realism, the state, and foreign policy*. Cambridge University Press.
MacDonald, P. K., & Parent, J. M. (2011). Graceful decline? The surprising success of great power retrenchment. *International Security, 35*(4), 7–44.

Mandelbaum, M. (1988). *The fate of nations: The search for national security in the nineteenth and twentieth centuries*. Cambridge University Press.

Mandelbaum, M. (2019). *The rise and fall of peace on earth*. Oxford University Press.

Marsh, K. (2012). Managing relative decline: A neoclassical realist analysis of the 2012 U.S. defense strategic guidance. *Contemporary Security Policy, 33*(3), 487–511.

Mearsheimer, J. J. (1990). Back to the future: Instability in Europe after the Cold War. *International Security, 15*(1), 5–56.

Mearsheimer, J. J. (2001). *The tragedy of great power politics*. W. W. Norton.

Mearsheimer, J. J. (2011). Imperial by design. *National Interest, 111*(Jan/Feb), 16–34.

Mearsheimer, J. J. (2014). Why the Ukraine crisis is the West's fault: The liberal delusions that provoked Putin. *Foreign Affairs, 93*(5), 77–89.

Mearsheimer, J. J., & Walt, S. M. (2007). *The Israel lobby and U.S. foreign policy.* Farrar, Straus and Giroux.

Meibauer, G. (2020). Interests, ideas, and the study of state behaviour in neoclassical realism. *Review of International Studies, 46*(1), 20–36.

Modelski, G. (1978). The long cycle of global politics and the nation-state. *Comparative Studies in Society and History, 20*(2), 214–235.

Modelski, G., & Thompson, W. R. (1996). *Leading sectors and world powers: The coevolution of global politics and economics*. University of South Carolina Press.

Montgomery, E. B. (2006). Breaking out of the security dilemma: Realism, reassurance, and the problem of uncertainty. *International Security, 31*(2), 151–185.

Morgenthau, H. J. (2006). *Politics among nations: The struggle for power and peace* (7th ed., revised by K. W. Thompson & D. Clinton). McGraw-Hill (Original work published 1948).

New York Times. (2016, July 21). *Transcript: Donald Trump on NATO, Turkey's coup attempt and the world*. https://www.nytimes.com/2016/07/22/us/politics/donald-trump-foreign-policy-interview.html

Niebuhr, R. (2005). *Moral man and immoral society: A study in ethics and politics.* Continuum (Original work published 1932).

Nieman, M. D., Martinez Machain, C., Chyzh, O., & Bell, S. R. (2021). An international game of risk: Troop placement and major power competition. *Journal of Politics, 83*(4), 1307–1321.

Nye, J. S., Jr. (2010). American and Chinese power after the financial crisis. *Washington Quarterly, 33*(4), 143–153.

O'Mahoney, A., Priebe, M., Frederick, B., Kavanagh, J., Lane, M., Johnston, T., Szayna, T. S., Hlavka, J. P., Watts, S., & Povlock, M. (2018). *U.S. presence and the incidence of conflict.* RAND Corporation.

Onea, T. (2012). Putting the 'Classical' in neoclassical realism: Neoclassical realist theories and US expansion in the post-Cold War. *International Relations, 26*(2), 139–164.

Onea, T. (2021). Neoclassical realism and grand strategy. In G. Meibauer, L. Desmaele, T. Onea, N. Kitchen, M. Foulon, A. Reichwein, & J. Sterling-Folker, (Eds.), Forum: Rethinking neoclassical realism at theory's end. *International Studies Review, 23*(1), 274–277.

Organski, A. F. K., & Kugler, J. (1980). *The war ledger*. University of Chicago Press.

O'Shea, P. (2019). Strategic narratives and US military bases in Japan: How 'deterrence' makes the Marine base on Okinawa 'indispensable.' *Media, War & Conflict, 12*(4), 450–467.

Pape, R. A. (2005a). Soft balancing against the United States. *International Security, 30*(1), 7–45.

Pape, R. A. (2005b). *Dying to win: The strategic logic of suicide terrorism*. Random House.

Posen, B. R. (1993). The security dilemma and ethnic conflict. *Survival, 35*(1), 27–47.

Rathbun, B. (2008). A rose by any other name: Neoclassical realism as the logical and necessary extension of structural realism. *Security Studies, 17*(2), 294–321.

Ripsman, N. M., Taliaferro, J. W., & Lobell, S. E. (2016). *Neoclassical realist theory of international politics*. Oxford University Press.

Rose, G. (1998). Neoclassical realism and theories of foreign policy. *World Politics, 51*(1), 144–172.

Schweller, R. L. (1996). Neorealism's status-quo bias: What security dilemma? *Security Studies, 5*(3), 90–121.

Schweller, R. (2018). Opposite but compatible nationalisms: A neoclassical realist approach to the future of US-China relations. *Chinese Journal of International Politics, 11*(1), 23–48.

Sestanovich, S. (2020, May). The brilliant incoherence of Trump's foreign policy. *The Atlantic*. https://www.theatlantic.com/magazine/archive/2017/05/the-brilliant-incoherence-of-trumps-foreign-policy/521430/

Sheafer, T., & Dvir-Gvirsman, S. (2010). The spoiler effect: Framing attitudes and expectations toward peace. *Journal of Peace Research, 47*(2), 205–215.

Snyder, G. H. (1984). The security dilemma in alliance politics. *World Politics, 36*(4), 461–495.

Snyder, G. H. (2002). Mearsheimer's world—Offensive realism and the struggle for security: A review essay. *International Security, 27*(1), 149–173.

Taliaferro, J. W., Lobell, S. E., & Ripsman, N. M. (2009). Introduction: Neoclassical realism, the state, and foreign policy. In S. E. Lobell, N. M. Ripsman, & Jeffrey W. Taliaferro (Eds.), *Neoclassical realism, the state, and foreign policy* (pp. 1–41). Cambridge University Press.

Taliaferro, J. W., Lobell, S. E., & Ripsman, N. M. (2018). Is peaceful change in world politics always desirable? A neoclassical realist perspective. *International Studies Review, 20*(2), 283–291.
Thompson, W. R. (1983). Uneven economic growth, systemic challenges, and global wars. *International Studies Quarterly, 27*(3), 341–355.
Walldorf, C. W., Jr., & Yeo, A. (2019). Domestic hurdles to a grand strategy of restraint. *Washington Quarterly, 42*(4), 43–56.
Walt, S. M. (2018). *The hell of good intentions: America's foreign policy elite and the decline of U.S. primacy*. Farrar, Straus and Giroux.
Waltz, K. N. (1979). *Theory of international politics*. McGraw-Hill.
Waltz, K. N. (1993). The emerging structure of international politics. *International Security, 18*(2), 44–79.
Westad, O. A. (2017). *The Cold War: A world history*. Basic Books.
Zakaria, F. (1998). *From wealth to power: The unusual origins of America's world role*. Princeton University Press.
Zakaria, F. (2008). *The Post-American world*. W. W. Norton.

CHAPTER 4

The System, the Psyche, and the Stayers

*

The preceding chapter laid out the first, primarily system-level explanation of why a negative outlook on the prospects for retrenchment consistently prevails in foreign-policy deliberations and decision-making. But this still leaves much of the actual filtering mechanism unaccounted for. Neoclassical realism does not really deal in any depth with the microfoundations of Waltz's first and second images; its focus and its strengths lie elsewhere.

And, clearly, there are other prominent schools of thought in international relations that have far more faith in the potential benevolence of the international system and its constituent structures and actors. The end of the Cold War spawned much optimism regarding the future of international politics (Fukuyama, 1989). Many expected wars and security dilemmas to become much less relevant on account of the spread of political and economic liberalism, which would constrain bellicose leaders, enhance economic interdependence, and strengthen trust among states (Bueno de Mesquita et al., 1999; Maoz & Russett, 1993; Mousseau et al., 2003). Others pointed to similar effects being produced by a steadily increasing number of liberal international institutions (Ikenberry, 2001). If the incentives and willingness to fight really had declined markedly, so would also security competition.

J. Jakobsen, *The Geopolitics of U.S. Overseas Troops and Withdrawal*, Palgrave Studies in International Relations,
https://doi.org/10.1007/978-3-030-94488-9_4

Indeed, as pointed out in Chapter 2, even many realists, in particular offshore balancers, draw the conclusion that partial retrenchment is unquestionably the optimal way to proceed for the United States. This is so even as they hold that the international system remains anarchic, highly competitive, inherently dangerous, and fundamentally "realist." System-level theories cannot say much about *why* it is that the foreign-policy community tends to take its cue from the gloomy descriptions and predictions of world affairs rather than from either the more optimistic ones or from the pessimistic ones that still advocate retrenchment. Neither has realism in its *classical* variant elaborated much on its standard assumption that human beings are hardwired into seeking power and domination over others. And the origins of the very human tendency to abhor losses, even small ones, and to interpret the reigning status quo as a natural state of affairs that should be preserved, even at great cost, are also mostly overlooked by prominent realist theories (Taliaferro, 2004a, p. 5). For a fuller understanding of how systemic stimuli translate into foreign-policy preferences and recommendations, I will, in this chapter, first turn to the *psychological* underpinnings of the "transmission belt" that have mostly been left underexplained by realist theory. Thereafter, the system level is reintroduced, primarily in the guise of the notorious domino theory.

Prospect Theory and International Security

In their 2018–2019 article "Bad World," published in *International Security*, Dominic Johnson and Dominic Tierney point to an apparent general puzzle in international politics: Even though state leaders, when judging themselves, tend to exhibit certain positive biases (such as overconfidence), negative phenomena and developments *in a state's external environment* seem to carry much more weight in the minds of decision-makers than do positive ones (Johnson & Tierney, 2018–2019). This observation and phenomenon is what Paul Rozin and Edward Royzman had previously referred to as a "negativity bias." They argue that this bias operates "across a wide range of domains, and in nonhuman as well as human animals," and it works to ensure "that in most situations, negative events are more salient, potent, dominant in combinations, and generally efficacious than positive events" (Rozin & Royzman, 2001, p. 297).

Bad Is Badder Than Good Is Good

The notion that "bad is stronger than good" (Baumeister et al., 2001) breaks with the expected-utility principle, which asserts that actors in international relations, and in most other realms, are basically rational. The rational-actor assumption holds that, when confronting choices, actors seek to maximize their expected utility by carefully considering the likely outcomes from different courses of action. They do so without displaying any consistent biases, such as showing an undue preference for the status quo, or over-weighting or underweighting the utility of alternatives that lead to negative outcomes.

However, rational behavior in the expected-utility-maximization sense hardly seems to be an overly dominant mode of behavior in international affairs. To the contrary, a plethora of oft-observed acts cannot easily be explained by rational-actor models, as they indicate that the good and the bad have an *asymmetrical* impact on perceptions, judgments, and decisions (Johnson & Tierney, 2018–2019, p. 96). We notice quite often, for example, that states and their leaders magnify the threats that originate from abroad, and that they act upon that magnification. For instance, when the superpower conflict ends, Washington quickly and almost inevitably identifies new "demons" that are judged to be at least as bad as was Soviet Communism. Furthermore, small-scale military interventions are often followed by mission creep; when mission creep sets in and the intervention proves a failure, the interventionist doubles down on its commitments; losses of territory or influence are not offset in the minds of leaders by similar gains in other places; costly wars are initiated for dubious reasons, but the subsequent road to a compromise peace is usually extremely arduous and sometimes outright impassable; the status quo, in whatever the topic area, is seen by most as perfectly natural, while its alternatives (*any* alternative) are deemed an affront to order, stability, and the "world community"; and withdrawing forward-deployed troops from any region or host nation is by many interpreted as the first step toward the collapse of an empire.

With respect to the external environment, therefore, people seem to be more sensitive to negative phenomena than to positive ones. These are inherently psychological mechanisms that quite likely originate in our evolutionary past (McDermott et al., 2008). At root, negativity biases should have been favored by natural selection as they prescribe decisions and behavior that *minimize the likelihood of experiencing very costly outcomes* (Haselton & Nettle, 2006; Johnson et al., 2013). They might well have worked to help our ancestors survive and reproduce (Johnson & Tierney, 2018–2019, p. 96). They are, in effect, nature's way of forcing upon humans (and animals) mandatory insurance—albeit with excessive premiums (as is surely also the case for many more mundane types of insurance!). Yet, our ingrained fear of (mostly harmless) spiders and snakes, and of heights and the dark—and the lengths to which we are prepared to go to shield ourselves from the unlikely risks of being invaded by the military of another state or killed by transnational terrorists—are not necessarily "a shortcoming in rational behavior;" instead, they signify "an adaptation of behavior to a complex, uncertain world. Biased mechanisms are not design defects of the human mind, but rather design features" (Haselton & Nettle, 2006, p. 59).

Prospect Theory, Loss Aversion, and the Status-Quo Bias

Over four decades ago, Daniel Kahneman and Amos Tversky set out to develop a theory of decision-making under conditions of risk that took just such design features into consideration. They called it prospect theory (Kahneman & Tversky, 1979). Contrasting markedly with the tenets of expected-utility theory, prospect theory's "fundamental ideas," which have been borne out by ample experimental empirical evidence, "are that reference points exist, and that losses loom larger than corresponding gains" (Kahneman, 2011, p. 297). In practical terms, the negative feelings incurred by people from having to, say, *relinquish* one hundred dollars are considerably stronger than the positive feelings experienced from *receiving* the same sum. Thus, the way outcomes from decisions are *framed*—specifically, whether they are placed in the realm of losses or of gains—matters greatly (Berejikian, 2002, p. 762). The framing process broadly concerns how people tend to make sense of an external environment—the world—that is normally quite ambiguous.

Two elements—loss aversion and status-quo (or reference-point) bias—form the core of prospect theory. They are closely linked to one another, and both are of great interest to us here. They play a fundamental role as cognitive short-cuts that help guide decision-makers' perceptions and understanding of a given situation, ultimately enabling them to make and to implement decisions in what is basically a murky world. First, *loss aversion* is simply the idea, expectation, and apparent empirical reality that "losses loom larger than corresponding gains" (Tversky & Kahneman, 1991, p. 1039). This goads people into undertaking risky action in order to try and recuperate what has been lost, while it dampens their risk acceptance when they are faced with the prospects of gaining something that were never theirs in the first place. This offers an explanation, for instance, of why animals as well as communities and states are usually willing to fight much harder over territory that is theirs (or has once been theirs) than over the territory of others. The very close sibling to the concept of loss aversion is the "endowment effect." The term was coined by Richard Thaler, and it describes how one attaches greater value to an object already in one's possession or endowment than one does to a similar object when it is not part of one's endowment. It is exactly this "underweighting of opportunity costs" that constitutes the endowment effect (Thaler, 1980, p. 44).

Second, the decisions of people and states are influenced by the *reference point* to which possible outcomes are compared. Specifically, a *status-quo bias*—"a preference for the current state" (Kahneman et al., 1991, p. 194)—commonly applies. The concept was originally introduced by the economists William Samuelson and Richard Zeckhauser; their experimental evidence suggested that people, in an even broader way than suggested by the loss-aversion mechanism, tend to prefer remaining at some existing state (Samuelson & Zeckhauser, 1988). The disadvantages of leaving the status quo normally overshadow the advantages. This is in no small part due to people becoming psychologically committed to the existing state of affairs, which causes them to overweight sunk costs, which in turn helps justify past commitments (Samuelson & Zeckhauser, 1988, p. 37). Moreover, the reference point, usually the status quo, easily attains normative qualities of its own. In one passage of Christopher Clark's *The Sleepwalkers*, a magnificent account of the confusion that besieged decision-makers in the run-up to the First World War, the author alludes to an Austrian public lawyer, Georg Jellinek, who lived and worked in the late nineteenth century. This lawyer, Clark (2012, p. 361) writes, had published an analysis of

> what he called 'the normative power of the factual'. By this he meant the tendency among human beings to assign normative authority to actually existing states of affairs. Human beings do this, he argued, because their perceptions of states of affairs are shaped by the forces exerted by those states of affairs. Trapped in this hermeneutic circularity, humans tend to

> gravitate quickly from the observation of what exists to the presumption that an existing state of affairs is normal and thus must embody a certain ethical necessity. When upheavals or disruptions occur, they quickly adapt to the new circumstances, assigning to them the same normative quality they had perceived in the prior order of things.

In an exquisite way, Clark here presents the two essential elements of the reference-point bias. As the preceding paragraph has clarified, the first is that people tend to prefer to remain at the status quo. To invoke yet another Austrian from times past, the great novelist Robert Musil and his immortal masterpiece *The Man Without Qualities*: the vast majority of people lack "a sense of possibility"; that is, they lack the ability to contemplate alternative states of affairs, viz., to "conceive of everything there might be just as well," which is a quality possessed only by those very few who "attach no more importance to what is than to what is not" (Musil, 2005 [1930], pp. 11–12).

Changes to the Reference Point

The second element of the reference-point bias is reflected by the last sentence of the quote from *The Sleepwalkers*; it concerns the possibility that "upheavals and disruptions" may create a new status quo or reference point, which then eventually (or rather quickly, as the Austrian lawyer surmised) comes to possess the same "ethical necessity" as the old one. People and states adapt, albeit often very reluctantly. For example, in 1990, the "ethical necessity" of Yugoslavia was not in doubt among Europeans (or Americans) living *outside of* that country. Yet, a few years and wars (and an abundance of suffering) later, one could hardly find *any* individual *anywhere* in Europe (or the United States) clamoring for the return of that multi-ethnic state whose "ethical necessity" had completely evaporated in the eyes of the world.

However, the reference point at any given time need not necessarily align with the current status quo. In the realm of international affairs, for instance, some countries—such as rising powers—may exhibit aspirations that help move *their* reference points away from the status quo and to what are as-yet-unrealized states of affairs where their power positions have been improved (Levy, 1992, p. 289). If a nation should eventually reach such a state, for then to go on and improve its position even further, one would expect it to "renormalize" for that additional gain relatively quickly, naturally adopting a bias for the new reference point (Jervis, 1992, p. 200). Contrarily, states are not likely to renormalize as quickly, or ever, to losses, whether they are real and current or merely hypothetical future ones. Hence, it is not particularly surprising to find that the U.S. foreign-policy establishment is permeated by notions of the virtues of continued U.S. primacy. Even as America's relative power is in steady (but still fairly slow) decline, most U.S. foreign-policy professionals are and remain ardent *Stayers*. This they are and remain in part for the reason that such a preference for continued primacy and deep international engagement rests on a very general, if mostly unspoken, belief in the "normative quality" and "ethical necessity" of the state of affairs as it has been for over seven decades.

Things were more complicated for Japanese leaders in the early 1940s. It is widely acknowledged that Japan's attack on the United States at Pearl Harbor in December 1941 was extremely hazardous. Still, it is also widely acknowledged that, for Tokyo, it was a choice between two glaring evils. The status quo, where Washington was effectively strangling Japan's economy, worked to the clear detriment of Japan's aspirations; its reference point was rather a smoothly-operating Greater East-Asia Co-Prosperity Sphere—a Japanese Empire—, which was mostly a *prospective* state of affairs that the nation was willing to take enormous risks to realize (Levy & Whyte, 1997). Japan, of course, had in the preceding years committed huge resources to the pursuit of territorial expansion. These resources soon became seen as sunk costs, which needed to be justified by way of spending even and ever more resources aimed at both securing the nation's previous gains and moving closer to its real level of aspiration. All this, as we know, took place within a context where Japan's overall geostrategic position was under severe pressure. Even so, retrenchment was shunned in preference for a policy that bordered on the suicidal. Although irrational on the face of it, Japan's audacious attack on the United States nevertheless reflects a relatively general phenomenon in international relations, plausibly accounted for by prospect theory, where "risks are taken in a bad situation in the hope of recovering past losses, turning the tide of battle, or recouping sunk costs" (McDermott, 2004, p. 300).

In the particular case of Japan, the reference point did not align with the status quo, making Tokyo's reference-point bias *destabilizing* to the larger system. In most circumstances, however, when decision problems are framed by states, the status quo functions as the accepted reference point. The way people and states tend to adapt to the murkiness of the world—by over-weighting losses, by overvaluing what one possesses, and by harboring a preference for things to remain as they are—does have certain order-maintaining effects (Levy, 1997, p. 91); the stability of international affairs, as Robert Jervis (2017, p. 97) puts it, is bolstered insofar "as many states are similarly prone to settle for the status quo rather than undertake risky actions that could bring gains if they succeed but lead to war if they do not."

Prospect Theory and the International Relations Literature

In that sense, the arguments of prospect theory align rather closely with the "defensive" strand of realist theory (Taliaferro, 2004a, Chapter 2). This is so even if a schism exists between these theories with respect to causal antecedents. Whereas prospect theory essentially attends to independent variables at the individual level of analysis, defensive realism holds that system-level factors are the main shaper of state behavior. Consequences and predictions are largely the same, though. Defensive realists stress that states first and foremost seek to *maintain* their place in the hierarchy of nations (Waltz, 1979, p. 126); that is, "[s]tates are defensive positionalists" (Grieco, 1990, p. 40) preoccupied with upholding their share of capabilities or power. Hence, they are in most instances status quo-seeking actors. They are also actors who generally place more weight on the negative signals that radiate from the international system than they do on the positive ones—they "fear each other even when information favoring accommodation is present" (Johnson & Tierney, 2018–2019, p. 98)—, and they are actors who worry much about *relative* gains and losses and less about *absolute* ones; they can even be willing to accept that their own resources and power diminish as long as they do not do so at a steeper rate than that suffered by their competitors.

Contrarily, prospect theory primarily explains *individual* decision-making, even if such is a reaction or an answer to pressures from, and the uncertainties of, the individual's external environment. It is not a given that *group* decision-making, such as that which normally undergirds the enactment of foreign policies, can be similarly explained. Still, there is much to suggest that groups act no different than individuals in this respect. Experimental evidence based on prospect theory's ideas and hypotheses suggests that group decisions are instead associated with an *amplification* of the mechanisms operating at the individual level (Whyte, 1993). A steadily increasing number of studies in political science and international relations have therefore constructed their analyses around the presumption that the "aggregate risk taking behavior of the group (…) should be largely consistent with prospect theory"; this means, among other things, that systemic "incentives filter through an elite decision-making process that systematically skews toward loss avoidance" (Taliaferro, 2004b, pp. 191, 195).

Robert Jervis (1970, 1976) contributed with several early publications, such as his first two books from the 1970s, wherein he convincingly applied psychological theories to explain foreign-policy behavior and international relations outcomes. Later reviews of the literature showed that the application of prospect theory more specifically had, with time, become a reasonably widespread practice in scholarship on world affairs (Boettcher III, 2004; Levy, 1992; Mercer, 2005; Vis & Kuijpers, 2018). For example, some highly famous international events have been investigated based on concepts and mechanisms derived from Kahneman and Tversky's original ideas and findings. Such events include Japan's attack on Pearl Harbor, as referred to above (Levy & Whyte, 1997); the 1962 Cuban missile crisis (Haas, 2001); the loss-averse behavior of Great Britain during the 1956 Suez Crisis (Richardson, 1992); and the change in how U.S. President Franklin Roosevelt *framed* the 1938 Munich Crisis to himself, which made him, in the short space of the crisis itself, go from an isolationist mode into believing that intervention into European affairs had become an absolute necessity (Farnham, 1992).

Others have used prospect theory to elucidate more general phenomena. Gary Schaub (2004), for example, applies the concepts of loss aversion and the endowment effect to account for why deterrence (which does not necessitate any action—it merely threatens action if the adversary crosses any red lines—, and which aims at securing the status quo or something in the possession of the deterrer) is considerably "easier" than compellence (which *does* demand action, and which is aimed at inflicting losses on the target). These mechanisms have also been used to explain the proclivity of the United States to deploy troops and to intervene militarily abroad; such deployments and interventions are more common when Washington aims to counter perceived threats and prospective losses to its geopolitical position than when it seeks to acquire geopolitical gains (Blechman & Kaplan, 1979; Nincic, 1997). Indeed, great powers in general, argues Jeffrey Taliaferro (2004a, 2004b), are prone to undertake risky diplomatic and military interventions even in relatively peripheral areas where their very core security interests are hardly in play. This they do in an attempt to avoid the losses to the country's power, prestige, or geopolitical interests that they—the senior officials of the great powers—reckon have already materialized or *might* do so if the country fails to take action. Moreover, commitments deepen as costs are sunk: the great power perseveres with its intervention with tenacity, but also at great cost to itself; and it might also escalate a failing intervention in order to recuperate what has earlier been lost.

Taliaferro's work explicitly combines defensive realism and prospect theory to produce what he calls the "balance-of-risk theory" (which we can subsume under neoclassical realist theory). Core features of this theory include its heavy emphasis on loss aversion and, relatedly, on decision-makers' inclination to take excessive risks when facing possible losses. More broadly, Talieferro insists that a negativity bias commonly applies, which causes foreign-policy elites to act as if moderate threats are really great threats, as if peripheral interests are really core interests, and as if things that *might* go wrong (without intervention) *will* go wrong (without intervention). Leaders of great powers, he writes, are

> prone to engage in worst-case thinking, heavily discount the future, and focus on the immediate necessity of avoiding the loss. Decision-makers will tend to treat negative but probable outcomes as if they were certain. Adverse developments in areas previously deemed peripheral to a great power's security will suddenly take on tremendous importance in elite deliberations. Central decision-makers are more likely to adhere to the so-called domino theory – a set of interconnected beliefs and assertions regarding the interdependence of strategic commitments, the relative prevalence of bandwagoning over balancing in alliance formation, the opportunistic nature of adversaries' actions, and the cumulative effect of conquests. (Taliaferro, 2004b, p. 201)

Threat Sensitivity and System-Level Dominoes

The quote from Taliaferro contains two elements of particular importance here. The first concerns the tendency of leaders and foreign-policy professionals to *inflate threats*. Kahneman and Tversky (1979, pp. 282–283) argue that events that are extremely unlikely are associated with rather unpredictable behavior: people tend either to ignore such possibilities or to overweight them. Either way, people are inclined to overweight the *utility* of outcomes relative to the *probabilities* of their occurring, which does not mesh well with expected-utility theory (Levy, 1997, pp. 91–92).

The international relations literature, furthermore, indicates that states very rarely *under*estimate threats, whereas they often *over*estimate them (Johnson & Tierney, 2018–2019, p. 102). Granted, this can be due to a mix of forces located at different levels of analysis. Domestic public opinion (and the media), for example, is likely to punish leaders harder if the latter (mistakenly) underrate severe threats to national security than if they overreact to threats that never materialize (Nincic, 1990, p. 374). A state's bureaucracy and foreign-policy community also have incentives to exaggerate threats since their budgets and prestige are linked to these threats being vigorously counteracted (Cavanaugh, 2007–2008, p. 567). At the individual level, as the preceding pages have suggested, distinct psychological forces exist to make people—leaders and ordinary citizens alike—prone to focus especially on, and to search for, negative information (Johnson & Tierney, 2018–2019, p. 106).

Threat Sensitivity in Games of Strategy

This negativity bias of individuals, though, rests on interpretations of events and developments at the *system* level. In international-security affairs, sometimes a state simply doesn't get a second chance if it underrates a low-probability, high-impact outcome. Under conditions of anarchy, this places severe strain on the scope for cooperation among states; there always exists a possibility that another state (or a non-state actor) can take advantage of one's own conciliatory (or gullible) posture. Mutually beneficial payoffs from cooperation can, in the long run, be harvested from an iterated Prisoners' Dilemma—the game that denotes a situation where cooperation is preferred to conflict, but where the *possibility* of being cheated by the other is a major concern—if, that is, the issue in question centers on non-security matters, such as trade. Even if cheating or defection by others might occur from time to time, such cheating does not normally represent an existential threat; hence, players often continue with these games relatively unperturbed even when they incur short-term losses. Issues of concern to one's core national-security interests, on the other hand, are less likely to be conceived of as the object of iterated games; some of them are instead interpreted as one-off games where one's adversary's cheating marks the end of the game. This leaves the cheating part with the spoils of victory and the conciliatory part with the "sucker's payoff" (Jervis, 1978).

That is to say, one-off Prisoners' Dilemmas incentivize defection or "cheating" by both (or all) parties. Defection is the dominant strategy, meaning that non-cooperation is always the preferred course of action since it leads to the *certain avoidance* of a player's very worst outcome (which yields the sucker's payoff). This also means that it is the *rational* choice of action when players (i.e., states) cannot make any fully binding commitments, which they, as a matter of assumption, cannot do under anarchy. The characteristics of the Prisoners' Dilemma are therefore an allegory of some of the tragedy of international affairs, where states are encouraged to assume the worst (of the others).

This is also, if only in part, the case for a second famous game-theoretic metaphor, namely Chicken. Here, however, the outcome associated with mutual defection is considered the worst outcome for both (all) parties (and not the second worst, as in a Prisoners' Dilemma). Regarding the rationality of making worst-case assumptions, therefore, Chicken comes with a twist. If both players assume the worst of the opponent, both will opt for a more conciliatory strategy *so as to be certain* to avoid the catastrophic outcome of mutual defection. But the twist lies in the fact that, in such a game, it pays to signal as strongly and credibly as possible that oneself is anything but conciliatory, for thus to increase the likelihood that the other will be intimidated into making concessions. Unlike the Prisoners' Dilemma, a game of Chicken does not have any unambiguous equilibrium outcome, which makes for games with much huffing and puffing and posturing. But *sensitivity to threats* is an even more important element of this game, as overconfidence in the benevolence of the other leads oneself (as well as the other!) into the abyss if one's judgment eventually proves faulty.

In a stylized and normative sense, therefore, these two game-theoretic metaphors, taken together, advise actors to overweight worst-case and underweight better- or best-case scenarios in matters of international security. Not surprisingly, real life is replete with examples of mechanisms that induce states to act to *avoid tragedy* (Brands & Edel, 2019); this is so even if it is often a matter of debate whether they belong to the realm of Chicken or a Prisoners' Dilemma. The United Kingdom, for example, assumed the very worst when Germany, in the late nineteenth century, announced its ambitions to construct a world-class naval fleet. This spurred a bitter arms race between the two that is widely considered to have contributed to the First World War. In the Sudetenland Crisis of 1938, it was rather the overly *conciliatory* posture by the United Kingdom and France which, at first, led them to suffer the sucker's payoff and thereafter goaded them into assuming the very worst of Nazi Germany's intentions. In the late 1950s, the United States were (wrongly) convinced of the existence of a huge "missile gap" in the Soviet Union's favor, which caused Washington to undertake a massive build-up of its own missile arsenal *so as to be certain* to deny the Soviets any nuclear first-strike capabilities. The 2003 U.S. invasion of Iraq was fundamentally premised on faulty beliefs. Bolstered by quite subjective interpretations of available intelligence, these beliefs were largely preconceived ones; they held that Iraq possessed weapons of mass destruction, had a fully functioning program for the development of nuclear arms, and abetted *jihadi* terrorists. The "War on Terror" itself elevated the threats from al-Qaeda and its affiliates to the status of existential perils, spawning a momentous effort by the United States to contain international terrorism. And, currently, relations between the United States and China are significantly impacted by mutual suspicions about the other's intentions. According to Henry Kissinger (2014, p. 229), one of the staple traits of this relationship is that "the military maneuvers and defense programs of the other," even in the case that "they are 'normal' – that is, composed of measures a country would reasonably take in defense of national interest as it is generally understood – (...) are interpreted in terms of worst-case scenarios."

The Domino Theory

The second key element of Taliaferro's quote concerns the prevalence of *domino-theory thinking*, to which all of the real-life examples in the preceding paragraph are connected. Loss aversion at the individual level can help explain why the possibility of *negative* chains of events being set in motion at the system or international level looms larger in the minds of leaders and public opinion than does the possibility of *positive* such chains of events. Setbacks in the international sphere, even relatively minor ones, are often construed as the beginning of something bad that will only grow worse; this is so especially since they imply, or are thought to imply at any rate, that one's overall reputation will suffer in the process (Levy, 1992, pp. 285–287). In the late 1970s, for example, the United States' "gain" of Somalia mattered less than the simultaneous "loss" of Ethiopia to the Soviet Union. In general, Moscow adding to its portfolio of allies (events that were perceived in Washington as losses) made a much deeper impression on U.S. leaders and public opinion than did the many instances in which the Soviet Union lost allies (which were considered gains in the United States) (Jervis, 1992, p. 192).

Notably, this difference in reactions to negative and positive dominoes need not primarily be a product of biases at the individual level; to the contrary, the chain reactions themselves belong squarely to the system level, usually being products of a complex system whose constituent parts are interrelated in very opaque and hard-to-predict ways. Using the germ as their basic model, Paul Rozin and Edward Royzman, who coined the negativity-bias concept, argue that negative events in the world as such are also normally more *contagious* than positive ones. “Minimal exposure to germs,” they write, “can lead to maximal effects, because germs (…) are self-replicating. (…) Negative events may inherently (…) be more contagious, generalize more to neighboring domains, and be more resistant to elimination” (Rozin & Royzman, 2001, pp. 314–315). Small initial outbreaks of new diseases, such as the swine flu or COVID-19, are thus subject to attempts at early, costly intervention by states that seek to contain them before they transform into unmanageable global pandemics.

Granted, germs do not per se belong to the international-political sphere. But great powers regularly use similar reasoning (either explicitly or implicitly) when they try to battle negative events in the realm of world affairs. They often tacitly employ the metaphor of dominoes, fearing that a failure to intervene early against the first piece of domino will ultimately lead to a whole row of them falling. This is a highly contentious metaphor, particularly in the United States. It is nonetheless a very common one. It is also one which tends to saturate the grand-strategic thinking and fuel the security worries of great powers and empires (Snyder, 1991, p. 3).

Linked to the name of former U.S. President Dwight D. Eisenhower, but really describing “phenomena and beliefs [that] are ancient,” the domino metaphor or theory describes “the expectation that a defeat or retreat on one issue or in one area of the world is likely to produce, through a variety of mechanisms, (…) further demands on the state by its adversaries and defections from its allies” (Jervis, 1991, p. 22). In terms of the more specific thematic focus of this book, dominoes can be said to fall insofar as a withdrawal of forward-deployed troops creates adverse geopolitical chain reactions—that is, one or more secondary effects of negative geopolitical change—that are not fully *predictably*, *inevitably*, or *directly* a result of the withdrawal itself. Thus, the metaphor centers on the possibility of the occurrence of what Michael Mandelbaum (1988, pp. 148–151) calls "adverse chain reactions" in international affairs. These are essentially events, or chains of events, that are located sometime and somewhere in the future: each time the domino theory is invoked, the events about which one worries are likely as yet only *possibilities*; they are *hypothetical* developments depicting a *yet-to-be-realized future*—a *prospective* future; but amassing resources and the will to contain these events only *after* they have become manifest does not suffice, because such chains of events, once set in motion, are soon rendered irreversible; hence, domino-theory thinking catalyzes states and their leaders into moving fast and early, and to act according to the dictates of the negativity bias and the insurance principle; and it pushes them to treat otherwise rather peripheral interests as if they are really core interests, since *not* doing so opens up the way for the peripheral pieces of dominoes to knock over other pieces, and then some more, and more still, until the chain reaction eventually affects dominoes that are more unambiguously part of one's core national interests.

Dominoes and Political Realism

Hypothetical adverse geopolitical chain reactions, or dominoes, belong to the international level. However, decisions to implement policies to battle them are only taken after signals flowing from the system are filtered through the minds of policymakers. Insofar as their decision-making is influenced by negativity biases, these policymakers will be more prone to settle for a better-safe-than-sorry policy than what is predicted by expected-utility theory. To wit, the likelihood of adverse chain reactions *actually* occurring is in many or most cases probably overstated—although how often they are overstated, and by how much, are questions that are virtually impossible to answer with any confidence considering that we are here operating in the domain of hypotheticals and counterfactuals. Neither is it really straightforward to make a normative case for or against interventions and influence maximization.

On the one hand, as Chapter 3 explicated, one essential element of realist international relations theory, common to virtually all of its variants (including "human-nature"-centered classical realism), is its emphasis on the risks and threats that flow from the mere fact that international politics, unlike domestic politics, is neither formally nor actually hierarchically arranged. States and their leaders are, as they should be, inclined to worry a great deal about what bad events *might* occur; adopting a more care-free posture may eventually prove a gravely regrettable choice. Realists say that the security dilemma is a tragedy, but their theories also indicate that it is wholly *understandable* that many states get caught up in one. Prisoners' Dilemmas are what they are—viz., structures or strategic situations wherein cooperative and conciliatory moves are at base *non*-rational. The main exceptions are situations where elaborate institutional mechanisms have been established, at which point the Prisoners' Dilemma ceases to be a Prisoners' Dilemma; it is now rather an "Assurance Game," which incidentally depicts a strategic situation that is not normally associated with international anarchy but instead with formal institutionalized hierarchy. And neoclassical realists, when describing the interventions and expansion and influence-seeking of states, tend also to convey that they *understand* the inclinations and the needs of these states, which inhabit a murky, anarchic world where one's core and peripheral interests can often only with great difficulty be disentangled. This is a world in which states, again often quite understandably, are inclined to project their power and to extend their influence as far as their material resources allow.

The specter of dominoes *possibly* falling plays an important, if sometimes unspoken role, in standard political-realist theory, as scholars try to understand why states behave like they do. There's a certain irony here, though, in that "the domino theory has been roundly criticized by many of the very analysts" who believe that international politics is a highly conflictual arena ridden by interstate competition over resources, security, power, influence, and "vacuums" or "shatterbelts" along the rimland of Eurasia (Owens, 1999, p. 70). "Anarchy," writes Mandelbaum (1988, p. 140), "makes chain reactions possible." And, at any given time, only a few of the world's states have the strength required to prevent external dominoes from falling. Strength, or relative power, also being a property of the system, we are here as well near the core of realist theory. Forestalling adverse chain reactions and falling dominoes requires power projection, pre-emptive expansion, the exertion of influence over others, and an inherent reluctance to withdraw from areas already included in one's sphere of influence. This is, again with Mandelbaum (1988, p. 140), the "pattern of international conduct of the strong," which is "thus a product of the structure of the international system and a particular distribution of power within it." This is also perfectly in line with the basic driving forces of (structural and neoclassical) realism.

Throughout the long third chapter of his 1988 book *The Fate of Nations*, Mandelbaum displays a profound ***understanding*** of why it is that strong nations (his text focuses on the United States) are beset by this urge to undertake costly projects abroad that may on the face of it look quite excessive. Evidently straying from the rational, in reality it is not necessarily, or indeed even normally, the case that expansive policies by a great power are based on wholly unsound reasons and reasonings. The way that the various parts of the international system are interconnected—which adds to the system's murkiness, we might add—is what pulls the powerful overseas. "One thing can lead to another," is a more elementary description of this feature of an international system in which problems "can gather momentum, like a boulder rolling down a hill," and where "[t]he person standing at the foot of the hill has an interest in what happens at the top" (Mandelbaum, 1988, pp. 137–138).

The Domino Skeptics

Many others, one should still point out, would argue to the contrary, that the boulder is in all but the most exceptional of circumstances either smaller or more secure—or both—than what Mandelbaum here suggests. Critics of the U.S. grand strategy of primacy contend that much of Washington's policy is based on misplaced alarmism where "threat inflators (...) construct elaborate and improbable sequences of events in order to convince Americans that faraway events might eventually cause them significant harm" (Walt, 2018, p. 150). The threat posed by Russia, for example, is much smaller than was the threat posed by the Soviet Union during the Cold War; Europe's economic strength and overall capabilities should more than suffice to keep any further expansionary ambitions of Moscow in check. The threat of failed states becoming safe havens for international terrorist groups unless the United States intervenes is also vastly overblown, many contend; the decision to combat every such group anywhere fails to pay heed to the significant differences among them, causing U.S. overcommitment in a "forever war against Jihadists" (Byman & McCants, 2017). International terrorism is not even close to representing any existential threat to the United States; U.S. policy is simply based on a "terrorism delusion" (Mueller & Stewart, 2012).

Similarly, critics lament, the status, power, and insidiousness of so-called "rogue states" have been unnecessarily elevated by a Washington that in the 1990s, for no sound reason, began to view them as first-rate military threats (Walt, 2018, p. 150). In reality, North Korea, Iran, Syria, Cuba, and—before they were attacked and invaded by the United States and its allies—Iraq and Libya neither posed nor pose any *serious* threats to other states, and certainly not to the United States. This would be so even should they harbor ambitions to develop nuclear weapons (which is or has been the case for all of them except Cuba, though only North Korea has actually succeeded in attaining such a capability).

Many instead point out that few if any threats tend to be as inflated as that of nuclear proliferation. The Cold War, whose main lessons seem to have been largely forgotten, demonstrated that it is perfectly possible to deter far more powerful "rogues" (such as the Soviet Union and Communist China) (Gavin, 2009–2010). This is so even when they have nuclear weapons; or, as some claim, even just *because* they have them, as such weapons markedly raise the threshold for starting conventional wars and, therefore, purportedly add a significant measure of stability to world and regional politics (Waltz, 2003 [1995]). Washington's "nuclear umbrella," stretched out to cover dozens of allies, is by and large superfluous; in all likelihood, these allies won't go nuclear even if U.S. troops are withdrawn and the umbrella is closed (Cha, 2001; Hughes, 2007). The debate about nuclear weapons is pervaded by references "to metaphors such as proliferation chains, cascades, dominoes, waves, avalanches, and tipping points"; but such metaphors are "inappropriate and misleading, as [they imply] a process of nuclear decision-making and a pace of nuclear weapons spread that are unlikely to transpire" (Potter & Mukhatzhanova, 2008, p. 159).

Critics of the domino theory also claim that the metaphor overstates how interdependent the world and its states really are. Whereas "the interdependence of dominoes lined up in a row is extreme," nations, although obviously interlinked, are simply "not as dependent on one another as the dominoes" (Shimko, 1994, p. 666). The mere choice to use such a metaphor, therefore, predisposes the analyst or the politician to exaggerate the likelihood that bad events will spark chain reactions. Chain reactions are inevitable when we are dealing with *actual* pieces of dominoes; but it is not so when we study and predict the behavior of states (or non-state actors). In fact, when the domino metaphor centers on the threat of smaller powers allying or aligning with larger ones (which the metaphor often, though not always, does), it essentially breaks with expectations derived from realism's balance-of-power theory (Jervis, 1991, p. 52). This theory foresees that, say, a rapidly rising major power will be met by widespread balancing behavior. Other states, fearing its rise, will seek to counter it rather than bandwagon with it. Put differently, any hypothesized chain reaction will likely be stopped once the first few dominoes have fallen.

Of course, states are never as well-behaved as theories would like them to be. As others have pointed out, jumping on the bandwagon, and thereby acting as a falling piece of domino, may under some circumstances be a perfectly reasonable choice as it makes it possible for a smaller state to share in the spoils of the ascending power's success—or it may at least work to appease it (Schweller, 1994; Walt, 1987, pp. 17–21).

Great Powers and the Domino Theory

Whether it is bandwagoning or balancing behavior that represents the dominant tendency in the world is really an empirical question. What is nonetheless clear, asserts Robert Jervis (1991, p. 52), is that Washington "has often acted as if bandwagoning were the dominant tendency." Yet, what is not as clear is whether this has been a prudent way of conducting foreign policy. Elsewhere, Jervis (1998, pp. 168–170) points to such examples as Cuba, whose 1959 revolution did not set off the avalanche of similar revolutions in Latin America that the United States feared it would, and U.S. failures in Lebanon in the 1980s, which did not cause allies to cut their ties to America. More infamously, the debacle that was the Vietnam War was followed neither by any region-wide communization of Southeast Asia (although Laos and Cambodia did become Communist as Washington was losing the war) nor by any full-scale collapse in U.S. prestige and credibility (Jervis, 1988, p. 676).

The Vietnam Domino

One could say that the Vietnam War, still a national trauma for the United States, gave the domino principle a bad name. It was President Eisenhower who, in April 1954, alluded to the profound national-security risks that were associated with the fall (to the Communist camp) of just a single domino, in this case Vietnam. Speaking to news reporters about the strategic importance of Indochina, Eisenhower invoked the

> broader considerations that might follow what you would call the 'falling domino' principle. You have a row of dominoes set up, you knock over the first one, and what will happen to the last one is the certainty that

> it will go over very quickly. So you could have a beginning of a disintegration that would have the most profound influences. (...) [W]hen we come to the possible sequence of events, the loss of Indochina, of Burma, of Thailand, of the Peninsula, and Indonesia following, now you begin to talk about areas that not only multiply the disadvantages that you would suffer through loss of materials, sources of materials, but now you are talking really about millions and millions and millions of people. Finally, the geographical position achieved thereby does many things. It turns the so-called island defensive chain of Japan, Formosa, of the Philippines and to the southward; it moves in to threaten Australia and New Zealand. (...) So, the possible consequences of the loss are just incalculable to the free world. (U.S. Department of State, Office of the Historian, 2021)

What was envisioned and feared was that a failure to intervene would cause a chain reaction that would eventually see the whole of Southeast Asia—and possibly also states beyond that region—ending up within the Soviet Union's (or China's) sphere of influence. As perceived by Washington, Vietnam was not only about Vietnam; it was about victory in the Cold War as such. Or, more precisely, it was about *not losing* the Cold War.

The war in Vietnam certainly constituted a multifaceted disaster for America. On the other hand, the very *withdrawal* of U.S. troops from the theater in the early 1970s was almost immediately followed by Communist takeovers in South Vietnam, Laos, and Cambodia; this lends at least some support to the very domino theory that underpinned the intervention in the first place. One of the most steadfast supporters of Washington's efforts in Vietnam was Lee Kuan Yew, prime minister of Singapore before, during, and after the war. He consistently argued that the independence of the whole region critically depended on the willingness of the United States to take on the enormous costs of intervening in what was seemingly the periphery of the Cold War (Kissinger, 2014, pp. 297–302). Furthermore, post-Vietnam debates in the United States centered strongly on the question of whether seeming peripheries should also be treated as *actual* peripheries by Washington, so as to avoid future Vietnams. The great-power historian Paul Kennedy (1988, p. 525) notes that this "disturbed existing allies, doubtless encouraged its foes, and caused wobbling neutrals to consider re-insuring themselves with the other side."

Empirically testing, and normatively evaluating, the domino theory are enterprises riddled with counterfactuality issues. Vietnam might have been the right war to fight for the United States; or it might have been the wrong one; or it might have been a mix. In any event, the war *did* happen, and the fact that most scholars would "now find the domino theory disastrously incorrect should not lead us to conclude it was not central to decision-makers" (Jervis, 2017, p. 18). Because it often was, and it often still is. This is in part due to human beings, including state leaders and foreign-policy establishments, being cognitively disposed to negative filtering of system-level information. The domino theory, as Jack Snyder (1991, p. 7) puts it, "is neither wholly true nor wholly false"; it is what it is, one might say, because the inclinations of humans are what they are. *And* because falling dominoes and adverse chain reactions *are* a real possibility in the anarchical international system. It is the very interconnectedness of the world that permits them to occur.

One of the founding fathers of geopolitical studies, Halford Mackinder, wrote a century ago about the world having become a "closed system." This "Post-Columbian Epoch," as he called it, differed from the world of the preceding "Columbian Epoch," which was characterized by openness and the existence of vast territories that had yet to be explored but which soon would be by adventurous (if also avaricious) mariners and other pioneers. In the new, tighter, and more interconnected system, "[e]very deed of humanity [would] henceforth be echoed and re-echoed in like manner round the world." That, Mackinder (1942 [1919], p. 23) said from the perspective of the year immediately following the First World War, would explain "why every considerable state was bound to be drawn into the recent war, if it lasted, as it did last, long enough."

Dominoes of Old

In fact, the specter of falling dominoes has driven great powers into extending their reach and intervening abroad throughout modern and ancient history alike. For example, the Peloponnesian Wars, which devastated the Greek city-states in the fifth century BC, can be conceived of as a series of efforts by the great powers at the time—Athens and Sparta—to forestall adverse chain reactions originating in the periphery of the system (Allison, 2017, pp. 35ff.). Two thousand years later, in the sixteenth century, the vast though geographically fragmented Habsburg Empire saw and reacted to an array of threats against its domain; rather than retrench from presumably less vital areas, and thereby avoid being overstretched, *all* of its possessions were deemed vital, necessitating substantial defensive efforts on multiple fronts (Kennedy, 1988, pp. 62ff.). The expansion of the Ottoman (Turkish) Empire had to be stopped; northern Germany had to be secured for Catholicism by a forward-deployed Imperial Army, as pressure mounted from ambitious Protestant princes supported by France and Sweden; territories in the Americas, and the riches that flowed from them, could certainly not be relinquished; and Habsburg possessions on the Italian Peninsula needed to be protected lest they create a vacuum that would be filled by the French (or the Ottomans, or both), which would, moreover, multiply future threats to the Empire. Finally, the Spanish Netherlands (roughly today's Belgium and Luxembourg) had to be defended virtually at all costs, again because the French (and also the United Provinces, i.e., today's Netherlands) stood ready to seize them if the Empire retrenched. Habsburg thinking was, according to Paul Kennedy (1988, p. 65), significantly influenced by

> the 'domino theory' – if the Netherlands were lost, so also would be the Habsburg cause in Germany, smaller possessions like Franche-Comté, perhaps even Italy. These were, of course, hypothetical arguments; but what is interesting is that the statesmen in Madrid, and their army commanders in Brussels, perceived an interconnected strategical whole, which would be shattered if any one of the parts fell.

The larger the empire, the more sensitive it is to the need to protect the "interconnected strategical whole." As its empire expanded in the nineteenth century, the British started observing, or imagining, possible threats and adverse chain reactions almost everywhere. This required further intervention, meaning further expansion, which added to the perceived threats and chain-reaction possibilities, which required further pre-emptive action, and so on. India, in particular, became the fulcrum of this type of reasoning. It was the most important piece of the British Empire; hence, the sea route to India needed to be protected; which meant that territories along the way, such as Egypt and South Africa, were vital to control and protect; which in turn also meant that London needed to shape events in adjacent areas (Mandelbaum, 1988, p. 151). The possible land routes to India created similar headaches for London. Following the Napoleonic Wars, fears that Russian encroachment into Central Asia, Iran, and Afghanistan was ultimately aimed at supplanting

Britain in India led Britain into undertaking its own expansionary efforts. This spurred the long-lasting "Great Game" of spying and expeditionary-force projection in High Asia between the two formidable empires (Hopkirk, 2006 [1990]).

That game lasted until an even more problematic foe appeared. German growth and ambitions in the late nineteenth and early twentieth centuries were viewed by London as no less than an existential threat. This was so because their quite possible endpoint *could*, or would, be German dominance on the Continent; German mastery of the world's oceans; and, finally, a German invasion of the Home Islands themselves. Ultimately, therefore, Britain decided to enter the First World War in 1914. The United Kingdom then repeated or reflected that decision when it entered the Second World War twenty-five years later. Again, a chain-reaction reasoning was applied, though now with the added fear that the Soviet Union's non-aggression pact with the Germans would spur Moscow into invading India, Afghanistan, and possibly also British-dominated Iran (Frankopan, 2015, pp. 365–366).

Neither did the marked decline of the British Empire in the post-war period lessen London's compulsion to think in terms of the domino theory. Most notorious is certainly the 1956 Suez Crisis, where Egypt's nationalization of the Suez Canal was interpreted in London almost as an apocalyptic event. If the United Kingdom decided to refrain from doing anything to reverse that (in their view) perfidious act, all other states in the region, British officials thought, would follow Cairo's example and move to nationalize their petroleum resources. In his best-selling book *The Silk Roads*, the historian Peter Frankopan (2015, p. 425) quotes a senior British diplomat at the time, who lucidly laid out the chain reaction that would presumably be set in motion if the Suez nationalization was left unanswered and Egypt eventually decided to shut the canal and stop the flow of oil:

> If Middle East oil is denied to [us] for a year or two, our gold reserves will disappear. If our gold reserves disappear the sterling area disintegrates. If the sterling area disintegrates and we have no reserves (...) I doubt whether we shall be able to pay for the bare minimum necessary for our defence. And a country that cannot provide for its defence is finished.

To wit, the United Kingdom went to war (in collaboration with France and Israel) with no less an objective than to save the (remains of its)

Empire, though ultimately with disastrous results for its prestige—and for the Empire as such. But declining empires are apt to think in domino terms; for external intervention to keep the first few dominoes from tipping over is often seen as essential if the decline is to be stemmed and the empire saved. This is so even if the end result is sometimes the opposite of what is hoped for. Another empire, the Soviet Union, soon suffered inglorious collapse after having invaded Afghanistan in late 1979. That invasion had all to do with reversing what was viewed in Moscow as adverse chain reactions within Afghanistan that would soon spill over into the Cold War more broadly. According to threat-sensitive Soviet officials, the United States stood ready to dominate Central Asia if Moscow failed to take appropriate pre-emptive action (Frankopan, 2015, pp. 464–465).

Dominoes in U.S. History

However, neither rising nor firmly-established great powers are known to shy away from invoking the domino principle either. For example, forestalling adverse chain reactions caused by the specter of German hegemony in Europe catalyzed the United States into intervening in both world wars. "Contagion" was the metaphor used by President Truman in 1947 to justify intervention in Greece and Turkey (Schweller, 1994, pp. 80–81); the president feared that a failure to replace the influence exerted by the near-insolvent British would in the end hand Moscow access to the Middle East, Africa, and Asia (Levering, 2016 [1982], p. 35)—and, *possibly*, to France and Italy as well (Mandelbaum, 1988, p. 145). Then, in what serves as a classic example of chain reactions that *may* ensue, in 1950 the United States was forced to watch a Moscow-supported North Korea invade South Korea soon after the United States had pulled out its troops from the latter.

Washington drew what it judged to be the appropriate lessons from Korea, and its grand strategy was, for much of the rest of the Cold War, ingrained with domino-theory reasoning. The United States invoked the domino principle when it, post-Korea, proceeded to construct an elaborate "hub-and-spokes" alliance system in East Asia (Cha, 2009–2010); it obviously invoked it in Vietnam as well; in post-revolutionary Cuba (Gaddis, 2005 [1982], p. 212); in Chile in the early 1970s (Westad, 2005, p. 201); and in Central America, in general (Slater, 1987, p. 106), and Nicaragua (Schweller, 1998, pp. 80–81) and El Salvador (Brands, 2016, pp. 79–80) in particular. Furthermore, the United States feared that dominoes would fall in faraway Angola, so it also intervened in that conflict (Jervis, 1991, p. 23). After the Soviet Union invaded Afghanistan, officials in Washington foresaw that, without U.S. action of some sort, the Soviets would soon extend its power even further, "stretching from the Horn of Africa, across the Red Sea, to the shores of the Indian Ocean," ultimately enabling the Russians to gain "control of the Gulf's oil" (Westad, 2017, p. 495). Thus began Washington's decade-long support for the *Mujahideen* fighting the Russian invaders.

Neither did the end of the Cold War and the onset of the unipolar moment lessen Washington's perceived need to act early and proactively to support wobbly dominoes, as Iraq, under the unruly Saddam Hussein, was the first to discover. According to the U.S. foreign-policy scholar Hal Brands (2016, p. 326), one of the key principles of U.S. statecraft in the post-Cold War period

> was that the United States should meet emerging threats early – before they became full-on systemic crises or existential dangers such as the Cold War-era Soviet Union, and perhaps before they emerged in the first place. The idea was central to the administration's emphasis on retaining alliances and forward deployments in key regions – a primary point of which was to keep new threats from arising – and even more so in its refusal to let Saddam become hegemonic in the Gulf. More broadly, U.S. officials repeatedly stated that although Washington could not solve every problem, it must engage proactively to address those issues that might jeopardize global stability and key American interests if allowed to fester.

Domino Types and the Credibility Quandary

Central to this thinking, as Brands points out, were *proactive foreign engagement* and the *stationing of U.S. troops abroad*. Three types of dominoes, or three types of chain reactions, each centering on a specific category of actors, were to be forestalled by what was mostly a continuation of U.S. foreign and security policies of old.

Dominoes Big, Small, and Smaller Still

First, what normally constitute the weightiest perils are those originating from global or regional *major powers*. These are the types of threats associated with Napoléons and Hitlers (and Saddams), where the expansion and success of a great power beget further expansion and further success

(Jervis, 1998, p. 166). From the United States' point of view, potential powerful competitors or rivals constitute the primary security concern; rows of dominoes are always present, but such rows will seem particularly "precarious when there is a force at the other end prepared to give [them] a push" (Mandelbaum, 1988, p. 141). In today's world, such worries imbue U.S. views on the rise of China, with respect to its assertiveness in the South and East China Seas and its economic expansion through the Belt and Road Initiative; Russia's intrusions into its "backyard" and beyond; and Iran's wars-by-proxy in the Middle East (The White House, 2017).

The threat from great-power dominoes seems particularly acute when there exist power "vacuums" in other regions, because "vacuums always get filled and (…) the principal issue is not whether, but by whom" (Kissinger, 1994, p. 426). This is also one of the reasons why more peripheral areas of the globe, inhabited by smaller states, sometimes attain an oversized importance. A hegemon in decay will find it increasingly difficult to retain its influence in the periphery, leaving the latter vulnerable to conflict and paving the way for great-power rivals to enter the void. Insofar as the declining hegemon aims to uphold its position, which is more likely than not, great-power conflict can commence in what are seemingly less vital regions (with the Balkans serving as a recurring historical example) (Levy, 1985, p. 347; Pollins, 1996, p. 108).

Second, the lack of any true rivals to the United States after the end of the Cold War meant that a handful of *smaller states* received much greater attention from U.S. policymakers, who now placed them at the top of the hierarchy of threats to the nation's security. The main problem for Washington was that some of these secondary powers had made, and were making, substantial efforts to develop biological, chemical, and—most worryingly—nuclear weapons. The 1998 Rumsfeld Commission (1998), a non-partisan commission tasked with assessing this threat, was clear in its conclusion that the threat to the United States "posed by these emerging capabilities is broader, more mature and evolving more rapidly than has been reported in estimates and reports by the Intelligence Community."

As perceived by Washington, there were several interrelated risks associated with such developments. The appearance of new and smaller nuclear-armed states would immediately alter regional balances of power and introduce a significant measure of instability into regional systems. Nuclear weapons would also embolden their possessors, increasing their scope for conducting more aggressive foreign policies (Edelman et al., 2011). Even mere *ambitions* to develop a nuclear-weapons capability would be destabilizing, regardless of whether they ever manifested in a real bomb; such ambitions, it was reckoned, would be met by much hostility from other states and would bring on the agenda the issue of pre-emptive attacks (Beardsley & Asal, 2009, p. 253).

Worse still for the United States, nuclear weapons are the eminent "power equalizer," which means that nuclear-armed adversaries would severely cripple U.S. freedom of maneuver in important regions (Glaser & Fetter, 2005, p. 89; Smith, 2006). Containing or militarily confronting the smaller state would be an enterprise fraught with risk. However, *not* containing or confronting it would also have some very serious repercussions for regional, global, and U.S. security. In an article published in *Security Studies* in 1997, Barry Posen toys with the idea of what *could* happen in the hypothetical case where Washington refrained from intervening against an aggressive and expansionist, *nuclear-armed* Iraq. A "failure to act to reverse Iraqi aggression," he writes, would

> increase the desire of status quo powers for their own nuclear weapons. (...) The course of world politics can be changed dramatically for the worse. (...) Effectively, the relative U.S. power position will be undermined. The perceived utility of nuclear weapons will be enhanced. Thus, the incidence of war in general will probably increase, at least in the near term. The probability of nuclear accidents and ecological disasters associated with crash, low-budget nuclear weapons programs will also increase. Most seriously, there will be a greater risk of regional nuclear war. (Posen, 1997, pp. 7–8)

Posen here points to several *possible* adverse chain reactions that *might* follow from one state's successful development of a nuclear weapon. This reflects what is a staple argument of the non-proliferation literature, as it is of U.S. non-proliferation policy debates. A significant part of the rationale for the United States' global engagement, and for its overseas military bases, is to counter just such looming chain reactions. A retrenching America that no longer acts as the lynchpin of global non-proliferation efforts, many fear, will provide the spark that *might*: (i) open the way for smaller, recalcitrant states to develop weapons of mass destruction, including nuclear arms; (ii) thereby set in motion a chain reaction that will see regional adversaries of the new nuclear-weapon state start developing similar weapons of their own; and (iii) signal to many of Washington's allies that the U.S. "nuclear umbrella" under which they have lived in seeming security for decades is seriously leaking or about to be closed for good, which *might* result in even more states deciding to go nuclear. The reasoning here is simply that nuclear proliferation *might* beget nuclear proliferation; the metaphors used are dominoes, chain reactions, or "proliferation cascades." Proliferation pessimism thus generally applies, a pessimism that helps justify and undergird continued U.S primacy (McDonough, 2006, pp. 33–34).

The third type of dominoes or chain reactions is associated with *non-state actors*, in particular *jihadist* groups. In the early 1990s, for example, many voices advocated for a continuation of the U.S. engagement in Afghanistan, even after the Soviet Union had withdrawn its troops. These voices cited the risk that "extremist governments would control Kabul in the future and that by withdrawing from the field, the United States was throwing away a chance to exercise a moderating influence" (Coll, 2004, pp. 238–239). These fears materialized relatively quickly. In 1996, the Taliban, an Islamic fundamentalist group, overtook the Afghan capital. The Taliban soon established the so-called Islamic Emirate of Afghanistan, an entity that lasted until U.S. (and British) forces removed it in late 2001.

Similar fears, of course, have multiplied in recent years, spurred by the successful establishment and subsequent expansion of a "caliphate" by the Islamic State in Iraq and Syria. This problem evidently called "for a concerted and forceful international response or it [would] metastasize" (Kissinger, 2014, pp. 297–298). The response *was* indeed forthcoming, and it did succeed in rolling back and containing the problem even if the metastasis concerns didn't end with the dissolution of the physical caliphate (Byman, 2019, p. 246). In fact, the October 2019 announcement by President Trump that U.S. forces would finally withdraw from Syria (though not from Iraq, and the Syrian withdrawal was soon partially reversed) sparked warnings from an array of analysts, politicians, and military officials about the very real likelihood of different types of dominoes falling as a consequence of the decision (Ryan, 2019). Not everyone was impressed by these dire warnings, though. As one critical observer remarked about the more general issue of U.S. withdrawals from potential terrorist hotbeds (though with special reference to Afghanistan): "By that logic, if we can even call it that, the U.S. troops should never leave any place where terrorists have once operated – or might in the future.

It is a recipe for war unbounded by either geography or time" (Preble, 2019).

Reputation, Credibility, and Dominoes

As this quote hints to, with more than an ounce of derision, forward-deployed U.S. troops are subject to self-perpetuating dynamics that work against withdrawal. These are dynamics closely connected to concerns about status and prestige, reputation and credibility. These are also timeless forces that exert great influence on states' foreign and security policies—and have done so throughout history (Morgenthau, 2006 [1948], Chapter 6). Again, it is the *belief* (or the fact) that the world and events taking place within it form an interconnected whole that provides the link between dominoes and one's reputation. This belief (or fact), writes Thomas Schelling (2008 [1966], p. 124), induces states to fight over their image, or their "face"; indeed, "this kind of face is one of the few things worth fighting over," as "'face' is merely the interdependence of a country's commitments."

The problem for the United States in Vietnam was not (only) Vietnam per se, but rather the demonstration that Vietnam offered more broadly of American resolve, of the country's determination to fulfill the commitments it had taken on (Gaddis, 2005 [1982], p. 211; Levering (2016 [1982], p. 91). The same was the case for Korea in 1950: it was of high relevance to affairs in Europe because of the "demonstration effect" a successful conquest by the Soviet-supported North of the South would have (Mandelbaum, 1988, p. 161). Hence, the United States chose to act, thereby effectively establishing a commitment to the protection of South Korea from which it has since been unable and/or unwilling to extricate itself. The Cold War was rife with similar cases of commitments, reputational concerns, and (the risk of) armed conflict. Berlin was long the epitome of these dynamics; but the United States' portfolio of commitments only grew further as the Cold War progressed. And, once a piece of territory had become part of this portfolio, it remained. For example, as events in Korea unfolded, Washington rapidly became more committed to the protection of "Nationalist China" (i.e., Formosa or Taiwan), which also encompassed the smaller, Taipei-controlled islands of Quemoy (Kinmen) and Matsu. This was not the most natural of commitments, and it was not one which the United States necessarily wanted to see extended into the rest of the Cold War and beyond. However, unilaterally abrogating it was perceived as an even worse option. Schelling (2008 [1966], p. 83, f.n. 24) puts it somewhat amusingly:

> Almost everyone in America, surely including the President and the Secretary of State, would have been relieved in the late 1950s if an earthquake or volcanic action had caused Quemoy to sink slowly beneath the surface of the sea. Evacuation would then not have been retreat, and an unsought commitment that had proved peculiarly susceptible to Communist China's manipulation would have been disposed of. Such is the intrinsic value of some territories that have to be defended!

What was the case for Vietnam, for Korea, for Formosa, and for Berlin was and is also the case more generally: even limited withdrawals or limited defeats matter to decision-makers, who "seem to believe that small losses will multiply via the domino effect" (Jervis, 2017, p. 87). A loss of face following retreat in armed conflict, or a failure to follow through with a threat, is potentially harmful in a wider way. But so is reneging on a promise to remain at the side of one's ally; this *has the potential* to render incredible all other promises (or commitments) one makes—or at least those of a similar nature (Jervis, 1991, p. 25). Retrenchment under pressure from external actors is a particularly problematic move, even if states are no more inclined to retrench voluntarily under less adversarial circumstances (Gilpin, 1981, p. 194).

Of course, as with dominoes and adverse chain reactions more broadly, concerns for one's reputation are sometimes overdone. Some say it is overdone in *most* cases and that the currency of credibility and resolve is overvalued by leaders who fail to acknowledge the importance of case-specific context (Huth, 1997, pp. 84–85). A state's overall prestige and reputation might well, objectively speaking, be fairly impervious to isolated events, which means that commitments are "more independent than interdependent" (Mercer, 1996, p. 9). Furthermore, the risk of smaller states bandwagoning with one's major-power rival might be overblown, which leads countries such as the United States "to squander resources in strategically meaningless conflicts (e.g., Vietnam) in order to reassure allies who are likely to remain loyal in any event" (Walt, 1985, p. 18).

Scholars, Policymakers, and Falling Dominoes

Yet, in the final analysis, what determine the course of action a state takes are the *subjective* perceptions about the strategic situation it faces (Ripsman et al., 2016, Chapters 1–3). It is patently clear that adverse chain reactions, including those that stem from a demonstrated lack of resolve or a failure to fulfill one's promise, are a real *possibility* in the realm of international politics, as most domino skeptics would also readily admit (Walt, 1985, p. 18). Then it matters less for scholarship if such concerns for one's reputation are *generally* exaggerated by great powers, as is likely the case if we take our cue from a combination of international relations realism and prospect theory. The bad things that *might* happen in world affairs are overweighted compared to the good things that may materialize. The failure to escape commitments from which one would ideally want to escape belongs to a category of tragic dynamics or outcomes in international politics (Schelling, 2008 [1966], pp. 63–66). This tragedy, though, just like the security dilemma and the competition for power and influence among the great powers, is nonetheless an understandable one. Or it should be, at any rate. But many analysts note, *and rightly so*, that fears are quite often exaggerated. Christopher Layne (1997, p. 115), who belongs to the domino skeptics, writes about the way offshore balancers (a label under which he also belongs) base their analyses and recommendations "on the belief that concrete vital interests should determine U.S. commitments (rather than credibility determining commitments and commitments, in turn, determining interests)."

One couldn't agree more! Yet still, big states and their leaders do not seem to act in this way; instead, and somewhat ironically, notes Robert Jervis (1982–1983, p. 12), reputational and domino concerns often "lead states to act and draw inferences in a pattern opposite from the one that we – and most other analysts – imply." In that sense, one may lament, states *misbehave* quite regularly. But states act as they act, and they believe what they believe; and these acts and beliefs are what we must explain and understand—with a solid dose of empathy, I want to add.

Two of the most acclaimed political realists in academia, Zbigniew Brzeziński and Henry Kissinger, were also renowned policymakers directly involved in some of the most famous and notorious, or notoriously famous, international-security issues of the Cold War. In their writings *and* in their policymaking and policy recommendations, they both highlighted greatly the risk of dominoes falling. As U.S. National Security

Advisor in the late 1970s and early 1980s, Brzeziński *acted* on the belief "that the Soviets invaded Afghanistan because they were 'emboldened by our lack of response over Ethiopia'" (quoted in Jervis, 1998, pp. 166–167). As a *scholar*, he was also adamant that U.S. resolve and credibility should not be taken lightly. The United States had, for instance, to keep their forward-deployed troops in South Korea, not only or even mostly because of Korea per se, but not least because it was "difficult to conceive of the Japanese continuing to rely on continued U.S. deployment on Japanese soil in the wake of an American abandonment of South Korea" (Brzeziński, 1997, p. 190).

Henry Kissinger is and was no different. In his capacity as U.S. National Security Advisor during both the height and the latter stages of the Vietnam War (when he simultaneously served as Secretary of State), he oversaw the very domino-based and credibility-driven U.S. policy that was at the time heavily criticized by most others. In his writings, he quite emphatically defends this policy that he himself greatly helped shape (e.g., Kissinger, 1994, pp. 645ff.). But he also defends more generally the view that the United States' credibility is of fundamental value in a conflict-ridden world in which adverse chain reactions *can* quite conceivably cause U.S. power and influence to wither. In *Diplomacy*, for example, Kissinger (1994, p. 491) writes about the quite sudden and surprising decision to intervene in what turned out to be a long and bloody war on the Korean Peninsula (he himself did not take part in that decision):

> In the final analysis, all of the belligerents in the Korean War drew important lessons from it. The American statesmen of the period deserve to be remembered for the vision with which they committed forces to a distant country which, only a few months earlier, they had declared to be irrelevant to American security. When the challenge came, they had the courage to reverse themselves because they understood that acquiescence to the communist occupation of Korea would undermine the American position in Asia, especially America's crucial relationship with Japan.

Kissinger has for half a century offered his readers an abundance of sound analysis, even if some parts of his work, in particular where he writes about Vietnam, can appear to be somewhat self-congratulatory (though perhaps unavoidably so). Kissinger, however, is a realist scholar (and practitioner), and he is a realist whose understanding of world affairs includes an acknowledgment—and here he oftentimes takes the perspective of a U.S. citizen—of the damaging potential of dominoes, especially when smaller dominoes are knocked over, or *appear* to be so, by the very biggest and most adversarial ones. A major portion of Kissinger's work (again both academically and in practice) deals with hypotheticals and counterfactuals. Such analyses are incredibly challenging. They are often also incredibly fun to do, and they are incredibly important. They also spark an incredible amount of policy and scholarly disagreement.

*

The Scottish historian Niall Ferguson (2003) asks where the world would have been without the British Empire, a counterfactual question that his brilliant book—*Empire: How Britain Made the Modern World*—attempts to answer (in a worse place, it pretty much concludes). Offshore balancers and primacists—the *Leavers* and the *Stayers*, respectively—whose analyses rather center on *America's* role in the world, focus on the same basic issues: where would the world—and America—be without the U.S. Empire, or with a significantly scaled-back version of empire? Where would the world and America be if America (partially) retrenched? In a worse place, say primacists. In a better place, say offshore balancers. These two camps mostly encompass realist scholars, which goes to show the breadth of this school of thought. It also goes to show that systemic signals are not easily interpreted or digested. The signals are murky because the world is murky. In that sense, policymakers who insist on magnifying the negative signals and diminishing the positive ones should perhaps be exonerated for choosing the better-safe-than-sorry option—the one that is usually associated with the status quo. And, whether they should be exonerated for this or not is really an evaluative or normative matter only of secondary importance herein. More important is the observation we make that these policymakers *do* (tend to) act under the seeming presumption that much or most of what *can* go wrong under retrenchment *will* go wrong under retrenchment. Taken as a whole, U.S. policymakers (and indeed the policymakers of great powers in general)—and on this there is little or no disagreement between primacists and offshore balancers—are steadfast believers in dominoes, in adverse chain reactions, and in the inherent conduciveness, righteousness, and "normative quality" or "ethical necessity" of the status quo. They are, in other words, *Stayers*.

References

Allison, G. (2017). *Destined for war: Can America and China escape Thucydides's trap?* Houghton Mifflin Harcourt.

Baumeister, R. F., Bratslavsky, E., Finkenauer, C., & Vohs, K. D. (2001). Bad is stronger than good. *Review of General Psychology, 5*(4), 323–370.

Beardsley, K., & Asal, V. (2009). Nuclear weapons as shields. *Conflict Management and Peace Science, 26*(3), 235–255.

Berejikian, J. D. (2002). Model building with prospect theory: A cognitive approach to international relations. *Political Psychology, 23*(4), 759–786.

Blechman, B. M., & Kaplan, S. S. (1979). U.S. military forces as a political instrument since World War II. *Political Science Quarterly, 94*(2), 193–209.

Boettcher III, W. A. (2004). The prospects for prospect theory: An empirical evaluation of international relations applications of framing and loss aversion. *Political Psychology, 25*(3), 331–362.

Brands, H. (2016). *Making the unipolar moment: U.S. foreign policy and the rise of the post-Cold War order*. Cornell University Press.

Brands, H., & Edel, C. (2019). *The lessons of tragedy: Statecraft and world order*. Yale University Press.

Brzeziński, Z. (1997). *The grand chessboard: American primacy and its geostrategic imperatives*. Basic Books.

Bueno de Mesquita, B., Morrow, J. D., Siverson, R. M., & Smith, A. (1999). An institutional explanation of the democratic peace. *American Political Science Review, 93*(4), 791–807.

Byman, D. (2019). *Road warriors: Foreign fighters in the armies of jihad*. Oxford University Press.

Byman, D., & McCants, W. (2017). Fight or flight: How to avoid a forever war against Jihadists. *Washington Quarterly, 40*(2), 67–77.

Cavanaugh, J. M. (2007–2008). From the "Red Juggernaut" to Iraqi WMD: Threat inflation and how it succeeds in the United States. *Political Science Quarterly, 122*(4), 555–584.

Cha, V. D. (2001). The second nuclear age: Proliferation pessimism versus sober optimism in South Asia and East Asia. *Journal of Strategic Studies, 24*(4), 79–120.

Cha, V. D. (2009–2010). Powerplay: Origins of the U.S. alliance system in Asia. *International Security, 34*(3), 158–196.

Clark, C. (2012). *The sleepwalkers: How Europe went to war in 1914*. Penguin.

Coll, S. (2004). *Ghost wars: The secret history of the CIA, Afghanistan, and Bin Laden, from the Soviet invasion to September 10, 2001.* Penguin.

Edelman, E. S., Krepinevich, A. F., & Montgomery, E. B. (2011). The dangers of a nuclear Iran: The limits of containment. *Foreign Affairs, 90*(1), 66–81.

Farnham, B. (1992). Roosevelt and the Munich crisis: Insights from prospect theory. *Political Psychology, 13*(2), 205–235.

Ferguson, N. (2003). *Empire: How Britain made the modern world.* Penguin.

Frankopan, P. (2015). *The silk roads: A new history of the world.* Bloomsbury.

Fukuyama, F. (1989). The end of history? *National Interest, 16*(Summer), 3–18.

Gaddis, J. L. (2005). *Strategies of containment: A critical appraisal of American national security policy during the Cold War.* Oxford University Press. (Original work published 1982.)

Gavin, F. J. (2009–2010). Same as it ever was: Nuclear alarmism, proliferation, and the Cold War. *International Security, 34*(3), 7–37.

Gilpin, R. (1981). *War and change in world politics.* Cambridge University Press.

Glaser, C. L., & Fetter, S. (2005). Counterforce revisited: Assessing the Nuclear Posture Review's new missions. *International Security, 30*(2), 84–126.

Grieco, J. M. (1990). *Cooperation among nations: Europe, America, and non-tariff barriers to trade.* Cornell University Press.

Haas, M. L. (2001). Prospect theory and the Cuban missile crisis. *International Studies Quarterly, 45*(2), 241–270.

Haselton, M. G., & Nettle, D. (2006). The paranoid optimist: An integrative evolutionary model of cognitive biases. *Personality and Social Psychology Review, 10*(1), 47–66.

Hopkirk, P. (2006). *The Great Game: On secret service in High Asia.* John Murray. (Original work published 1990.)

Hughes, L. (2007). Why Japan will not go nuclear (yet): International and domestic constraints on the nuclearization of Japan. *International Security, 31*(4), 67–96.

Huth, P. K. (1997). Reputations and deterrence: A theoretical and empirical assessment. *Security Studies, 7*(1), 72–99.

Ikenberry, G. J. (2001). *After victory: Institutions, strategic restraint, and the rebuilding of order after major wars.* Princeton University Press.

Jervis, R. (1970). *The logic of images in international relations.* Princeton University Press.

Jervis, R. (1976). *Perception and misperception in international politics.* Princeton University Press.

Jervis, R. (1978). Cooperation under the security dilemma. *World Politics, 30*(2), 167–214.

Jervis, R. (1982–1983). Deterrence and perception. *International Security, 7*(3), 3–30.

Jervis, R. (1988). War and misperception. *Journal of Interdisciplinary History, 18*(4), 675–700.
Jervis, R. (1991). Domino beliefs and strategic behavior. In R. Jervis & J. Snyder (Eds.), *Dominoes and bandwagons: Strategic beliefs and great power competition in the Eurasian Rimland* (pp. 20–50). Oxford University Press.
Jervis, R. (1992). Political implications of loss aversion. *Political Psychology, 13*(2), 187–204.
Jervis, R. (1998). *System effects: Complexity in political and social life*. Princeton University Press.
Jervis, R. (2017). *How statesmen think: The psychology of international politics*. Princeton University Press.
Johnson, D. D. P., & Tierney, D. (2018–2019). Bad world: The negativity bias in international politics. *International Security, 43*(3), 96–140.
Johnson, D. D. P., Blumstein, D. T., Fowler, J. H., & Haselton, M. G. (2013). The evolution of error: Error management, cognitive constraints, and adaptive decision-making biases. *Trends in Ecology & Evolution, 28*(8), 474–481.
Kahneman, D. (2011). *Thinking, fast and slow*. Farrar, Straus and Giroux.
Kahneman, D., & Tversky, A. (1979). Prospect theory: An analysis of decision under risk. *Econometrica, 47*(2), 263–291.
Kahneman, D., Knetsch, J. L., & Thaler, R. H. (1991). Anomalies: The endowment effect, loss aversion, and status quo bias. *Journal of Economic Perspectives, 5*(1), 193–206.
Kennedy, P. (1988). *The rise and fall of the great powers*. Random House.
Kissinger, H. (1994). *Diplomacy*. Simon and Schuster.
Kissinger, H. (2014). *World order*. Penguin.
Layne, C. (1997). From preponderance to offshore balancing: America's future grand strategy. *International Security, 22*(1), 86–124.
Levering, R. B. (2016). *The Cold War: A post-Cold War history*. Wiley-Blackwell. (Original work published 1982.)
Levy, A. S., & Whyte, G. (1997). A cross-cultural exploration of the reference dependence of crucial group decisions under risk: Japan's 1941 decision for war. *Journal of Conflict Resolution, 41*(6), 792–813.
Levy, J. S. (1985). Theories of general war. *World Politics, 37*(3), 344–374.
Levy, J. S. (1992). Prospect theory and international relations: Theoretical applications and analytical problems. *Political Psychology, 13*(2), 283–310.
Levy, J. S. (1997). Prospect theory, rational choice, and international relations. *International Studies Quarterly, 41*(1), 87–112.
Mackinder, H. J. (1942). *Democratic ideals and reality*. National Defense University Press. (Original work published 1919.)
Mandelbaum, M. (1988). *The fate of nations: The search for national security in the nineteenth and twentieth centuries*. Cambridge University Press.

Maoz, Z., & Russett, B. (1993). Normative and structural causes of democratic peace, 1946–1986. *American Political Science Review, 87*(3), 624–638.

McDermott, R. (2004). Prospect theory in political science: Gains and losses from the first decade. *Political Psychology, 25*(2), 289–312.

McDermott, R., Fowler, J. H., & Smirnov, O. (2008). On the evolutionary origin of prospect theory preferences. *Journal of Politics, 70*(2), 335–350.

McDonough, D. S. (2006). *Nuclear superiority: The 'new triad' and the evolution of nuclear strategy*. Adelphi Paper 383. International Institute for Strategic Studies.

Mercer, J. (1996). *Reputation and international politics*. Cornell University Press.

Mercer, J. (2005). Prospect theory and political science. *Annual Review of Political Science, 8*(1), 1–21.

Morgenthau, H. J. (2006). *Politics among nations: The struggle for power and peace* (7th ed., revised by K. W. Thompson & D. Clinton). McGraw-Hill. (Original work published 1948.)

Mousseau, M., Hegre, H., & Oneal, J. R. (2003). How the wealth of nations conditions the liberal peace. *European Journal of International Relations, 9*(2), 277–314.

Mueller, J., & Stewart, M. G. (2012). The terrorism delusion: America's overwrought response to September 11. *International Security, 37*(1), 81–110.

Musil, R. (2005). *The Man Without Qualities* (S. Wilkins & B. Pike, Trans.). Picador. (Original work published 1930.)

Nincic, M. (1990). U.S. Soviet policy and the electoral connection. *World Politics, 42*(3), 370–396.

Nincic, M. (1997). Loss aversion and the domestic context of military intervention. *Political Research Quarterly, 50*(1), 97–120.

Owens, M. T. (1999). In defense of classical geopolitics. *Naval War College Review, 52*(4), 59–76.

Pollins, B. M. (1996). Global political order, economic change, and armed conflict: Coevolving systems and the use of force. *American Political Science Review, 90*(1), 103–117.

Posen, B. R. (1997). U.S. security policy in a nuclear-armed world or: What if Iraq had had nuclear weapons? *Security Studies, 6*(3), 1–31.

Potter, W. C., & Mukhatzhanova, G. (2008). Divining nuclear intentions: A review essay. *International Security, 33*(1), 139–169.

Preble, C. A. (2019, March 1). Ending the war in Afghanistan. *National Interest*. https://nationalinterest.org/blog/middle-east-watch/ending-war-afghanistan-46157

Richardson, L. (1992). Avoiding and incurring losses: Decision-making in the Suez crisis. *International Journal, 47*(2), 370–401.

Ripsman, N. M., Taliaferro, J. W., & Lobell, S. E. (2016). *Neoclassical realist theory of international politics*. Oxford University Press.

Rozin, P., & Royzman, E. B. (2001). Negativity bias, negativity dominance, and contagion. *Personality and Social Psychology Review, 5*(4), 296–320.

Rumsfeld Commission. (1998, July 31). *The Rumsfeld Commission Report*. Congressional Record. Retrieved August 12, 2021, from https://fas.org/irp/congress/1998_cr/s980731-rumsfeld.htm

Ryan, M. (2019, October 20). 'They are livid': Trump's withdrawal from Syria prompts rare public criticism from current, former military officials. *Washington Post*. https://www.washingtonpost.com/world/national-security/they-are-livid-trumps-withdrawal-from-syria-prompts-rare-public-criticism-from-current-former-military-officials/2019/10/19/d9455f08-f1ba-11e9-8693-f487e46784aa_story.html

Samuelson, W., & Zeckhauser, R. (1988). Status quo bias in decision making. *Journal of Risk and Uncertainty, 1*(1), 7–59.

Schaub, G. Jr. (2004). Deterrence, compellence, and prospect theory. *Political Psychology, 25*(3), 389–411.

Schelling, T. C. (2008). *Arms and influence*. Yale University Press. (Original work published 1966.)

Schweller, R. L. (1994). Bandwagoning for profit: Bringing the revisionist state back in. *International Security, 19*(1), 72–107.

Schweller, R. L. (1998). *Deadly imbalances: Tripolarity and Hitler's strategy of world conquest*. Columbia University Press.

Shimko, K. L. (1994). Metaphors and foreign policy decision making. *Political Psychology, 15*(4), 655–671.

Slater, J. (1987). Dominos in Central America: Will they fall? Does it matter? *International Security, 12*(2), 105–134.

Smith, D. D. (2006). *Deterring America: Rogue states and the proliferation of weapons of mass destruction*. Cambridge University Press.

Snyder, J. (1991). Introduction. In R. Jervis & J. Snyder (Eds.), *Dominoes and bandwagons: Strategic beliefs and great power competition in the Eurasian Rimland* (pp. 3–19). Oxford University Press.

Taliaferro, J. W. (2004a). *Balancing risks: Great power intervention in the periphery*. Cornell University Press.

Taliaferro, J. W. (2004b). Power politics and the balance of risk: Hypotheses on great power intervention in the periphery. *Political Psychology, 25*(2), 177–211.

Thaler, R. (1980). Toward a positive theory of consumer choice. *Journal of Economic Behavior & Organization, 1*(1), 39–60.

Tversky, A., & Kahneman, D. (1991). Loss aversion in riskless choice: A reference-dependent model. *Quarterly Journal of Economics, 106*(4), 1039–1061.

The White House. (2017, December). *National security strategy of the United States of America*. Retrieved April 23, 2020, from https://trumpwhitehouse.archives.gov/wp-content/uploads/2017/12/NSS-Final-12-18-2017-0905.pdf

U.S. Department of State, Office of the Historian. (2021). *Foreign relations of the United States, 1952–1954, Indochina, Volume XIII, Part 1*. Retrieved March 13, 2020, from https://history.state.gov/historicaldocuments/frus1952-54v13p1/d716

Vis, B., & Kuijpers, D. (2018). Prospect theory and foreign policy decision-making: Underexposed issues, advancements, and ways forward. *Contemporary Security Policy, 39*(4), 575–589.

Walt, S. M. (1985). Alliance formation and the balance of world power. *International Security, 9*(4), 3–43.

Walt, S. M. (1987). *The origins of alliances*. Cornell University Press.

Walt, S. M. (2018). *The Hell of good intentions: America's foreign policy elite and the decline of U.S. primacy*. Farrar, Straus and Giroux.

Waltz, K. N. (1979). *Theory of international politics*. McGraw-Hill.

Waltz, K. N. (2003). More may be better. In S. D. Sagan & K. N. Waltz (Eds.), *The spread of nuclear weapons: A debate renewed* (pp. 3–45). W. W. Norton. (Original work published 1995.)

Westad, O. A. (2005). *The global Cold War: Third World interventions and the making of our times*. Cambridge University Press.

Westad, O. A. (2017). *The Cold War: A world history*. Basic Books.

Whyte, G. (1993). Escalating commitment in individual and group decision making: A prospect theory approach. *Organizational Behavior and Human Decision Processes, 54*(3), 430–455.

CHAPTER 5

The Geopolitical Logic of U.S. Overseas Troops

*

The preceding chapter revolved around the broad argument that the fuzzy, hard-to-interpret characteristics of the anarchic international system, and of many or most events in world affairs, *combined with* the very human tendency to be sensitive to threats, to abhor losses more than gains, and to prefer the familiarity of things as they are, incentivize a type of behavior that is fairly cautious if one faces prospective gains but considerably less so if losses are looming. This is a type of behavior that corresponds with a *drive to shape and to control the external environment*, to influence it and secure it, so that no big or medium-sized boulder starts rolling down the hillside. The shaping and the controlling are done by the big states, by those whose power resources allow them to extend their reach abroad and overseas. Great powers expand as their resources expand. And they don't normally retrench unless their power resources are dwindling at a rate which makes retrenchment in the short or medium term all but inevitable. If relative power doesn't change much, and in the absence of any new significant vacuums that need to be filled by someone, the expectation is one of status quo-seeking behavior.

Of course, dominoes can start falling anywhere. And not even the greatest powers can possibly react to any and every such threat. Likewise, not any and every falling domino can, or indeed should, be

J. Jakobsen, *The Geopolitics of U.S. Overseas Troops and Withdrawal*, Palgrave Studies in International Relations, https://doi.org/10.1007/978-3-030-94488-9_5

judged as a major threat; not all dominoes should be treated equally. Thomas Schelling has done more than perhaps any other scholar to direct our attention toward the nebulous dynamics of games of strategy, wherein material resources and psychological factors (such as demonstrated resolve and credibly conveyed commitments) conjointly produce outcomes whose effects often extend far beyond the seeming boundaries of the particular game itself. But Schelling was unquestionably cognizant of the need to avoid overdoing chain-reaction thinking and credibility obsession. With reference to the U.S. Formosa Resolution of 1955 (which commanded the administration to counteract any significant threat to the island), he writes that "[i]f Congress passed such a resolution for every small piece of the world that it would like the Soviets to leave alone, it would cheapen the currency [of U.S. prestige and reputation]. A nation has limited resources, so to speak, in the things it can get exceptionally concerned about" (Schelling, 2008 [1966], p. 51).

The material resources of the United States are still vast, but they are also still limited. To a greater or lesser extent, Washington does *prioritize* among dominoes; some are judged to deserve and to require securing by America, whereas others are more liable to being neglected by it. The better part of this chapter will account for some of the key factors underpinning such a prioritization. In this, I draw greatly on classical geopolitical theory—on Halford Mackinder and Nicholas Spykman, and on Alfred Thayer Mahan—whose analyses combine geography and power and history to explain which regions of the world matter (or don't), and *why* they matter (or don't). Those regions that matter, per Washington's viewpoint, have generally, and for decades, been splattered with U.S. military bases and forward-deployed troops.

Overseas Troops as Domino Management

However, this chapter begins by elucidating the main functions that these bases and troops have, or are thought to have, in the United States' efforts to manage threats originating from overseas. In theory, U.S. military bases abroad, and the attendant forward-deployed troops, serve to forestall most or all of the various kinds of possible adverse geopolitical chain reactions, or falling dominoes, that were reviewed in the preceding chapter. They do so on account of both physical and psychological mechanisms. These bases enable the United States to fight faraway enemies, creating the material and logistical foundation necessary for the projection of power. Yet, most of the time, the forces stationed at U.S. overseas bases are *not* engaging any adversary in physical battle; their presence is rather meant to *deter* enemies and to *reassure* allies (as well as international economic actors) about Washington's commitment to their security and to the world and regional order. Ultimately, however, these mechanisms and objectives rest on the willingness of the United States to fight if the need arises.

Deterrence and reassurance—along with the occasional war-fighting—for the general purpose of upholding the status quo are the broad functions of the U.S. overseas presence. Within these wide categories exist numerous more specific missions with which the troops are explicitly or implicitly tasked. These missions include dissuading states from engaging in military and security competition and arms races; counteracting the spread of nuclear, biological, and chemical weapons; battling terrorist groups and thereby protecting Americans from future terrorist attacks; filling geopolitical vacuums before they are filled by rival states; preventing trade in narcotics and other illegal commodities; thwarting attempts by regional major powers to expand; keeping a lid on regional conflicts; responding quickly to natural and humanitarian disasters; increasing other nations' faith in the long-term viability of democracy and market economy; raising the projected costs for authoritarian regimes of human-rights abuses; undergirding international law, institutions, and norms; ensuring that vital sea lanes are kept open; bolstering the confidence of international investors and traders that geopolitical risks will remain low; and, in general, upholding (or extending) the influence of the United States.

U.S. Troops for Deterrence

Actual war-fighting aside, these forward-deployed forces usually fulfill a *deterrent* purpose by signaling to regional powers—larger and smaller ones alike—that attempts to alter the status quo will trigger a U.S. response that is bound to inflict intolerable costs on the aggressor. Deterrence, in other words, involves a threat that is contingent on the behavior of the target: the threatener promises that: "'If you do x I shall do y to you.' If the opponent expects the costs of y to be greater than the benefits of x, he will refrain from doing y; he is deterred" (George & Smoke, 1974, p. 48).

Effective deterrence hinges on the target behaving reasonably rationally in a formal sense: if the target's expected utility of challenging the status quo is lower than the expected utility of refraining from making such a challenge, it will likely choose the latter (Huth, 1999, pp. 28–29). These calculations, in turn, depend on the target's estimation of the *credibility* of its opponent's threat. Credibility, for its part, rests on a combination of *capabilities* and *resolve* or *will*. This is one of the enduring mechanisms or truths in international security. Two centuries ago, the Prussian general and military theorist Carl von Clausewitz (2007 [1832], p. 16) asserted that "[i]f you want to overcome your enemy you must match your effort against his power of resistance, which can be expressed as the product of two inseparable factors, viz. *the total means at his disposal and the strength of his will*" (italics in the original).

In the case of the United States, it is clear that, in matters concerning the defense of the U.S. homeland, this combination automatically lends enormous credibility to any U.S. threat to reciprocate against an enemy attack. Things get far more complicated, though, when the United States engages in *extended deterrence*; that is when Washington threatens to use force against a state in order to deter it from attacking a U.S. ally. Protecting others, as opposed to oneself, is automatically associated with less credibility. This is in large part due to the question marks that will inevitably appear with respect to the threatener's resolve or willpower when the issue does not directly concern one's own territorial security. If push really comes to shove, why would the threatener really be willing to fight for another nation? Why would U.S. authorities let U.S. soldiers die for the benefit of others? Indeed, for many countries (though not the United States), empirical data suggest that the majority of the population is unwilling even to fight for *their own* country (Jakobsen & Jakobsen, 2019).

Extended-deterrence credibility can especially be questioned when the threatener has chosen not to station any forward-deployed troops within the borders of, or near to, the ally that may be targeted. A lack of a physical presence certainly hampers (perceptions of) the threatener's resolve; and the lack of sufficient (or any) capabilities near the potential theater of armed conflict dramatically decreases the threatener's ability to forestall a *fait accompli*. For example, the 1930s saw several such examples where questionable resolve on the part of France and the United Kingdom (and certainly also the United States) combined with a lack of military means in or near territories that mattered, which opened the way for Nazi Germany to expand.

On both these accounts—capabilities and will or resolve—, forward-deployed troops can help change both the perceptions and the reality of the threatener's credibility. In particular, empirical studies have found that the presence of heavy ground forces works to bolster U.S. extended-deterrence credibility (Frederick et al., 2020; O'Mahoney et al., 2018). Overseas military bases, and surely the large ones, send some very specific signals that military forces stationed in the U.S. homeland can't. A key rationale for stationing troops abroad, for example, is to have them function as a "tripwire" that signals to adversaries that an attack on the U.S. ally will be treated as an attack on the United States itself. This increases the likelihood of a U.S. military response, thereby lowering the likelihood of the adversary attacking in the first place (Schelling, 2008 [1966], p. 119). This very logic underpinned the decision to deploy huge amounts of U.S. troops in Western Europe early in the Cold War. It was Thomas Schelling (1980 [1969]) who originally presented the tripwire logic; he argued that U.S. assurances to its allies were significantly strengthened by the forward-deployed troops, which were "plainly visible" and "manifestly connected up to the machinery of war" (Schelling, 2008 [1966], p. 99). In fact, troop numbers did not have to be overly huge to serve such a role; the main point of a tripwire is to increase the credibility of one's deterrent by placing *some* troops in "harm's way" (Schelling, 1980 [1969], p. 192). This continues to be a major function of many of the U.S. overseas troop contingents.

U.S. Troops for Reassurance and Order

Deterring adversaries through tripwires, moreover, is tightly connected to a *reassurance* mechanism that is directed toward allies and partners of the United States. A physical presence produces an added psychological effect since it amounts to a costly signal by the United States that it is committed to the security and stability of host countries, and of regions at large (Aklin & Kern, 2019; Braithwaite & Kucik, 2018). This was, as much as anything else, the rationale behind the initial discussions in the fall of 1948 of a North-Atlantic Treaty and a more permanent stationing of U.S. troops in Western Europe. This would demonstrate that Washington had ample staying power, and the Europeans gradually "started to believe that the United States would not withdraw from the continent anytime soon" (Westad, 2017, pp. 117–8).

More generally, forward-deployed U.S. troops crucially undergird world and regional orders by credibly conveying that Washington will continue to act as the lynchpin in their support. The troops, writes Kent Calder (2007, pp. 8–9), act to reinforce "common expectations of the political-economic future," which also points to their "long-term logic." Insofar as the United States can retain its dominant position and its active engagement in world affairs, this would, for most of the states in the world, represent a highly cherished continuation of the existing state of affairs.

A particular variant of (structural) political realism—*hegemonic stability theory*—stresses that clarity of world leadership, effectively meaning something at least resembling unipolarity, brings about a host of positive outcomes. For several decades following the Second World War, a major portion of the globe, namely that which was allied or aligned to the United States, could conduct their preferred business—rebuilding their countries and societies from the preceding calamity; focusing much on butter and less on guns; increasing foreign trade and investment—while not having to live in constant fear that interstate war could break out at any moment (though for some, such fear was certainly still an intermittent plague). Eventually, the end of the Cold War functioned as the midwife of a structural systemic change that saw a significant expansion of the area subsumed under and organized by U.S. leadership. This rapid shift from bipolarity to unipolarity, according to adherents of hegemonic stability theory, worked to bolster peace and order even further, constraining security competition and thereby rendering possible a period of unprecedented, nearly globe-spanning commercial expansion (Wohlforth, 1999).

This realist-affiliated theory contends that a transformation from pure anarchy, or from multipolarity or bipolarity, into the quasi-authority or quasi-hierarchy that is associated with a global hegemon, works to relieve the global and regional order problems that have been so common in history. This is particularly so if: (i) the hegemon is generally committed to *liberal* political and economic institutions, which normally align closely with the self-interests of the economically dominant actor (Krasner, 1976); and (ii) the hegemon bolsters the short- and long-term credibility of such commitments by demonstrating a willingness to project its military power abroad, by way of overseas bases and troops (Posen, 2003, p. 46).

As for the economic side of things: When the historian Niall Ferguson (2003, p. xx) pithily asks, "can you have globalization without gunboats?," his answer is that there are very sound reasons to doubt it. The international trade and investment system consist of actors that are sensitive to geopolitical risk; they therefore tend to value highly the market signals that are provided by U.S. military preponderance and force projection (Aklin & Kern, 2019; Drezner, 2013; Jones & Kane, 2012). Well over a century ago, it was the naval power of the British Empire that ultimately reassured businesses and investors that their capital could flow safely to other regions (Ferguson, 2003, pp. 368–9; Reynolds, 2000, p. 28). In more recent times, investors' faith in the system has been, as it still is, tightly linked to the role of the United States in the world. America's naval power, like that of its British predecessor, reinforces perceptions that sea lanes will be kept open and, therefore, that trade will continue to flow smoothly and that complex supply chains will not be interrupted (Drezner, 2013, pp. 70–71; Mandelbaum, 2019, p. 47). This comes in addition to land-based U.S. troops, which help create both the reality and perceptions of a stable investment environment. This, in turn, makes businesses more confident that they can safely sink their capital abroad (Biglaiser & DeRouen, 2007). In the absence of such confidence, global trade and investment could soon enter a self-perpetuating downward spiral, to the detriment of the world—and U.S.—prosperity.

In other words, U.S. military bases and troops abroad function to counteract potential adverse geopolitical chain reactions. They do so by way of both deterrence and reassurance mechanisms, signaling to adversaries and allies alike that the United States is credibly invested in supporting the status quo *for the long-term*. The reverse side of the coin, of course, is that a burgeoning *withdrawal* process *risks* setting in motion some of the very chain reactions that the troops were there to forestall in the first place. Even very limited and geographically circumscribed moves in the direction of retrenchment *can, potentially*, cause a first few dominoes to fall. For example, if we anticipate one of the cases that will be reviewed later, the October 2019 announcement by President Trump that the United States would pull out of Syria immediately created an opening for Turkey to invade the north-eastern part of that war-ridden country. This represented the starting shot of a qualitatively new phase in the Syrian civil war, which, according to most observers, would see both Russia and the Syrian regime increase their relative power as well as *possibly* emboldening the Islamic State (DeYoung et al., 2019).

President Trump soon proceeded to reverse the Syrian pullout (albeit only partially) This is therefore a useful illustration of (some of) the system-level constraints that beset inclinations or plans to withdraw soldiers. The expected pattern of U.S. forward-deployed troops—in terms of numbers as well as geographical distribution—is, as this book argues, quite heavily skewed toward the status quo.

Geopolitics and Prized Pieces of Geography

The preceding sections have emphasized that the overseas bases and troops are linked to a geopolitical logic that, as seen from Washington's perspective, highlights the indispensable role of the United States in ensuring that the status quo upholds and stability reigns in key regions. Forward-deployed U.S. troops are, so to speak, "domino managers" in the purported service of the United States and (again from Washington's perspective) of overseas regions—and ostensibly of the world at large.

But domino management in this sense presupposes access to suitable locations many of which are critical in order for Washington to be able to project power within and sometimes beyond regions of great importance. So, the question that also needs to be posed is: which regions, and which locations, *are* of great importance? No state is, or has ever been, omnipotent; every great power throughout history has had to prioritize among regions and locations with regard to where its focus and its energies should be directed. Not all possible geopolitical chain reactions are important enough, given the inevitable scarcity of resources that inflicts even the greatest of powers, to warrant the payment of a considerable "insurance" premium; and power cannot possibly be projected anywhere and always. A state's foreign policy, in other words, needs a *geographical direction*: places must be assigned strategic value, and a proper *geo*strategy must be formulated.

Geography, Geopolitics, and Geostrategy

Geostrategy is a variable. In theory, it can change very quickly. For example, a new government may decide to upend its nation's foreign-policy priorities, withdrawing its activities from one place and redirecting them toward another. An empire can retrench if its leaders so decide. Contrarily, a small, insular state can suddenly determine that its interests are better served if it becomes actively engaged in regional or extra-regional power politics. In practice, however, and as previous chapters have argued, a state's geostrategy cannot be unduly decoupled from the realities of absolute and relative power. Neither can it be too far removed from the realities of *geography*—if the state in question is to be secure and prosperous.

Power *and* geography matter; and political power plus geographic space is essentially *geopolitics*, a reality and a field of study that centers on "the political and strategic relevance of geography to the pursuit of international power" (Owens, 2015, p. 464). As seen from the viewpoint of any given nation, a place, a country, or a region is of geopolitical interest insofar as events and developments there have the potential to affect its national interests—notably its relative power position—in a profound way, either negatively or positively. To *manage* its geopolitical interests, any major or great power pursues a related *geostrategy* that is intended to uphold or improve that country's relative power position, and to support or further its national interest, vis-à-vis its rivals. The geostrategy provides guidance as to *where* the state should expend its energies; viz, over which places, countries, or regions it should strive to exert power (understood as influence).

When we expect the projection of military power, such as the construction and maintenance of overseas bases and the stationing of troops abroad, to exhibit a great deal of spatial and temporal continuity, this speaks to the enduring properties of geography. In Nicholas Spykman's lovely phrase: "Ministers come and ministers go, even dictators die, but mountain ranges stand unperturbed." The "geographic characteristics of states," he further writes, "are relatively unchanging and unchangeable," which means that "the geographic demands of those states will remain the same for centuries" (Spykman, 1938, p. 29). Russian leaders, for example, have for ages been immensely concerned about that part of the Great European Plain which runs north and east of the Carpathian Mountains; this is the flat landscape through which its rivals in the west have routed numerous invasions (and through which Russia has routed *its* invasions in the opposite direction). Russian leaders are still hugely concerned about this piece of geography, as they are of the Black Sea and its outlets through the Bosporus and the Dardanelles, which have also been Russian obsessions for centuries. The recent and ongoing conflict in and over Ukraine vitally concerns both of these geographically based matters.

The obsession of Imperial Great Britain was not only the Empire itself; of equal importance was to ensure a balance of power on the European continent, under the centuries-long assumption that a European hegemon would ultimately pose an existential risk to the Empire *and* to the Home Islands themselves. The latter is separated from the Continent by the narrow English Channel; the Channel has always been a barrier to invaders, but it has also always been a potentially surmountable such. Geography can cause security, and it can cause vulnerability; and the same piece of geography can cause both at the same time. Of course, in London they obviously do not any longer fret much or at all over the possibility of a cross-Channel invasion. The Empire is effectively no more; Western Europe has long since become peaceful; and the only entity vaguely resembling a European hegemon is the hugely bureaucratic and legalistic, non-militarized European Union.

That is to say, the *implications* or *effects* of enduring geographical features may change over the long haul. Western Europe might be peaceful in part because of a confluence of normative changes, tightly interlinked economies, and an institutional framework that massively incentivizes interstate cooperation. These features help render today's English Channel a geopolitically unexciting body of water—as unexciting as the historically vital Straits of Messina and Otranto have long since become. Yet, the geography of Western Europe has also lost some of its geostrategic relevance because changes in the balance of power have commingled with alterations to routes of communications to shift the center of geopolitical gravity away from Europe.

Islands and Ports and Sea Lanes of Communication

In bygone eras, the tiny archipelago of Malta was a highly coveted prize: for the Ottoman Empire, the "infidel rock" represented the last stepping stone of the maritime gateway to Catholic Rome. Contrarily, for Rome, for Emperor Charles V, and for the Knights Hospitaller (who were handed Malta in 1530 by Charles in his capacity as King of Sicily), it was the last line of defense for the Habsburg Empire and Western (Catholic) Christendom. Hence the immense importance of the Great Siege of Malta in 1565, where the islands' defenders succeeded in holding on to their prized asset by the skin of their teeth (Crowley, 2008, p. 43, Part II). Later, for Great Britain, Malta provided a crucial base for its ships *en route* through the Suez Canal and the Red Sea and to India. But for Britain as well, if Malta were to fall into the hands of an adversary—as it briefly did during the Napoleonic Wars and almost did during the Second World War—this would pose an existential threat to the entire Empire. Malta, thus, needed to be defended at virtually all costs, if one were British, or Maltese, or a knight bent on defending Catholic Christianity, or Pope Pius IV, or Charles V, or Philip II of Spain; or it needed to be conquered at virtually all costs, if one were Sultan Suleiman the Magnificent, or Napoléon, or Mussolini, or Hitler. Since the Phoenicians settled on the islands almost three thousand years ago, Malta has been "at the center of every invasion, migration, and trading enterprise in the history of the [Mediterranean] sea" (Crowley, 2008, p. 92). Nowadays, Malta is simply a prime tourist destination and a host of the European gambling industry. It is every bit as beautiful (and barren) as before, but it is now a peaceful island bereft of any geostrategic importance of note.

The historic roles of the Mediterranean and of Malta, one might say, have been supplanted by oceans and islands further to the east on the map. As the world's economic gravity and main trade routes have moved eastward, fueled not least by China's immense economic-growth rates, the Indian Ocean and the South and East China Seas (and also the South Pacific) have become increasingly more important—economically and, by implication, geostrategically. The rise of China has ignited a sea-lane-centered contest, which is admittedly still only in its adolescent phase, between the incumbent American hegemon and a serious Chinese challenger, with a few other regional major powers, such as India, also maneuvering for influence.

Islands and ports along the route from East Asia to the Persian Gulf and the Red Sea, respectively, have relatively suddenly appeared or reappeared as geopolitical loci. For some of these places, this is more or less history repeating itself, or history continuing. Sri Lanka, Malacca, and Singapore have centuries of experience of being pawns in the rivalry of great powers. For others, such as the Maldives and the ports of Gwadar (Pakistan) and Kyaukpyu (Myanmar), their rise to true geopolitical prominence represents something relatively new. The islands, reefs, and shoals of the South China Sea, for their part, have seen a fair bit of interstate quarreling before. Yet, they have certainly not seen it on quite the scale that prevails today. What are mostly tiny banks of sand with rather peculiar names—such as Subi Reef, Fiery Cross Reef, Mischief Reef (the most aptly-named of the lot!), Taiping Island, Johnson Reef, and Woody Island—are now constituent parts of the real-world geopolitical drama. As geographical features, they are enduring entities (although some have changed in size and shape due to land reclamation). Their renewed importance, though, is due to the very mix between shifts in the balance of power and changes to routes of communication and trade patterns, which are the two key elements that can alter the relative strategic values of different pieces of geography (Kelly, 2016, p. 127).

Classical Geopolitics and the Permanent Interests of a Sea Power

But much nonetheless remain the same, such as the general importance of sea lanes of communication. Writing in the late nineteenth century, one very famous American admiral, Alfred Thayer Mahan (1890), was adamant that sea power represented the key to prosperity and security for a *potential* great nation such as the United States. This was so even if the United States was traditionally, and by inclination and culture, primarily a land-focused power (Lambert, 2018, pp. 2ff.). Pointing to the spectacular rise and further growth of the British Empire, but also to the attendant relative failure of the more continentally focused France, Mahan contended that the country that he himself served, which had theretofore been primarily preoccupied with internal matters, needed to take up the mantle from the British and start constructing a world-class navy. History rested on and moved, and the fates of great nations were shaped, based on: (i) the production of goods, much of which needed to be traded; (ii) shipping, which effectuated trade; (iii) naval power, to protect shipping; and (iv) colonies, or bases, favorably scattered along the main sea lanes (Mahan, 1890, pp. 27–28).

In times of peace, Mahan noted, friendly ports controlled by other nations suffice as resting places for one's own ships. Yet, peace cannot be guaranteed. Britain's rise, of course, took place in an era in which

> there prevailed on the sea a lawlessness the very memory of which is now almost lost, and [where] the days of settled peace between maritime nations were few and far between. Thus arose the demand for stations along the road, like the Cape of Good Hope, St. Helena, and Mauritius, not primarily for trade, but for defence and war; the demand for the

possession of posts like Gibraltar, Malta, Louisburg, at the entrance of the Gulf of St. Lawrence, – posts whose value was chiefly strategic, though not necessarily wholly so. (Mahan, 1890, p. 28)

The United States, Mahan believed, needed the same global infrastructure: resting places, naval bases, coaling stations, and control over canals and maritime choke points; and all these locations would require protection by a significantly strengthened U.S. Navy. Military force would be needed in tandem with the commercial activities of a merchant fleet. Naval strength, and eventual naval superiority, developed alongside a global network of key locations or nodes, would ensure that U.S. war ships avoided ending up as "land birds, unable to fly far away from their own shores." U.S.-controlled bases along the major sea lanes, places where the ships could "coal and repair, would be one of the first duties of a government proposing to itself the development of the power of the nation at sea" (Mahan, 1890, p. 83).

Halford Mackinder and the Heartland

Published in 1890, Mahan's book—*The Influence of Sea Power Upon History*—was the first work of the fairly novel field of geopolitics that provided quite explicit geographically centered guidance for U.S. foreign policymakers. Granted, at first it influenced navalists in other countries, particularly Britain, to a much larger degree (Kennedy, 2017 [1976], pp. 182, 192). But as it turned out, it would later, albeit quite a while later, offer at least a partial explanation of the gist of *U.S.* geostrategy. Mahan's work, however, was soon complemented by the most renowned geopolitician of them all, Halford Mackinder.

A British geographer, Mackinder did not lay out any blueprint for *American* geostrategy per se; but his most famous works arguably accounted well for the main geopolitical issues and problems in his own time and well beyond. Moreover, Mackinder's arguments and theorizing help us understand why U.S. geostrategy evolved as it did from the Second World War onward, even if they didn't necessarily *directly* influence the decisions of U.S. policymakers. Mackinder still has numerous avid readers and academic heirs who emphasize that his writings "pointed to an enduring geostrategic reality of the first importance"; in particular, his supporters assert rather unassumingly, "he got the really big things right enough" (Gray, 2004, pp. 9, 12).

The things he got right enough included not least pinpointing whence the threats to the sea powers of the day would originate. The "geographical pivot of history," as was the title of his 1904 paper, was "that vast area of Euro-Asia which is inaccessible to ships, but in antiquity lay open to the horse-riding nomads, and is to-day about to be covered with a network of railways" (Mackinder, 1904, p. 434). He later relabeled this the Heartland (of Eurasia), arguing that whichever country controlled it would assume no less than a position of potential world domination. The way he delimited the Heartland changed over the decades; in "The Round World and the Winning of the Peace," an essay published in the middle of the Second World War, he had eventually ended up equaling the Heartland to the entire area of the Soviet Union west of the Yenisey River (Mackinder, 1943, pp. 598–599). Moreover, providing the gateway to the Heartland was East Europe, a concept introduced in 1919 by Mackinder (1942 [1919], p. 139). This denoted an area the control over which had spurred, and would continue to spur, intensive great-power contests. The quite famous dictum of Mackinder (1942 [1919], p. 150) with respect to these two vast pieces of territory (and to the even larger "World-Island," consisting of Eurasia and Africa) was:

Who rules East Europe commands the Heartland:
Who rules the Heartland commands the World-Island:
Who rules the World-Island commands the World.

The nation, or coalition, controlling the Heartland would secure "the greatest natural fortress on earth" (Mackinder, 1943, p. 601). This was a fortress in the sense that it shielded its possessor from attacks by the sea powers belonging to the "outer crescent," which encompassed, among others, Great Britain, the United States, and Japan (Mackinder, 1904, pp. 434, 436). The Heartland provided strategic depth in abundance (Mackinder, 1943, p. 599). With continued modernization, and with the development of new lines of communication (not least railways), and with the harvesting of all the natural resources that this territory so amply offered (Mackinder, 1943, pp. 599–601), its proprietor would eventually be able to make use of its "vast continental resources for fleet-building" (Mackinder, 1904, p. 436). This would lead to it posing a formidable and existential land-power-turned-sea-power threat to the original sea powers of the world.

Mackinder and the Geostrategy of Sea Powers

The main implication of Mackinder's reasoning for the sea powers of the outer crescent was, in a way, simple: no single great power (e.g., Germany or Russia, or, sometime in the future, China), or combination of great powers (i.e., Germany *allied to* Russia or, more relevant today, Russia allied to China), could be allowed to dominate East Europe (Mackinder, 1942 [1919], pp. 149ff.). Or, in broader terms, no Eurasian land power could be allowed to move within reach of hegemony. From the sea powers' point of view, whenever (or, ideally, well before) such a scenario started to attain some plausibility, this should work as a signal for them to act. As the contemporary scholar of geopolitics Colin Gray (2004, p. 16) writes, the pre-eminent maritime state of the day had actually "practiced sea power for the better part of three centuries before Mahan explained what they had been doing. In like manner," Gray continues, Mackinder (and Spykman, more about whom shortly) "provided a grander scale of explanation for what had been the golden rule of English, then British, statecraft for four hundred years: keep continental power divided." Great Britain had for ages pursued archetypical balance-of-power policies, which also speaks to the relation between political realism and classical geopolitics.

This, more than anything else, had characterized Britain's geostrategy in Europe in the eighteenth century, both before and during the French Revolutionary and Napoleonic Wars. More related still to Mackinder's thesis and warnings was Britain's behavior in the nineteenth century and at the start of the twentieth, a period during which considerable energy was spent by London attempting to block Russian expansion in Southeast Europe (i.e., the Turkish Straits), in Central and South Asia (during the "Great Game"), and in East Asia (hence the 1902 Anglo-Japanese Alliance) (Hall, 1955, pp. 110–111). Then, of course, came the *ententes* with France, and with Russia itself, in the pre-World War I period. Eventually, the jockeying for position by the great powers of Europe transformed into the actuality of a world war—but one in which the hegemonic threat stemmed from Germany, a hybrid land-sea power, as opposed to Russia. In that war, another of the powers of the outer crescent—the United States—joined the fight against the German-led alliance, albeit it did so rather late in the conflict (and the U.S. Navy mostly served to ferry soldiers across the Atlantic). Late or not, it thereby assumed just the role that Mackinder advised for a state of its kind and geographical location.

Then, as we know, the whole scenario was for all intents and purposes repeated two decades later (although the Second World War saw Imperial Japan, which was in any case as much or more of a land power by virtue of its continental territorial ambitions, switching to Germany's side). And in that war, the U.S. Navy certainly played a key role.

The two world wars of the first half of the twentieth century, and the Cold War of the second half, generally pitted one or more predominantly land-based powers against a coalition of states whose core was made up of powers belonging to the maritime outer crescent (with the Soviet Union between June 1941 and September 1945 being the primary exception). It is true, of course, that in the world wars, the main aggressor—Germany—did not control any of the Heartland when the fighting commenced, and it was not a pure land power. But it did seek, in both wars, to expand into huge tracts of the Heartland, which spawned a plethora of horrific battles for control over the pivot area.

Of great import was also the fight for control over what Mackinder referred to as the *inner* or *marginal crescent* (1904). These were territories made up of combinations of land and sea powers (including Germany itself). They encompassed huge land areas, but vitally also *ocean-facing* ones; they contained the majority of the world's population; and they included "most of continental Europe west of Russia, the Maghreb, the Middle East, and continental South, Southeast, and East Asia" (Owens, 2015, p. 471). For Mackinder (1904, p. 436), they were territories of eternal conflict; for they were territories the control over which by the pivot would lead to it having "the empire of the world" within reach.

This meant that the predominantly sea-based powers of the outer crescent would have to devise a geostrategy directed toward forestalling such a possibility. Already in 1904, Mackinder (1904, p. 436) had noted that this threat should work to

> throw France into alliance with the over-sea powers, and France, Italy, Egypt, India, and Corea would become so many bridge heads where the outside navies would support armies to compel the pivot allies to deploy land forces and prevent them from concentrating their whole strength on fleets.

Mackinder held firmly on to these ideas, writing about the outer-crescent powers' need for bridgeheads conveniently located at the inner crescent, whence amphibious sea power could be deployed as the crucial container of land power (Mackinder, 1943, pp. 601–602). By now, writing in 1943, he had introduced an additional geographical concept—the *Midland Ocean*, which was the North Atlantic—whose constituent powers should form an association revolving around "a bridgehead in France, a moated aerodrome in Britain, and a reserve of trained manpower, agriculture and industries in the eastern United States and Canada" (Mackinder, 1943, p. 604). What Mackinder here proposed, and foreshadowed (though at the time with the threat from Germany in mind), was in effect not much less than the North Atlantic Treaty Organization itself. His vision of bridgeheads and bases and power projection by insular maritime states in territories of the marginal crescent was, as we know, six years later to become a virtual permanent reality in the form of a U.S.-centered and U.S.-dominated North-Atlantic Alliance—though now with the Soviet Union as the pivotal, Heartland-based threat to be contained.

Nicholas Spykman and Bases for the Balance of Power

For all his focus on the Pivot and the Heartland, Mackinder also foreshadowed another concept, that of the *Rimland*. Building on the ideas of both Mahan and Mackinder, the Dutch-born American geopolitician and international relations realist Nicholas Spykman argued that the true center of power politics and international conflict was not the Heartland, but instead what he called the Rimland of Eurasia, which was Mackinder's inner marginal crescent relabeled. The great wars had not really constituted "a core binary contest between landpower and seapower" (Gray, 2015, p. 887), but rather one between mixed land-sea alliances. They centered on the overriding issue of who would control this vast, wealthy, and populous rimland area, which represented an existential threat, if controlled by one's adversary, and a marvelous geostrategic opportunity, if controlled by (or allied to) oneself.

Who would rule the world, and who would be hindered from ruling the world, depended on the balance of power in the Eurasian littoral. This was a contest from which the United States could ill afford to abstain, Spykman argued in *America's Strategy in World Politics*. Published in 1942, against the backdrop of Nazi Germany's onslaught on the Soviet Union, and of Pearl Harbor, the book unapologetically advised American interventionism, American world-spanning balance-of-power policies, and "the constant improvement of [America's] own relative power position" (2017 [1942], pp. 21–22). Isolationism might still have been an option, but it would really not do at all; absent U.S. control over vital sea lanes, and over what had by now become an equally vital airspace, America's adversaries—whether based in the Rimland or the Heartland—stood to gain such control at the expense of the United States.

Control over sea lanes and airspace, in turn, hinged not only on bridgeheads in Continental Europe or Asia, but also on a *worldwide oceanic network of military bases*, a portion of which had already been acquired by the United States by the time Spykman wrote his book. For example, key with respect to command of the sea lanes in the North Atlantic were early U.S. wartime deals with the governments of Canada and the United Kingdom (not least for port access in New Brunswick and Newfoundland, respectively), and with Denmark (for naval and air bases in Greenland), as was the joint U.S.-U.K. occupation of Iceland (Spykman, 2017 [1942], p. 189). Further south in the Atlantic, Spykman saw the Azores and the Cape Verde Islands as possible additional stepping stones for Washington. Closer to the U.S. homeland, Bermuda, Cuba (Guantánamo Bay), Puerto Rico, and the Bahamas linked up with Newfoundland in the north to create an almost impregnable protective shield in the Western Atlantic (Spykman, 2017 [1942], pp. 427–32). An array of bases scattered around the "American Mediterranean" (i.e., the Caribbean) were needed—and they were duly acquired in the initial stages of the war—as the area "was dangerously exposed to threats coming from the southeast" (Spykman, 2017 [1942], p. 433).

It was likewise in the Pacific Ocean. The immense vastness of this body of water, combined with the threat posed by Japan, made the existing American possessions (the Aleutian Islands, Hawaii, Midway, Wake, Johnson, and Samoa, along with the Philippines much farther to the west) inadequate as protectors of U.S. supply lines (Spykman, 2017 [1942], p. 192). Later in the war, the supply-line problem decreased considerably, as the United States fought its way steadily closer to the Japanese home islands, taking possession of ever more Pacific stepping stones along the way. These were pieces of real estate that held incalculable value, and even more so in the burgeoning era of aerial warfare. Before the war, Spykman had noted and foreseen that, whereas naval powers had certainly required bases and stations on "[o]ffshore islands and narrow peninsular points near weak powers," the onset of the era of air power would "mean that many now worthless and unclaimed bits of territory will become highly desirable" (Spykman, 1938, p. 48). The age of colonialism might have been in terminal decline, but great-power competition over precious bits of territory was certainly not.

This was even more so, Spykman (2017 [1942], p. 165) asserted, because the world, and notably the Rimland, or the "great coastal regions of the world," were now "interdependent not only economically but also politically. Oceans are no barriers; they are routes for the thrusts of sea power as well as highways of commerce." The Second World War had forcefully hammered home this point, which carried fairly seamlessly over to the Cold War. That grand-scale conflict witnessed the maintenance and expansion of the military-base network that the United States had acquired from the late nineteenth century and, especially, during the Second World War.

Again, Mackinder's notion applied: As one scholar wrote from the perspective of 1955, Germany might have been comprehensively beaten, but the threat remained that "various shifting alliances and combinations of power would (...) rotate around the pivot state"—that is to say, the Soviet Union—, which could conceivably "add the necessary oceanic frontage" to its enormous land territory (Hall, 1955, p. 110). Or the pivot could, at some future point, be represented by China, substituting for Russia as Germany had previously done. Mackinder (1904, p. 437) rather presciently hinted to such a possibility in the closing words of his 1904 essay. For now, though, with the Cold War looming and then exploding into life in the Korean War, it was Moscow that needed to be contained. Containment, though, vitally centered on Spykman's Rimland; that is to say, on the construction of "'rimland dykes'"—viz., military bases—stretching "along the Russian periphery that either would force a Soviet mellowing and submission or an internal implosion" (Kelly, 2016, p. 53).

U.S. Cold War geostrategy did not, then, solely reflect the ideas of an even more famous geopolitician, George Kennan, who is by many or most regarded as the father of Washington's containment policy; U.S. Cold War geostrategy was also Mackinder plus Spykman. But it was Admiral Mahan as well, for without sufficient sea power, and without the necessary ocean-based infrastructure—the indispensable island nodes—there was no way to reach the Rimland to put pressure on the Heartland power(s). A contemporary geopolitician, Phil Kelly (2016, p. 115), is very close to the mark when he writes that "[t]he balance between this sea power versus land power debate seems best resolved by Spykman's rimland, which refines these contrasts to include both sea and land orientations within his middle location as a place for major international happenings."

A Network of Bases on (Rim)land and at Sea

Hence, the Heartland *and* the Rimland *and* the sea lanes of the world's oceans were vital. One is therefore tempted to say that the United States should and would be everywhere. With respect to the normative contents of their writings, therefore, these representatives of classical geopolitics—Mahan, Mackinder, and Spykman—would, had they lived today, undoubtedly have been *Stayers* and not *Leavers*. This was certainly a fitting characteristic of Spykman, who essentially recommended *on*-shore balancing in Eurasia along with "flexible circumferential seapower" (Gray, 2015, p. 884). The Eurasian supercontinent, and the major and great powers inhabiting that vast portion of the earth's landmass, shouldn't be left to their own fate; it was in the *permanent* interest of the United States to see to it that no potential hegemon arose there, that power was fairly evenly distributed in Eurasia, and that any burgeoning geopolitical chain reaction was nipped in its bud by the projection of sufficient U.S. military power. What developed in stages during the Cold War was the actual realization of such ideas: the United States constructed a military-base network centered on the rimlands of Eurasia. Western Europe, East Asia, and, eventually, the Persian Gulf were singled out as the three main locations where the United States would be intervene by way of military alliances and the forward deployment of U.S. troops.

However, sea power, or naval dominance, would be a necessary pendant to forces onshore. Stopping Eurasian dominoes from being pushed over by any great power came to be seen as a core U.S. interest. This also hinged on U.S. control over key *communication lines*—in particular, the oceanic ones—which thereby also gained the status of a core and *permanent* U.S. national-security interest. The contemporary geopolitical analyst Jakub Grygiel (2006, p. 27) writes:

> The bulk of military power is still projected via land or sea. (...) [S]upplies are shipped via sea, land, and air routes, underscoring the importance of safe lines of communication. (...) A state that controls lines of communication has full strategic independence. It does not have to rely on the goodwill and protection of other states to access the resources it needs, project power where it wants, and maintain commercial relations with

> whom it wants. When a state does not have control over the routes linking it with the source of resources and other strategic locations, it falls under the influence of the power in charge of those lines of communication.

Two points apply. First, the logic that Grygiel here presents is basically the same as that which permeates international politics and international security more generally: under anarchy, trust among states is in short supply; worst-case scenario thinking directs much of states' security policies; and the exertion of influence is to a large extent motived by a perceived need to forestall similar moves by one's competitors. A great naval power, which the United States has been for decades, is hugely incentivized to exert as much control as possible over important sea lanes; if it does so, it gains the opportunity to deny such control and free passage to others. Free passage through the world's sea lanes is normally not denied anyone in peacetime, no matter what power rules the waves; denial is usually reserved for times of conflict, which (from the perspective of more harmonious times) "leaves the illusion of a free maritime space" (Grygiel, 2006, p. 28). That the United States insists on serving as an allegedly benevolent guarantor of safe passage and freedom of navigation, a role that is firmly based on the material reality of its naval power, sometimes masks the fact that this would give it enormous leverage should conflict break out. This fact, moreover, significantly affects its leverage in peacetime as well, since the resolution of even quite innocuous security, diplomatic, or commercial issues are shaped, to a degree, by the vague yet always looming shadow of military conflict.

The second point of note is that sea power and true control over communication lines are not possible if one only controls, or has secure access to, a small handful of nodes along the routes of relevance. A *network* of an array of geographically diversified nodes—the whole gamut of Mahan's oceanic infrastructure—is required. It is perhaps a bit simplistic, but it nonetheless approaches the truth, to state that having access to only a few bases is as useless as having access to none—if *global* power is sought. This implies that a naval power that withdraws from, or is forced to give up control over, a few of the nodes along an important oceanic highway risks seeing its entire naval-power edifice unravel. (The same issues are also of relevance to attendant needs for long-distance *air* mobility.) That is to say, global powers need a widespread *global* network of military bases. This applies nearly as much to land as to sea (or air). On land, it is true, the global power can be more selective with respect to what particular regions or sub-regions are of interest; bases can be more concentrated geographically. On the world oceans, though, one would need secure access to sundry nodes along the whole length of the sea lanes.

The main point, in any case, is that these bases must be perceived as *interdependent* ones; their usefulness can only be properly assessed when they are viewed as a geostrategic whole. These are nodes that are by necessity connected to a number of other nodes, in regional or global networks (Erickson et al., 2010, p. 215). Furthermore, these *en route* systems should also contain resiliency or redundancy measures: bases should be sufficiently numerous and dispersed so as to decrease both the consequences of a sudden lack of access to particular bases and the effects of enemy attacks on, or threats to, one or a few of the facilities (Lostumbo et al., 2013, pp. 44ff., 115). For example, the U.S. Air Force maintains the necessary infrastructure to support several routes over the Pacific and Atlantic Oceans (Pettyjohn & Vick, 2013, pp. 30ff.). As for the Pacific, it includes a northern route (linking Alaska to Japan) and a southern route (from Hawaii to Japan or Southeast Asia, via Guam). The Atlantic encompasses a northern and a mid-Atlantic route (with endpoints in the U.K./Germany and Spain, respectively), in addition to a southern one (via the Caribbean and Ascension Island and to Africa).

It was in no small part the *lack of* such global infrastructure which hampered the great-power ambitions of Kaiser Wilhelm's Germany and Tsarist Russia (the latter of which was famously routed by Japan in the war of 1904–05 primarily due to the mix between sheer geographic distance to the war theater and a complete lack of control over the sea lanes leading there) (Grygiel, 2006, pp. 27–28). Contrarily, the British Empire was upheld in large part due to the very *existence of* such an infrastructure, as was the case for other sea-based empires in previous eras (Modelski, 1978). London was quite conscious about the interdependence of the various parts, each of which had a role—often a crucial role—to play in the defense of the Empire and of British hegemony. Writes the historian John Darwin (2009, pp. 3–4):

> British ability to provide naval protection or to send reinforcements to the scene of a conflict would have been very limited without the resources the imperial system supplied. It was strategic control of the Cape Colony (...) that secured the gateway to Asia from European waters. The prime function of Egypt, occupied by the British in 1882, was to preserve British

> use of the Suez Canal and protect the 'Clapham Junction' of imperial communications. Malta, Aden, Ceylon, Singapore, Hong Kong, Esquimalt (on Vancouver Island), the Falklands and Halifax, Nova Scotia, formed the network of bases from which the Royal Navy patrolled the world's sea-lanes.

Protecting its precious "stepping stones – the oases that linked together to connect back to London" (Frankopan, 2015, p. 280)—was a task of immense importance to foreign-policymakers in London. The stepping stones, however, while providing the "'keys' which locked up the globe" for Britain (Kennedy, 2017 [1976], p. 154), inevitably also created additional insecurities; for if they served to protect the Empire, they also needed protection themselves. Today, of course, the British network of bases is a paltry one even if it is still, in a sense, "global." On a more positive note, the United Kingdom's perceptions of insecurities have lessened in tandem with the (near-complete) dissolution of the Empire. The United States has long since inherited Britain's imperial role *and* its perceptions of insecurities. The vast majority of the world's stepping stones, at sea and onshore, belong to America, handing it enormous power, and often giving it enormous anxiety. This is so not least since the prospect of *losing* stepping stones—even smaller ones—is perceived by many as a grave threat to the entire interdependent network, and hence to U.S. hegemony at large.

Expectations for the Empirical Analysis

With regard to empirical patterns, which I will investigate in the subsequent three chapters, what expectations can we derive from the first two parts of the book? In general terms, we can expect three results from an analysis of: (a) developments in U.S. troop numbers over the span of the Trump presidency; and (b) arguments and narratives spurred by and aimed at counteracting threats, proposals, or plans to withdraw troops from various locations. First, we can expect that *any* retrenchment suggestion from the president (sometimes, but not always, firmly supported by his administration) immediately and consistently incites an array of protesting voices. These are the voices of the broader foreign-policy community, whose members warn against the negative geopolitical chain reactions—the falling dominoes—that ostensibly *could* ensue from the implementation of such a suggestion. Second, we can expect that in these situations, the voices, arguments, and narratives of the *Stayers* powerfully overwhelm the voices, arguments, and narratives of the *Leavers*—in terms of numbers as well as the attention acquired. Third, we can therefore expect that these two groups have an asymmetrical impact on policy (notwithstanding that the president in the period of relevance himself was a *Leaver*), which most probably results in the numbers and geographical distribution of U.S. troops proving to exhibit a great measure of continuity.

The better part of book so far has accounted for the key driving forces behind such expectations. Specifically, four broad categories or dimensions, each harboring multiple interlinked causal pathways, combine to help keep U.S. policy on the same track as it has been for over seven decades. These dimensions are: power, history, psychology, and geography.

Power

First, a core assumption of political realism, in all its variants, is that *power* matters enormously. Material power, and in particular *relative* material power, shapes interests, threat perceptions, and the amount of international influence sought, and ultimately acquired and upheld, by any state. The United States is, with regard to its relative economic and military power, in steady but nonetheless fairly slow decline. There is still a long way to go before the multipolarity label can reasonably be applied to describe the *global* distribution of power. The implication of this is that retrenchment, in the foreseeable future, is not at all inevitable. (But neither is it likely that we will see a major expansion of U.S. global force projection, given the prevailing and projected trends in relative power.)

We can therefore expect, first, that overall U.S. troop numbers generally reflect continuity. The reality of the distribution or balance of power is not the only reason for this expectation—history, psychology, and geography work in the same direction—, but it is a fundamental one nonetheless. Second, we can expect that the many and consistent threats, proposals, and plans by the U.S. executive to withdraw troops work to incite a plethora of stark warnings from the broader foreign-policy community, most of whose members are *Stayers*.

The narratives of the *Stayers*, we would surmise, draw heavily upon system-level reasoning connected to perceived realities of relative power. That is to say, we can expect the *Stayers*, among other things, to highlight the still-dominant role of the United States; the virtues of U.S. hegemony and leadership (which require global force projection for deterrence and reassurance purposes); the potent threat posed by a few assertive global and regional major powers, which need to be deterred, balanced, or contained; the risk that geopolitical "vacuums" will be filled by adversaries unless the United States actively projects *its* (military) power into the relevant locations; and the negative impact that even limited troop withdrawals would have for U.S. reputation and credibility as a whole. With overseas military bases and troops forming the material backbone, and most potent symbol, of Washington's global influence, as U.S. relative power continues to decline, these bases and troops will likely be among the *last* elements of that power to be voluntarily relinquished.

Such arguments, we expect, will dominate over the arguments posed by the *Leavers*, which are far fewer in numbers. They will stress that a mix of factors makes retrenchment a preferred policy option in most cases. *Leavers* will argue that the costs of global and regional leadership and order-keeping are much larger than what is allowed for by U.S. economic resources. They will contend that in nearly all regions (with East Asia being the one possible exception), regional allies of America can themselves easily undertake, and finance, any balancing and containment directed against menacing states. They will also emphasize that it is only natural to withdraw troops from numerous locations (notably, the Greater Middle East) where the United States has suffered very painful, largely self-inflicted geopolitical wounds; these include failed military interventions that have spurred multiple unintended consequences, and they include cases where security dilemmas have been generated due to the U.S. military presence.

History and Psychology

History is a second important driver of continuity, which we also expect will be prominently reflected in the warnings, arguments, and narratives of the *Stayers*. In part, this involves the sunk costs of the military bases hosting U.S. troops; the value of the existing basing infrastructure will play a significant role in these narratives, even as the *Leavers* will stress that upholding this globe-spanning military infrastructure is unjustifiably costly.

But history also combines with *psychology* to increase the likelihood of continuity. As argued in Chapter 4, humans are inclined to abhor losses and to prefer to remain at a self-defined, though inherently "natural," reference point. With respect to U.S. global engagement and projection of military power, this reference point is overwhelmingly the status quo. Hence, the arguments and narratives of the bulk of the foreign-policy elite of the United States will, or so we can expect, exhibit an unmistakable status quo bias. This need not be explicitly expressed; but we can still expect arguments and narratives at least to refer tacitly to the long history of U.S. leadership, order-keeping, network of alliances, and overseas military presence. Retrenchment, *Stayers* would say, will work to unravel deeply rooted structures that have long since attained the status of something fundamentally natural, something akin to an "ethical necessity" harboring "normative qualities" of its own. The *Stayers* would take for granted that the alternative—*any* alternative—to the prevailing order would be worse, likely far worse, than what has carefully been constructed by America, often in cooperation with its allies, over the course of more than seven decades.

A generally-working, very influential *negativity bias* will work to reinforce such mechanisms. As noted, we expect the arguments and narratives of the *Stayers* to be significantly shaped by loss aversion and a status quo bias. More broadly, we can expect the *Stayers* (a group that also encompasses many prominent media organizations) to frame the consequences of retrenchment in worst-case-scenario terms. Proposed troop withdrawals, large and small ones alike, will spur a plethora of warnings about the various negative geopolitical chain reactions, or falling dominoes, they will ultimately cause. Depending on context-specific factors, there will be warnings about dominoes of different kinds: rising major powers expanding their influence, threatening to establish regional hegemonies; smaller "rogue"-type states redoubling their efforts to develop weapons of mass destruction; and international and domestic terrorist organizations gaining territorial control in failed states, which will then act as safe havens whence attacks against the West can be planned and executed.

In contrast, the arguments and narratives of the *Leavers* will tend to drown when faced with the *Stayers*' warnings about all the things that *can*, or rather *will*, go wrong if America scales back its global military presence. The *Leavers*, we can suspect, will attempt to argue that such negative chain reactions are highly unlikely to manifest, and that the arguments of the *Stayers* really amount to scaremongering. In effect, the *Leavers* will want to convince others that, in many cases, it makes perfect sense to scale back U.S. activities. However, the *Leavers* will have an incredibly difficult job convincing the *Stayers* to accept that the United States needs to re-evaluate its ingrained preference for the status quo and get ready to renormalize to a new, more modest geopolitical reference point.

Geography

The fourth and last of the major drivers behind the expected continuity is *geography*. We would expect the arguments and narratives of the *Leavers* to emphasize, or at least indicate, that the United States is eminently secure behind its two surrounding moats—the Atlantic and the Pacific Oceans—and that it now needs to focus overwhelmingly on domestic matters rather than on quixotic foreign-policy issues that are purportedly of little real relevance to U.S. national security. Yet again, we would expect these arguments and narratives to be superseded by the voices of the *Stayers*.

Here as well, the *Stayers* might not be overly explicit in their references to the enduring interests that America takes, or *should* take at any rate, in what happens in Eurasia. Their narratives will nonetheless reflect the theories of Mackinder, Spykman, and Mahan. The *Stayers* will perceive a continued threat from the actual or potential Heartland powers, which is Russia and China, respectively. These threats need to be deterred, balanced, and contained, the *Stayers* would say, and the United States is the only country commanding the capabilities needed to do so. U.S. troops must remain, therefore, along the rimlands of almost the whole of Eurasia. This includes Europe, the Persian Gulf and the Greater Middle East, and Southeast and Northeast Asia. The *Stayers* would argue that U.S. troop numbers should increase in the Asia–Pacific region, but they would strongly object to any real downsizing of troops elsewhere. U.S. troops might be somewhat more expendable in other regions of less perceived importance or urgency, such as most of Africa and the Americas. Still, the main point of the Stayers is that the United States should remain where they are—and where they have been for decades. Bases and troops should also remain along the permanently vital sea lanes of communication, the *Stayers* would contend. The islands and ports, *all* the islands and ports, to which the United States has secure access—in the Pacific, in the Atlantic, and in the Indian Ocean—form a coherent whole that is key to the whole infrastructure that supports U.S. global influence. Even a loss of a few of these locations *may*, or so we can expect the *Stayers* to argue, ultimately make the whole construction unravel.

*

We could perhaps now state that the *Stayers* believe—or, my *expectation* is that they believe—that every U.S. overseas military base, and every U.S. forward-deployed soldier, is as valuable as the next one; they should all be treated equally, that is. We should expect this to be reflected in the narratives and arguments of the *Stayers*. When suggestions or plans are made to withdraw troops *from somewhere or from anywhere*, the *Stayers* will react forcefully. They will certainly do so when the withdrawal suggestion concerns a main operating base hosting thousands of U.S. troops. But they will also do so, I expect, when the projected withdrawal concerns but a handful of troops from a location only peripheral to what can reasonably be deemed a core national interest of the United States.

The pullout of a few soldiers, *Stayers* would argue, signals something that is of much, much greater significance than the mere numbers themselves can possibly indicate: it signals that America is a waning power, a power-in-retreat. In that sense, the *whole* structure of U.S. global presence, resting on the very material reality of the vast network of military bases and troops, is of vital importance—or is perceived to be so. The *Stayers* might not explicitly argue that either the entire structure remains in place, or one might as well disassemble it all; but this is often what is implied by the arguments and narratives that are put forth. Faced with such scenarios, the *Leavers* will have a very hard time indeed making the United States withdraw from any overseas location at all.

References

Aklin, M., & Kern, A. (2019). Moral hazard and financial crises: Evidence from American troop deployments. *International Studies Quarterly, 63*(1), 15–29.

Biglaiser, G., & DeRouen, K. Jr. (2007). Following the flag: Troop deployment and U.S. foreign direct investment. *International Studies Quarterly*, *51*(4), 835–854.

Braithwaite, A., & Kucik, J. (2018). Does the presence of foreign troops affect stability in the host country? *Foreign Policy Analysis, 14*(4), 536–560.

Calder, K. E. (2007). *Embattled garrisons: Comparative base politics and American globalism*. Princeton University Press.

Crowley, R. (2008). *Empires of the sea: The Siege of Malta, the Battle of Lepanto, and the contest for the center of the world*. Random House.

Darwin, J. (2009). *The empire project: The rise and fall of the British world-system, 1830–1970*. Cambridge University Press.

DeYoung, K., Dawsey, J., & Fahim, K. (2019, October 17). In Turkey's president, Trump seems to have found a soul mate. *Washington Post*. https://www.washingtonpost.com/national-security/in-turkeys-president-trump-seems-to-have-found-a-soul-mate/2019/10/17/7fea3ada-eeb5-11e9-89eb-ec56cd414732_story.html

Drezner, D. W. (2013). Military primacy doesn't pay (nearly as much as you think). *International Security, 38*(1), 52–79.

Erickson, A. S., Walter III, L. C., & Mikolay, J. D. (2010). Diego Garcia and the United States' emerging Indian Ocean strategy. *Asian Security, 6*(3), 214–237.

Ferguson, N. (2003). *Empire: How Britain made the modern world*. Penguin.

Frankopan, P. (2015). *The silk roads: A new history of the world*. Bloomsbury.

Frederick, B., Watts, S., Lane, M., Doll, A., Rhoades, A. L., & Smith, M. L. (2020). *Understanding the deterrent impact of U.S. overseas forces*. RAND Corporation.

George, A. L., & Smoke, R. (1974). *Deterrence in American foreign policy: Theory and practice*. Columbia University Press.

Gray, C. (2004). In defence of the Heartland: Sir Halford Mackinder and his critics a hundred years on. *Comparative Strategy, 23*(1), 9–25.

Gray, C. S. (2015). Nicholas John Spykman, the balance of power, and international order. *Journal of Strategic Studies, 38*(6), 873–897.

Grygiel, J. J. (2006). *Great powers and geopolitical change*. Johns Hopkins University Press.

Hall, A. R. (1955). Mackinder and the course of events. *Annals of the Association of American Geographers, 45*(2), 109–126.

Huth, P. K. (1999). Deterrence and international conflict: Empirical findings and theoretical debates. *Annual Review of Political Science, 2*(1), 25–48.

Jakobsen, J., & Jakobsen, T. G. (2019). Tripwires and free-riders: Do forward-deployed U.S. troops reduce the willingness of host-country citizens to fight for their country? *Contemporary Security Policy, 40*(2), 135–164.

Jones, G., & Kane, T. (2012). U.S. troops and foreign economic growth. *Defence and Peace Economics, 23*(3), 225–249.

Kelly, P. (2016). *Classical geopolitics: A new analytical model*. Stanford University Press.

Kennedy, P. (2017). *The rise and fall of British naval mastery*. Penguin. (Original work published 1976.)

Krasner, S. D. (1976). State power and the structure of international trade. *World Politics, 28*(3), 317–347.

Lambert, A. (2018). *Seapower states: Maritime culture, continental empires and the conflict that made the modern world*. Yale University Press.
Lostumbo, M. J., McNerney, M. J., Peltz, E., Eaton, D., Frelinger, D. R., Greenfield, V. A., Halliday, J., Mills, P., Nardulli, B. R., & Pettyjohn, S. L. (2013). *Overseas basing of U.S. military forces: An assessment of relative costs and strategic benefits*. RAND Corporation.
Mackinder, H. J. (1904). The geographical pivot of history. *The Geographical Journal, 170*(4), 421–444.
Mackinder, H. J. (1942). *Democratic ideals and reality*. National Defense University Press. (Original work published 1919.)
Mackinder, H. J. (1943). The round world and the winning of the peace. *Foreign Affairs, 21*(4), 595–605.
Mahan, A. T. (1890). *The influence of sea power upon history, 1660–1783*. Little, Brown.
Mandelbaum, M. (2019). *The rise and fall of peace on earth*. Oxford University Press.
Modelski, G. (1978). The long cycle of global politics and the nation-state. *Comparative Studies in Society and History, 20*(2), 214–235.
O'Mahoney, A., Priebe, M., Frederick, B., Kavanagh, J., Lane, M., Johnston, T., Szayna, T. S., Hlavka, J. P., Watts, S., & Povlock, M. (2018). *U.S. presence and the incidence of conflict*. RAND Corporation.
Owens, M. T. (2015). In defense of classical geopolitics. *Orbis, 59*(4), 463–478.
Pettyjohn, S. L., & Vick, A. J. (2013). *The posture triangle: A new framework for U.S. Air Force global presence*. RAND Corporation.
Posen, B. R. (2003). Command of the commons: The military foundation of U.S. hegemony. *International Security, 28*(1), 5–46.
Reynolds, D. (2000). *Britannia overruled: British policy and world power in the twentieth century*. Routledge.
Schelling, T. C. (1980). *The strategy of conflict*. Harvard University Press. (Original work published 1960.)
Schelling, T. C. (2008). *Arms and influence*. Yale University Press. (Original work published 1966.)
Spykman, N. J. (1938). Geography and foreign policy, I. *American Political Science Review, 32*(1), 28–50.
Spykman, N. J. (2017). *America's strategy in world politics: The United States and the balance of power*. Routledge. (Original work published 1942.)
von Clausewitz, C. (2007). *On War* (M. E. Howard & P. Paret, Trans.). Oxford University Press. (Original work published 1832.)
Westad, O. A. (2017). *The Cold War: A world history*. Basic Books.
Wohlforth, W. C. (1999). The stability of a unipolar world. *International Security, 24*(1), 5–41.

CHAPTER 6

U.S. Overseas Troops: Empirical Patterns, 2017–2021

*

The current and the next chapter will put the theory of great-power persistence to the test. There, I analyze empirical patterns and developments in the numbers and geographical distribution of U.S. military bases and troops from the end of the Obama administration to the end of the Trump administration. This chapter presents an overview of world and regional trends, while Chapter 7 studies more closely developments and cases of attempted withdrawal within regions of particular interest.

Domino Narratives

The empirical data consist of numbers and narratives. The numbers tell their own important story, even if they cannot give us a complete picture of the extent to which continuity has dominated over change, or vice versa, regarding the U.S.'s overseas military-force posture. Most importantly, the numbers cannot on their own assist us much in *explaining why* continuity, or change, has characterized deployment decisions in specific contexts; identifying the causal chains requires an investigation into the *narratives* surrounding a variety of cases. These narratives are in most instances brought to public attention by international or national media, which therefore represent an invaluable trove of empirical information

J. Jakobsen, *The Geopolitics of U.S. Overseas Troops and Withdrawal*, Palgrave Studies in International Relations,
https://doi.org/10.1007/978-3-030-94488-9_6

for this project. The narratives consist of a mix of arguments, rationales, allusions to the purpose of the troops, criticism (often of the president), and even ideologies of sorts. The narratives might sometimes be self-serving; more often, though, they are likely constructed with purported U.S. national-security interests in mind. The narratives regularly take their cue from systemic developments of the hypothetical sort, and they tend to revolve around the issue of falling dominoes—that is, of geopolitical chain reactions, in particular negative ones.

The narratives are constructed by various members of the broader foreign-policy community, the "domestic guardians of primacy." I have earlier accounted for who these guardians are; suffice it to repeat briefly here that they are numerous and widespread (and very well described by Stephen Walt [2018]), and they are overwhelmingly *Stayers*. Among other groups, the guardians consist of the U.S. Congress; the U.S. military and intelligence services; lobby groups; think tanks; diplomats; commentators and analysts; university scholars; consultants; and much of the media. These voices are often supported by quite similar ones in those host countries that are directly affected by U.S. decisions on bases and troops (and sometimes voices from countries that are not directly affected weigh in as well).

These are the voices who construct and tout the narratives that I surmise constitute quite a solid barrier against real change. One challenge here, of course, is that the narratives cannot normally *on their own* cause continuity (or change, for that matter). At least we cannot automatically and effortlessly create or recreate a causal link between them and actual policy decisions. In a few cases we can, though. It is certainly easiest to do so when members of the U.S. Congress, which is formally authorized to make decisions on the *funding* of troop withdrawals, act to block the president's push for retrenchment. This was what the Congress did in 2020, with respect to the U.S. troops in Germany and Afghanistan (though with little success in the latter case) (Zengerle, 2020). In both cases, domino concerns weighed heavily on the decision-making process. Such a direct link can also easily be established when numbers are backed up by an admission by the president that his position on a particular withdrawal issue changed because he received some strong advice from one or more *Stayers*. This happened when President Trump ordered a troop surge, rather than a drawdown, in Afghanistan in 2017 (Mehta & Copp, 2017). And sometimes information comes to light that shows that members of the foreign-policy community by and large take the matter into their own hands, effectively making the decision on troop numbers themselves. This was apparently what happened in Syria in 2020, when the U.S. envoy to the country, together with his team, hid from the president and senior leaders the true number of U.S. troops deployed there (Williams, 2020). This they ostensibly did to ensure that the *actual* number stayed above what was judged by them to be necessary to avoid setting in motion some very damaging geopolitical chain reactions.

Yet, in most cases, the evidence of *causal* relationships is less unambiguous. We cannot then know for certain that the narratives of the *Stayers* of the foreign-policy community have had real and independent effects on policy decisions. In such instances, we need to thread carefully, and we need to tease out the links as best as we can—if there really *are* links, that is—using a mix of numbers and narratives in the process. Vitally, we also need to be cognizant of, and to control, the role played by the domestic forces that I reviewed in the last main section of Chapter 2. In some of the cases of proposed or possible withdrawal that failed to manifest, it could be that the main cause is instead to be found in the very whimsical or incoherent nature of Trump and his team; or it could be down to the president primarily using withdrawal threats to enhance his bargaining power on other issues; or it could be that U.S. public opinion did not really fancy withdrawal, circumscribing the room for maneuver of a president who sought re-election; or it could be that sheer institutional inertia and bureaucratic constraints were the main cause of continuity; or it might be that pressure from vested economic interests made the president think twice about pulling out the troops.

A mix of forces is certainly in play in some of the cases of interest. Even so, what we also do know is that threats, proposals, and plans to withdraw U.S. troops from virtually any location incite debates that are usually heavily one-sided, and therefore not really debates at all. This creates an environment wherein actually implementing a decision to withdraw troops becomes immeasurably more difficult than what would have been the case if silence had prevailed instead. To withdraw troops, in such an atmosphere, also involves accepting that one—meaning the president and his administration—will have to suffer the attendant political and reputational costs, which are bound to be quite substantial. This is particularly so when the matter at hand—because this is the message relayed by the guardians of primacy—concerns a core U.S. national-security interest, such as the international balance of power; the possible metastasizing of transnational terrorism; the spread of authoritarianism; the potential blocking of trade routes; the proliferation of weapons of mass destruction; or any other *potentially* serious threat to America. These are issues and warnings that cannot be taken lightly by any president or administration, no matter how unconventional and controversial he and it is, even in cases where threats to U.S. national security are grossly inflated by their propagators.

The Numbers (and Their Defects)

For our purposes here, the time period covered by the empirical analysis represents an ideal case—or at least as ideal as is realistically possible. This is so for two main reasons. First, Donald Trump assumed the presidency after the previous decade had witnessed steadily growing *perceptions* that U.S. relative power was in serious decline, and that the unipolar moment had passed. The 2008–2009 financial crisis came to be seen as one potent symbol of such a change, as did the quagmire-like, practically unwinnable U.S. wars in Iraq and Afghanistan. The behavior of other major powers also constituted a portent of major shifts in the balance of power. Some emphasized that China's newfound foreign-policy assertiveness coincided with the financial crisis (Christensen, 2011). Others have noted that Russia's hybrid war against the West really commenced as early as in 2007 (Orenstein, 2019). This was the very same year as when Russian President Vladimir Putin, at a security conference in Munich, Germany, had castigated the United States and unipolarity alike in what was widely perceived, at the time and in hindsight, as a highly ominous speech (Shanker & Landler, 2007). It seemed clear that Russia and China (and, in the Middle East, Iran) would do what it took to help hasten the onset of multipolarity (Cooley & Nexon, 2020, p. 81). Subsequent years gave further indications of this. Russia's invasion of Georgia (2008), and of the Crimea and eastern Ukraine (2014); China's unabashed expansion in the South China Sea; and Iran's suspected nuclear-weapons ambitions and projection of power in the Middle East—all these events and developments raised serious questions about the standing of the United States in these regions, and in the world at large. The overriding issue seemed to be whether U.S. global military engagement was at all sustainable.

Second, the election of Donald Trump, who often raised such questions himself, meant that the United States was to be led, for at least four years, by a self-avowed *Leaver*. This was an unprecedented situation for a nation and a people that, if I once more lean on Sebastian

Schmidt's pithy yet fitting comment, believed that "having military bases all over the globe was not any more out-of-the-ordinary than drinking coffee in the morning" (Schmidt, 2020, p. ix). By 2017, however, it seemed clear that the status quo *was* going to be seriously questioned, following a decade of increasing geopolitical tensions in all those regions where U.S. military bases and troops had been present for generations. It was now up to President Donald Trump to make good on his election-campaign promises that his nation would scale back its global activities. One is indeed tempted to ask: If America wasn't going to retrench under *his* tenure, would it ever retrench at all?

The Data

The numbers presented and analyzed draw on data from the Pentagon's Defense Manpower Data Center (DMDC) (U.S. Department of Defense, 2021). The data represent a snapshot of the number of active-duty forward-deployed U.S. troops in any location as of the end of March (Q1), June (Q2), September (Q3), and December (Q4). This yields a total of 17 data points, running from December 31, 2016 (representing the last quarter of the Obama administration), to December 31, 2020 (representing the last quarter of the Trump administration). While most of the analyses end on the last day of 2020, wherever it is necessary I also consider developments in the following few months. The new president, Joe Biden, was inaugurated on January 20, 2021. There is nothing to indicate that U.S. troop numbers changed in any perceptible way in the three weeks in-between (with Afghanistan and Iraq constituting single-country exceptions, an issue to which I will soon return); whereas ending the data series on March 31 would risk conflating any changes made during the very last stages of the Trump era with changes made in the early months of the Biden administration. I will, in any case, devote ample space in the concluding chapter to developments in the first year of Biden's term.

The DMDC data encompass all *active-duty* service members in the Army, Navy, Marine Corps, Air Force, and Coast Guard "who are permanently assigned for duty" at the locations in question. The Pentagon explicitly excludes from these numbers service members "on temporary duty" and those supporting "contingency operations." As for the "permanent" troops, the alternative to using active-duty troop numbers would be to add to them the number of National Guards and reserve troops (with civilian staff at U.S. bases also coming in addition). This would lead to significantly higher numbers for a small handful of U.S. overseas territories: Puerto Rico, the U.S. Virgin Islands, American Samoa, and Guam. All other states and territories only have insignificant numbers of reserves and National Guards.

Most countries hosting U.S. troops do so voluntarily. However, sometimes the troops were, at least originally, part of an invasion force or deployed in the immediate aftermath of a war (Martinez Machain & Morgan, 2013, p. 103). While the deployment rationales might differ between these two types of cases, I nevertheless include both in my analyses. In part, I do so because the line separating war deployments from other types of missions is often quite blurry. For example, troop numbers in the broader Middle East have fluctuated for the last three decades, and sometimes greatly so, depending on U.S. policy decisions with respect to the wars in Iraq (twice) and Afghanistan. During the height of the latest Iraqi war, a large share of the troops that were actually *in* Iraq were assigned by the DMDC as being stationed at nearby U.S. bases in the United Arab Emirates and Kuwait (O'Mahoney et al., 2018, pp. 9–10). In addition, cases of actual war-fighting are rare; this only includes Afghanistan, Iraq, and Syria—as per the Pentagon's classification—and U.S. troop numbers have been fairly modest in all three countries during Trump's tenure. These numbers have also been kept formally secret by the Pentagon, a matter to which I will return shortly.

Caveats

While most of the country- and territory-specific numbers should be quite accurate, there are reasons to doubt their reliability in more than a handful of cases. Part of the problem is that the Pentagon's numbers exclude some specific categories of U.S. troops. Temporary staff and soldiers on contingency missions are not counted, and neither are troops stationed afloat; the latter are registered as located at their home port even should that be thousands of miles from where they *actually* are at the time of counting (Kane, 2006). Among those that are not counted are, for instance, rotational troops that are part of NATO's deterrence mission vis-à-vis Russia (Operation Atlantic Resolve) in European states such as Poland, Romania, and Bulgaria (Emmott, 2017; Gigova, 2017). For example, around 4500 U.S. troops rotate through Poland as part of the deterrence efforts against Russia (Gearan & Sonne, 2019), but the

DMDC only report the number 162 (for 2020_Q4). The around 600 U.S. troops in Kosovo are not included either (Reuters, 2020c). Likewise, the official number of U.S. troops in Sub-Saharan Africa measured only in the seven hundreds for all quarters in 2020. Although the actual number of U.S. troops on that continent at any given moment is fairly meager too, it is at least eight or nine times larger than what the DMDC tally suggests (Browne, 2020). For example, there are several thousand U.S. service members stationed at Camp Lemonnier, a U.S. main operating base in Djibouti, and the vast majority of them are not included in the official statistics (Bearak, 2019). DMDC numbers for nearby Somalia (and for Kenya) have also excluded most of the U.S. troops actually present there (Stewart & Houreld, 2020); whereas, in North Africa, some 400 U.S. peacekeeping troops stationed in Egypt's Sinai Peninsula fall more clearly and understandably outside of the DMDC classification (Times of Israel, 2020).

This definitional and classification problem might be further compounded if the current administration follows the advice of Mark Esper, Trump's defense secretary between July 2019 and November 2020. His suggestion was that the United States boost the use of rotational deployments in lieu of "permanently" stationed troops, purportedly to create "strategic flexibility" and to decrease the U. S. military footprint (Burns, 2020). In any case, the information relayed by the DMDC is certainly not complete if the objective is to get both a sense of the total number of U.S. military-service personnel *actually* located abroad (total numbers are significantly underestimated) and an unambiguous picture of whether the Trump tenure has seen change or continuity. The rather shadowy nature of some of the numbers, coupled with some relatively evasive replies from the U.S. administration on these issues, prompted *The Washington Post*, in October 2020, to warn that they intended to sue the Pentagon for not fulfilling its obligations under the Freedom of Information Act (Brannen & Goodman, 2020).

A related part of the problem is the secrecy surrounding some U.S. bases and physical installations. This issue is well-acknowledged by those who analyze the yearly U.S. Base Structure Reports, which provide numbers on property owned and leased by the U.S. military abroad (Herrera & Cicchini, 2013, pp. 141–142; Stravers & El Kurd, 2020). For instance, the last such report that has been made public, for Fiscal Year 2018 (which reports U.S. assets as of the end of September 2017), fails to include quite a few facilities (and attendant U.S. troops) that are known to exist (Stravers & El Kurd, 2020; Turse, 2019; U.S. Department of Defense, 2018). Such "secret" or officially unreported facilities can be found in several regions, although there are reasons to believe that underreporting is particularly salient with respect to the Middle East and Africa. Having said that, the DMDC data are far more comprehensive than the corresponding data in the Base Structure Reports (which also include troop numbers, though its list of countries is limited).

While I sympathize with the initiative of *The Washington Post*, I don't intend myself to sue anyone over this. Instead, I will proceed with the analysis using the numbers that *are* publicized. Three things make me certain that most of the potential problems can be rectified. First, the *Post*'s main issue was, understandably, with the lack of publicized troop numbers for Afghanistan, Iraq, and Syria. These I needed to estimate myself for each quarter, though in this endeavor I was invaluably assisted by reporting done by that very same newspaper (and quite a few others). Second, the rest of the country-specific numbers, even if they are on average underestimated, nonetheless exhibit substantial continuity from 2016_Q4 to 2020_Q4. That is to say, there is little reason to believe that the Pentagon has changed its definitions, or the way that troop numbers are counted, over the span of this period. Third, despite that the tables and figures that will be presented later only report the DMDC data (again with the exceptions of Afghanistan, Iraq, and Syria), I supplement and cross-check these data with information from an array of additional sources. This information is analyzed and presented throughout the two empirical chapters in order to gain a fuller, and hopefully almost complete, picture of the numbers and geographical locations—and the *change* in and *continuity* of the numbers and geographical locations—of U.S. forward-deployed troops.

The Missing Bits: Syria, Iraq, and Afghanistan

As of the end of 2017, the Pentagon no longer publishes troops data for Syria, Iraq, and Afghanistan, ostensibly because such data could potentially benefit U.S. adversaries (Zenko, 2018, p. 12). Such an argument notwithstanding, U.S. officials quite often refer to troop numbers anyway—in interviews, press briefings, and the like. As a result, media reports of troop levels in these three countries tend to be quite consistent, and I use these secondary sources as the basis for estimating the missing data.

With respect to Syria, for example, the last official number released by the Pentagon, for September 2017, was 1547 U.S. troops. This marked another step of a small troop "surge" by the Trump administration (as reflected by comparison with data for previous quarters), which apparently reached its apex near the end of 2017. From that point on, and until the third quarter of 2019, I use 2000 as an approximation. This is based on media reports that consistently refer to this number from the beginning of 2018 until after the first-quarter troops data for 2019 were published for the rest of the world (March 31) (e.g., Clarke & Tabatabai, 2018; DeYoung et al., 2019; Gibbons-Neff, 2018; Sly & Morello, 2018). President Trump's December 2018 announcement that he had ordered U.S. troops to leave Syria immediately (DeYoung, 2018) did not, in fact, lead to any sudden withdrawal. Rather, the first phase of the pull out of forces only commenced around the end of March and the beginning of April 2019 (Schmitt, 2019). By the time of the second quarter of that year (July 31), troop levels had fallen to around 1000, with the same number also circulating by 2019_Q3 (September 30) (Ali, 2019; Myers, 2019; Seldin, 2019). By mid-December 2019, the reported tally was down to 600 (Reuters, 2019c). Thereafter, troop numbers seemed to stabilize (Detsch, 2020). After an apparent small reduction in the second quarter of 2020, the United States soon sent around 100 additional service members to Syria to counter what Washington alleged was Russian aggression, bringing the estimated number up to 600 again (Ward, 2020). After that, there has, to my knowledge, been no report of any adjustment in numbers. To the contrary, the U.S. envoy to Syria, Ambassador James Jeffrey, claimed in late October that he expected no significant change in the near-term future (Christou, 2020). It seems clear, then, that the number of U.S. troops stationed in Syria has been quite stable since the end of 2019. Stability aside, exactly *how many* troops the United States has on the ground is somewhat more uncertain, with some recent sources saying 900 (Cloud, 2021; Williams, 2020). The secrecy of the Pentagon does not permit us to have full confidence in either of these numbers. I therefore choose to follow the bulk of the early reporting, which has also been more consistent over time, and I henceforth use the more conservative tally of 600.

Iraq is a second country whose troops data are shrouded in secrecy by the Pentagon. However, from 2017_Q4, other sources consistently cite "more than 5,000 American troops" (Ali, 2019), usually specifying this further as amounting to 5200 (Al Holland, 2018; Jazeera, 2020;

Taylor, 2019). This is also the number used in this analysis from 2017_Q4 to 2020_Q3, as no sources, again to my knowledge, have reported any significantly different number between those two points in time. Things seemed to change in the third and, in particular, the fourth quarter of 2020, with signals coming from Washington in late August and early September that the U.S. contingent in Iraq would soon be reduced (Reuters, 2020e, 2020f). A gradual reduction did commence in the fourth quarter (Davison, 2020), and it seemed to continue in the last few weeks of the Trump presidency. By mid-January, official(!) statements by the Pentagon declared that the number was now down to 2500 (Garamone, 2021). This followed a few weeks of gradual reductions (Al Jazeera, 2021), which makes it a reasonable conjecture that the troop level at the year's end stood at around 2750, which is the number used by me for 2020_Q4.

As for Afghanistan, after the completion of the limited "troop surge" by November 2017 (Moore, 2017), numbers were for a long time relatively consistently reported as 14,000 (e.g., BBC, 2019b; Jaffe & Ryan, 2018; Lamothe et al., 2018; Sediqi & Torode, 2019). The caveat here are the reports in October 2019, based on statements from top U.S. officials, that Washington had "quietly withdrawn" some 2000 troops during that year (Gibbons-Neff & Mashal, 2019). Not knowing exactly when these troops left Afghanistan, I therefore use the number 14,000 from 2017_Q4 until and including 2019_Q2, and 12,000 for 2019_Q3 and Q4. Then, in late February 2020, the United States and its main Afghan adversary, the Taliban, reached a peace deal under whose terms Washington pledged to reduce its forces in Afghanistan to 8600 within 135 days, with a full withdrawal of U.S. troops being finalized within a year (DeYoung, 2020). The pull out evidently started in early March (BBC, 2020; Lawrence, 2020). News sources suggest that the withdrawal proceeded fairly steadily in the subsequent weeks, approximating the planned average of a daily 25-troop reduction (Atwood & Browne, 2020; Lee & Kibe, 2020). For the quarter ending on 31 March 2020, I therefore use the number 11,250, which corresponds to the phased withdrawal plans. The gradual reduction seemed to continue thereafter; and by the end of June, General Frank McKenzie, head of the command responsible for Middle East and Central Asia, the United States Central Command (CENTCOM), claimed that the stated target of bringing the troop number down to the projected level within the specified four-and-a-half months had now been reached (Mashal, 2020). I therefore use the number 8600 for 2020_Q2, as other sources do (Cloud & Wilkinson, 2020; Reuters, 2020d). After that, a steady drawdown seems to have taken place, with U.S. officials citing the goal of reducing the number of troops to 3000 or less by early 2021 (Kube & Lee, 2020). That objective was realized, according to the Pentagon itself, which reported the number of 2500 on January 15 (Garamone, 2021). This allegedly followed a few weeks where military commanders were rushing to withdraw some 1500 troops (Al Jazeera, 2021) from a reported number of 4000 in mid-December (Seck, 2020). It therefore seems reasonable to set the number of U.S. troops in Afghanistan to 3500 for the end of 2020. This is so even if reports surfacing in March 2021 suggested that the true number in the first few weeks of the Biden was indeed still 3500. These reports were based on claims by U.S., European, and Afghan officials that some Special Operations forces had been left out of the records, and that Afghanistan in addition still hosted many "temporary and transitioning units" that were not counted in the official statistics (Gibbons-Neff et al., 2021). As with Syria, however, I choose to follow the more conservative estimates, which have also attained a quasi-official status, while acknowledging that they are most probably quite consciously understated by the U.S. Department of Defense.

U.S. Troops in the Middle East

The Middle East, in general, poses a few challenges beyond these three states. Part of the issue, as mentioned earlier, concerns the possible and probable secrecy—and the concomitant underestimation of troop numbers—surrounding some of the U.S. military installations. Also, U.S. officials acknowledge that troops on "sensitive missions" are not included in official counts (Gold, 2018). Another problematic issue seems to be that the data only include troops stationed ashore. In October 2019, U.S. Special Representative for Iran, Brian Hook, claimed that the United States had, since May that year, sent an additional 14,000 troops to the Middle East, "as an investment in regional security" following the growing tensions between Iran, on the one hand, and Washington and its regional allies on the other (U.S. Department of State, 2019). This number is not, however, reflected by the DMDC data; these actually show a small *decline* of forces in the Middle East and North Africa from 2019_Q1 (16,992) to 2019_Q2 (15,916); a further small decline from the latter and to 2019_Q3 (15,664); and again to 2019_Q4; before numbers increased to 17,712 by the end of March 2020. Many of the troops that were added were reportedly stationed afloat (Barnes & Schmitt, 2019); hence they fall outside the definition of forward-deployed troops used by the Pentagon. It is nonetheless reasonable to surmise that some of these forces are stationed ashore. In addition, reports suggest that the 14,000 troops that were added were supplemented by both air-defense personnel and bombers (Reuters, 2019a).

What is weightier, though, is that many of these service members were considered by the Pentagon as being on contingency or temporary missions and so were not included in the tallies (Glaser, 2017, p. 2). This fits with the long-standing U.S. policy of maintaining in the Middle East a "'presence without permanence'" (Henry, 2006, p. 27)—which perhaps could more accurately be described as a policy of "permanent presence of temporary troops." Whatever the (mix of) reasons for these less-than-convincing official numbers, it seems patently clear that the tally of U.S. troops for this region is seriously understated, and that the United States persistently keeps somewhere between 45,000 and 65,000 troops around the Persian Gulf alone (Schmitt, 2020).

We need to bear in mind such caveats especially when analyzing regional patterns. However, apart from estimating the "officially" missing numbers for Afghanistan, Syria, and Iraq, I choose not to make any additional alterations to the data. Doing otherwise would be a major task, and the added value would certainly be debatable; such an endeavor could just as well distort the true picture of U.S. troops. Besides, as I mentioned above, these issues are in part alleviated by the fact that the underestimation does not seem to vary systematically over the relevant time period. In addition, I draw on several other sources as well, so as to get a more accurate picture of troop numbers in this and in other regions.

Global and Regional Patterns

Figure 2.1 in Chapter 2, a line diagram that depicted data on total U.S. overseas troop numbers from 1950 to 2021, indicated a slow long-term decline. This trend really started when the Korean War ended in 1953, although it was punctuated by a major upsurge during the Vietnam debacle, and a more modest but still significant increase following the 2003 U.S. invasion of Iraq. In this section, however, I leave the Cold War era fully behind, instead of focusing exclusively on developments in the post-Cold War period. In particular, the presentation and analysis will center on the years—or quarters, rather—of the Trump administration.

U.S. Troops and Power After the Cold War

Figure 6.1 takes the start of the post-Cold War period as its point of departure, comparing the development in troop numbers to trends in the relative power position of the United States from 1992—the base year, which is given an index number of 100—to 2020. Developments in the relative power (as-resources) of the United States are measured in two different ways, both of which are admittedly rough approximations. The first line, which draws on data from the World Bank (2021), measures the United States' share of world gross domestic product (GDP), measured in constant 2015 U.S.$ using official exchange rates. The second line shows developments in its share of total world military expenditures, with underlying data from the Stockholm International Peace Research Institute (SIPRI, 2021). The third line tracks developments in U.S. overseas troop numbers (Kane, 2004; U.S. Department of Defense, 2021).

Fig. 6.1 Index of U.S. power resources and U.S. overseas troops 1992–2020

Figure 6.1 indicates that the United States is in gradual decline, and that it has been for a few years. Yet, neither the GDP nor the military-expenditure measure exhibits any *dramatic* decline. The former starts to drop, albeit very slowly, from its apex in 1999, but it is still 86 in 2020. This means that the United States still commands nearly one quarter (or 23.6%) of world GDP. If we also consider that the United States, according to another measure, now accounts for nearly half of the world's business activity (The Economist, 2021), there is little or nothing to indicate that the relative strength of America's economy is in any way *rapidly* and *unequivocally* disintegrating. The drop was more substantial, though, for military expenditures: close to 100 as late as in 2010, the index thereafter declined at a somewhat higher tempo than the GDP measure, falling below 77 in 2017. This decline can in part be explained by the increased military spending of China in recent years. However, U.S. military budgets were also reduced in *absolute* terms between 2010 and 2017. The subsequent three years, however, saw a substantial uptick in U.S. military expenditures, both in an absolute and a relative sense. Indeed, in 2020 the index stood at 86.8, with the United States accounting for a whopping 39.3% of total world military spending. One should note that this is almost exactly the same share as it commanded in 1999—that is to say, during the height of unipolarity but before the major wars in Afghanistan and Iraq.

Taken as a whole, therefore, Fig. 6.1 does not point to a contraction of U.S. power resources at a level that would *inevitably* effectuate any massive withdrawal of U.S. overseas troops. Consider also that the military budget, according to SIPRI, constitutes 3.7% of U.S. GDP. Granted, this is a high share when compared with other advanced economies. Even so, it is lower than it was during the height of the wars in Afghanistan and Iraq (4.9% in 2009), and it is *far* lower than it was for most of the Cold War, where it ranged between 4.9% (during the height of *détente* in 1978) and 13.9% (in 1952). This indicates that the United States still has some room for further increases in its military spending. Insofar as the command of power resources greatly helps shape military-base posture and troop deployments, which this book reckons it does, the United States therefore still has the material wherewithal to uphold, and even to bolster, its global military infrastructure. These numbers, in other words, suggest that the United States does not *need* to retrench. Moreover, the marked fluctuations in troop numbers in this period, over and above what we may call a "base level" of sorts, seem to happen fairly independently of power resources or power potential; instead, the fluctuations are linked to dramatic international developments. This was true for the troop withdrawals in the years immediately following the end of the Cold War. It was also the case for the substantial increase that took place after 9/11, which was augmented further by the 2003 invasion of Iraq. After the 2007 "troop surge" in Iraq, however, numbers started falling again, and quite rapidly so under the two administrations of Barack Obama, even if the geographical distribution of these troops remained quite constant.

When Donald Trump entered the White House in January 2017, the index value stood at about 66 for the second year in a row—its lowest point not only since the end of the Cold War, but for generations. According to these numbers, the increase in the index value in the first year of the Trump administration (to 71.4) proved to be temporary. By the end of his tenure, it was back down to a lower value (59.5) than what President Trump inherited from his predecessor. As I have indicated earlier, though, these numbers only tell one part of the story about changes and continuities in troop deployments. The rest of this chapter, and the next, will hopefully help complete the picture.

Total Troop Numbers Under Trump

The kernel of the preceding section is that U.S. economic and military resources, whether measured in absolute or relative terms, do not by themselves compel any current or near-term revolutionary change in the U.S. overseas military-force posture. This means that other drivers of any eventual decrease or increase—or of continuity—must be identified. However, we first need to confront further the more obvious issue of whether there really *has* been any significant change since Trump became president, according to the official numbers.

Figures 6.2 and 6.3 explore global developments from the fourth quarter of 2016 (December 31) to the end of 2020 (December 31). Whereas Fig. 6.2 only includes data where host countries are known, Fig. 6.3 adds to these numbers overseas troops at "unknown" locations. This accounts for the discrepancy with respect to developments from the start until the end of this period. Figure 6.2 shows that total troop numbers were practically exactly the same at these two points in time: the Trump era began with 171,133 U.S. troops abroad, and it ended with 171,049 (but note that this latter number does not account for

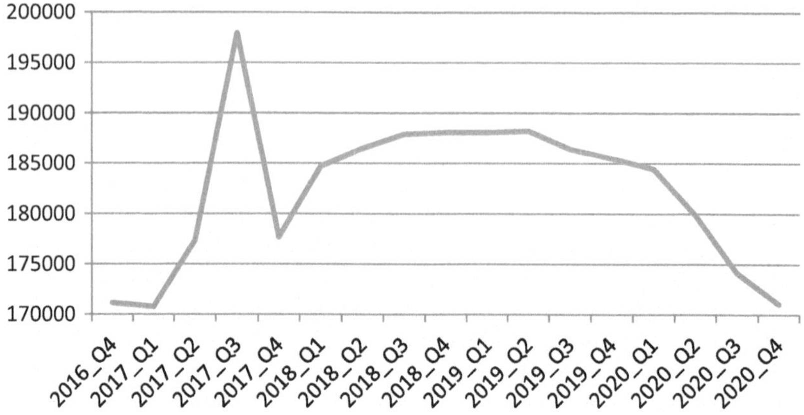

Fig. 6.2 U.S. overseas troops, world total, 2016(Q4)–2020(Q4)

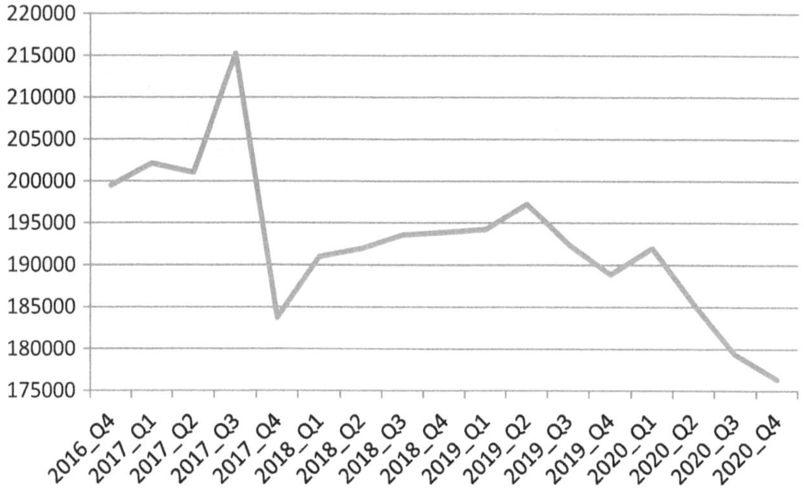

Fig. 6.3 U.S. overseas troops, world total, 2016(Q4)–2020(Q4), including "unknown" locations

the around 1250 troops that, according to my estimation, were withdrawn from Afghanistan (1000) and Iraq (250) in the last three weeks of his term in office). Figure 6.3, though, shows a *decrease* of some 23,000 troops (or 11.6%). The latter figure probably yields a more useful picture of the status of *global* deployment numbers, save for the very important caveat, discussed above, about missing data for key regions and countries: numbers of U.S. overseas military personnel classified as active-duty troops have apparently declined over Trump's presidency, albeit only modestly so.

A second noteworthy observation is the significant increase in troop numbers that took place almost immediately after Trump had taken office—that is, from Q2 to Q3 of 2017. That reflected both a decision by the president to intensify the fight against the Taliban in Afghanistan (with troop numbers there rising by more than 3000) and increased war efforts against the Islamic State around the Persian Gulf (Haltiwanger, 2017). Afghanistan apart, official troop numbers rose most sharply in the Middle East. A subsequent withdrawal of troops—or, more precisely, a withdrawal of troops formally designated as "permanently assigned"—from these same countries, which coincided with the crumbling of the bulk of the Islamic State's self-proclaimed caliphate, accounts for most of the drop in troop numbers from 2017_Q3 to 2017_Q4. Thereafter, official troop numbers picked up slightly; they then stabilized somewhat at just below 200,000 for several quarters; before they once again started falling during the latter months of the Trump administration.

Regional Troop Numbers Under Trump

If we now turn to regional troop numbers, what is noteworthy is, first, that Asia Pacific and Europe, which are the two largest host regions (again, according to the DMDC data, which effectively disregard most troops in the Middle East), have both witnessed an increase in U.S. troops over the period. For Europe, though, that increase is very modest, amounting instead to near-perfect overall *continuity*: troop numbers rose from 62,101 to 62,248 (+0.2%). Three of the five largest host countries—Italy, the United Kingdom, and Belgium—have received additional troops in the period, whereas the remaining two—Germany and Spain—have seen only a few of the troops leave (a reduction of 0.9 and 9.7%, respectively). Greece and the Netherlands are still important hosts, while the figure suggests a decrease in troop numbers for Kosovo, Romania, and Hungary. However, Kosovo continues to host between 500 and 700 U.S. troops that are part of NATO's Kosovo Force (KFOR) and so are not included in the DMDC tally (Bytyci, 2020; Lekic, 2020). Romania, for its part, has in recent years gained an elevated status in U.S. security and defense thinking; even though troop numbers have fallen, the United States has, inter alia, invested heavily in upgrading its surveillance and reconnaissance capabilities in the country, thus boosting its military presence in the vicinity of the Black Sea (Associated Press, 2021).

The increase is pronounced in the Asia–Pacific region: Facing both a nuclear-weapons capable North Korea and, even weightier, a rising China, the United States has *added* some 21,500 troops to this region from 2016_Q4, which amounts to an increase of 32%. The biggest hosts have seen the largest absolute increase. Japan's numbers have risen from 39,345 to 56,474 (+43.5%); South Korea's from 23,468 to 25,430 (+8.4%), and Guam's from 3831 to 6172 (+61.1%; some 2600 National Guards/reserves come in addition). Moreover, Australia has assumed a key role as host for the 2500 strong U.S. Marine Rotational Force-Darwin (Harkins, 2020), even if it actually varies from quarter to quarter whether or not the DMDC data count these troops as being "permanently assigned."

As noted earlier, numbers for the Middle East are most likely far less precise and comprehensive with respect to the actual number of U.S. military personnel, including those on "temporary" missions, deployed in the region. Official numbers suggest a total decline of nearly 14,000 troops (–51.6%). This involves ostensible reductions in all the major hosts. These apparent reductions do not, however, correspond well with what seems to be the reality on the ground. For example, I noted earlier that, according to widely circulated reports that were also confirmed by U.S. officials, the United States *added* around 14,000 troops in the Middle East between May and October 2019, after tensions with Iran started escalating (Reuters, 2019a; U.S. Department of State, 2019). Many of these troops were stationed afloat—and still more were regarded as "non-permanent"—and these were therefore not included in the Pentagon's numbers. In any event, the region reportedly saw even further additions of U.S. troops after 2019_Q3, following two specific developments. The first was the September 14, 2019, drone attack against Saudi Aramco plants in Saudi Arabia, which the United States blamed on Iran (Rapp-Hooper, 2019; Reuters, 2019b). The second was the late December and early January escalation of tensions between the United States, on the one hand, and Iran and Iraq-based Shia militias backed by Iran, on the other (McKay, 2020; Reuters, 2020a). The former event fails to impact the official numbers for 31 December 2019, which instead show a slight *decrease*.

In other words, it is hard to read too much into the official figures, even if we can be reasonably confident that the numbers for some—such as Afghanistan, Iraq, and Turkey—are fairly precise. A January 8, 2020, report from *Reuters* (2020b), which draws on estimations made by various research institutions, suggests much higher troop numbers for the main hosts than what are officially presented by the Pentagon. This includes Bahrain (with an estimated 5000 U.S. troops), Qatar (10,000), Saudi Arabia (2500), Kuwait (14,500), and the United Arab Emirates (5500). The grand total for these five countries is 37,500, which is far above the official Pentagon data for 2019_Q4 (6975 for the same group of states). Thousands of troops in states such as Turkey and Jordan come in addition. Yet another tally, published at the same time by *The Economist* (2020), yields the number of 41,000 U.S. troops in these five states combined; and in addition come over 25,000 U.S. troops stationed elsewhere in the region (if we include Afghanistan). In other words, troop numbers—that is, the combination of permanent and temporary, ashore and afloat—in the Middle East seem to have stayed the same under the Trump administration, even if a few thousand eventually left Iraq and Afghanistan. The DMDC reports a paltry 9880 U.S. troops in the Middle East at the end of 2020, with an estimated 6850 additional troops residing in Afghanistan, Iraq, and Syria. The total—16,730—is far less than the more realistic claim that there are between 45,000 and 65,000 U.S. troops stationed in the region at any given time (Schmitt, 2020).

As for Africa, the Pentagon's data show a decline that is large in percentage terms but low with respect to absolute numbers (from 2505 to 747). Here, too, there are reasons to be cautious when interpreting the numbers and trends. On the one hand, we know that numbers for this region are comparatively low, as they always have been (Kane, 2006, p. 7). On the other hand, the data only list one country with over 100 U.S. troops in 2020_Q4 (Djibouti with 121, mostly encompassing U.S. Marines), which does not mesh well with generally accepted assessments from other sources (Bearak, 2019). Reports also suggest that the United States has a substantial number of smaller military facilities spread around the African continent, most of which are tasked with counterterrorism operations (Turse, 2017, 2018). The estimation by David Vine (2015,

p. 313), that Washington's military "presence on the continent has grown from a few hundred to (...) likely between seven thousand and eleven thousand US troops," reflects the reality better than the Pentagon's numbers. Other, recent reports suggest that the United States has a total of 4400 troops in East Africa and 1400 in West Africa (Paquette & Warrick, 2020).

The Americas

That leaves the Americas. In one sense, this is the most important region for Washington because it represents the United States' "backyard." But it is also, in an equally important sense, the *least* vital region, which is why I do not conduct a more detailed analysis of it in the next chapter. For a start, there are no regional rivals or "peer competitors" to the United States here, and no Eurasian great power has made any serious attempts to establish a military foothold in the Western Hemisphere since the end of the Cold War. Neither does this region contain any state harboring nuclear-weapons ambitions; there are hardly any authoritarian regimes to be found (Cuba, Venezuela, and Nicaragua are the notable exceptions); and *jihadi* terrorists are very few and far between. The overwhelming power of the United States in its own near-abroad, not least as it manifests in naval superiority, means that it need not consistently demonstrate its power. The mere common knowledge of U.S. superiority would pretty much suffice; there is thus little need to amass huge forces at a plethora of main operating bases around the region as overt symbols of Washington's commitment to deterrence and reassurance. Considering also the closeness of Caribbean littoral states to bases in the United States itself, it is not surprising that troop numbers are comparatively low.

Deterrence and reassurance may well be implied more than openly proclaimed, but it is unlikely that this hampers their credibility and effectiveness. The United States can therefore afford to design other, "lesser" missions for its forces in the hemisphere. U.S. military facilities in Latin America and the Caribbean are mostly tasked with troops training, weapons testing, and various "police"-type work such as counterinsurgency warfare, anti-narcotics operations, intelligence collection, and humanitarian assistance (Lindsay-Poland, 2009, pp. 73–74). For such reasons, and due to widespread local opposition to a U.S. military presence in many countries, the region (outside of the U.S. mainland, that is) contains relatively few main operating bases, but a much vaster network of smaller installations or "quasi-bases" (Bitar, 2016).

Indeed, U.S. forces in the region have always been spread rather thinly (Kane, 2006, p. 10). The major staging areas for U.S. troops are, per December 31, 2020, Puerto Rico and the U.S. Virgin Islands (11,638 and 690 troops, respectively, if we include National Guards and reserves), Cuba (i.e., U.S.-administered Guantánamo Bay, with 728 troops belonging to the U.S. Army, Navy, and Marines), and Honduras (350 troops). It is no coincidence that these states and territories are the main locations for Washington. Honduras is in the middle of Central America, so forces based on the Soto Cano Air Base are able to serve missions in the whole of the sub-region (South, 2019). Guantánamo Bay, for its part, is centrally located in the Caribbean, and it has provided the starting point for numerous U.S. military interventions in the region. As usefully located in the Caribbean is Puerto Rico, whose status as a formal U.S. territory has made it only natural for a range of U.S. administrations to construct sizable military installations on the island. These troop-heavy main bases are supplemented by a range of other military installations dispersed around the Americas, including Canada, the Bahamas, the Dutch islands of Aruba and Curaçao, El Salvador, Costa Rica, and Peru (U.S. Department of Defense, 2018).

The official number of active-duty U.S. troops in the Americas (outside of the U.S. mainland) has decreased by 16.6% over the period under study here. However, when we include National Guards and reserves, numbers have actually *increased* by 30%, almost solely courtesy of additional forces in Puerto Rico. As is the case for numerous bases in other regions, key military facilities in the Americas—such as in Cuba and Puerto Rico—look set for an *expansion* in the near term (Associated Press, 2018; Fox, 2018; Rosenberg, 2017). All the while another key location—Greenland, an autonomous territory within the Kingdom of Denmark—, which geographically if not politically is situated in the Americas, is also being touted as a key to U.S. geostrategy in the years ahead. Even if President Trump's alleged wish to buy Greenland was unceremoniously rebuffed as an "absurd" suggestion by the Prime Minister of Denmark (BBC, 2019a; Salama et al., 2019), reports hold that Washington is seeking to make significant strategic investments on this huge Arctic island (Gronholt-Pedersen, 2019). This comes along with a strengthening of diplomatic and economic ties between the United States and Greenland (and Copenhagen), which are all efforts designed primarily to counter Russian and Chinese military and economic expansion in the Arctic (Morello, 2020).

*

Along with a presentation and critical review of the data, this chapter has given an overview of developments in the numbers and geographical distribution of U.S. overseas troops between the end of 2016 and until the end of President Trump's term in office. This is far from telling the complete story, however. Of particular interest is the process involving actual threats, proposals, or plans to withdraw troops; that is to say, I center much of the text to come on the tension between such proposals and the counter-arguments—or the narratives—espoused by the many *Stayers* of the foreign-policy community, who tend to work quite incessantly to uphold, and sometimes to bolster even further, U.S. force projection around the world. The next chapter will thus dig deeper into a variety of cases that illustrate some of the formidable barriers that the *Leavers* among Americans must overcome if actual retrenchment is ever, or at least in the foreseeable future, to become a reality.

References

Ali, I. (2019, October 20). *All U.S. troops withdrawing from Syria expected to go to western Iraq: Pentagon chief*. Reuters. https://www.reuters.com/article/us-syria-security-usa/all-u-s-troops-withdrawing-from-syria-expected-to-go-to-western-iraq-pentagon-chief-idUSKBN1WZ01S

Al Jazeera. (2020, March 12). *Pentagon threatens retaliatory strike after deadly attack in Iraq*. https://www.aljazeera.com/news/2020/03/pentagon-threatens-retaliatory-strike-deadly-attack-iraq-200312151654278.html

Al Jazeera. (2021, January 15). *US downsizes its troops in Afghanistan to 2,500*. https://www.aljazeera.com/news/2021/1/15/us-meets-its-goal-of-downsizing-in-afghanistan-to-2500-troops

Associated Press. (2018, July 3). *US Army to invest $62M in Puerto Rico in upcoming years*. https://apnews.com/cbcb094003ca4ddd8d3e471871de0701/US-Army-to-invest-$62M-in-Puerto-Rico-in-upcoming-years

Associated Press. (2021, January 5). *US Air Force deploys airmen, drones to base in Romania*. https://apnews.com/article/romania-air-force-black-sea-moscow-serbia-96b26b6a473191b72e83f1f463ca11c2

Atwood, K., & Browne, R. (2020, April 30). *US troop drawdown in Afghanistan running ahead of schedule*. CNN. https://edition.cnn.com/2020/04/30/politics/us-troops-afghanistan-drawdown/index.html

Barnes, J. E., & Schmitt, E. (2019, December 4). Iran is secretly moving missiles into Iraq, U.S. officials say. *New York Times*. https://www.nytimes.com/2019/12/04/us/politics/iran-missiles-iraq.html

BBC. (2019a, August 21). *Greenland: Trump criticises 'nasty' Denmark over cancelled visit*. https://www.bbc.com/news/world-us-canada-49423968

BBC. (2019b, September 3). *Afghanistan war: US-Taliban deal would see 5,400 troops withdraw*. https://www.bbc.com/news/world-asia-49559493

BBC. (2020, March 9). *Afghanistan conflict: US begins withdrawing troops*. https://www.bbc.com/news/world-asia-51810498

Bearak, M. (2019, December 30). In strategic Djibouti, a microcosm of China's growing foothold in Africa. *Washington Post*. https://www.washingtonpost.com/world/africa/in-strategic-djibouti-a-microcosm-of-chinas-growing-foothold-in-africa/2019/12/29/a6e664ea-beab-11e9-a8b0-7ed8a0d5dc5d_story.html?arc404=true

Bitar, S. E. (2016). *US military bases, quasi-bases, and domestic politics in Latin America*. Palgrave Macmillan.

Brannen, K., & Goodman, R. (2020, October 7). We're suing the Pentagon to find out where U.S. troops are deployed. *Washington Post*. https://www.washingtonpost.com/outlook/2020/10/07/were-suing-pentagon-find-out-where-us-troops-are-deployed/

Browne, R. (2020, February 2). *Top US general in Africa warns of growing threats as Trump administration weighs cutting troop numbers*. CNN. https://edition.cnn.com/2020/02/02/politics/africa-threats-trump-troops/index.html

Burns, R. (2020, July 21). Esper says US considering troop 'adjustments' in South Korea. *Washington Post*. https://www.washingtonpost.com/world/national-security/esper-says-us-considering-troop-adjustments-in-south-korea/2020/07/21/9c016ed6-cb5b-11ea-99b0-8426e26d203b_story.html

Bytyci, F. (2020, March 11). *Kosovo fears U.S. may pull troops out over Serbia tariff dispute*. Reuters. https://www.reuters.com/article/us-kosovo-usa-idUSKBN20Y1LE

Christensen, T. J. (2011). The advantages of an assertive China: Responding to Beijing's abrasive diplomacy. *Foreign Affairs, 90*(2), 54–67.

Christou, W. (2020, November 2). *US Envoy to Syria: 'No change in troop presence in Syria' whether Biden or Trump wins*. Syria Direct. https://syriadirect.org/us-envoy-to-syria-no-change-in-troop-presence-in-syria-whether-biden-or-trump-wins/

Clarke, C. P., & Tabatabai, A. (2018, October 16). America's indefinite endgame in Syria. *The Atlantic*. https://www.theatlantic.com/international/archive/2018/10/bolton-pledges-american-troops-syria-iran/573121/

Cloud, D. S. (2021, March 12). Inside U.S. troops' stronghold in Syria, a question of how long Biden will keep them there. *Los Angeles Times*. https://www.latimes.com/politics/story/2021-03-12/us-troops-syria-civil-war-biden

Cloud, D. S., & Wilkinson, T. (2020, July 2). White House tentatively agrees to leave some troops in Afghanistan past U.S. election. *Los Angeles Times*. https://www.latimes.com/politics/story/2020-07-02/afghanistan-trump-withdrawal-on-track

Cooley, A., & Nexon, D. (2020). *Exit from hegemony: The unraveling of the American global order*. Oxford University Press.

Davison, J. (2020, October 11). *Iraqi militias say they have halted anti-U.S. attacks*. Reuters. https://www.reuters.com/article/us-iraq-militias/iraqi-militias-say-they-have-halted-anti-u-s-attacks-idUSKBN26W0IN

Detsch, J. (2020, March 19). *Congress pushes for Syria funding boost*. Al-Monitor. https://www.al-monitor.com/pulse/originals/2020/03/congress-push-syria-funding-boost.html

DeYoung, K. (2018, December 20). With Syria decision, Trump's foreign policy doctrine emerges. *Washington Post*. https://www.washingtonpost.com/world/national-security/with-syria-decision-trumps-foreign-policy-doctrine-emerges/2018/12/20/ad055810-0487-11e9-9122-82e98f91ee6f_story.html?utm_term=.abbea1cfd4ec

DeYoung, K. (2020, March 1). Peace deal with Taliban sets timetable for U.S. exit but no long-term solutions for Afghans. *Washington Post*. https://www.washingtonpost.com/national-security/afghanistan-peace-deal-taliban/2020/02/29/c2d2ae14-5af2-11ea-9000-f3cffee23036_story.html

DeYoung, K., Dawsey, J., & Fahim, K. (2019, October 17). In Turkey's president, Trump seems to have found a soul mate. *Washington Post*. https://www.washingtonpost.com/national-security/in-turkeys-president-trump-seems-to-have-found-a-soul-mate/2019/10/17/7fea3ada-eeb5-11e9-89eb-ec56cd414732_story.html

Emmott, R. (2017, October 9). *NATO launches Black Sea force as latest counter to Russia*. Reuters. https://www.reuters.com/article/us-russia-nato/nato-launches-black-sea-force-as-latest-counter-to-russia-idUSKBN1CE0MJ

Fox, B. (2018, June 7). US military plans for future at Guantanamo because of Trump. *Washington Times*. https://www.washingtontimes.com/news/2018/jun/7/us-military-plans-as-if-guantanamo-wont-close-for-/

Garamone, J. (2021, January 15). *U.S. completes troop-level drawdown in Afghanistan, Iraq*. U.S. Department of Defense, DOD News. Retrieved September 15, 2021, from https://www.defense.gov/Explore/News/Article/Article/2473884/us-completes-troop-level-drawdown-in-afghanistan-iraq/

Gearan, A., & Sonne, P. (2019, June 13). Trump elevates Poland at the expense of Germany with troop decision. *Washington Post*. https://www.washingtonpost.com/politics/trump-elevates-poland-at-the-expense-of-germany-with-troop-decision/2019/06/12/335cf8a0-8d36-11e9-adf3-f70f78c156e8_story.html

Gibbons-Neff, T. (2018, April 5). U.S. forces in Syria: Building more outposts, pulling up lawn chairs. *New York Times*. https://www.nytimes.com/2018/04/05/magazine/syria-us-forces-manbij.html

Gibbons-Neff, T., & Mashal, M. (2019, October 21). U.S. is quietly reducing its troop force in Afghanistan. *New York Times*. https://www.nytimes.com/2019/10/21/world/asia/afghanistan-troop-reduction.html

Gibbons-Neff, T., Cooper, H., & Schmitt, E. (2021, March 14). U.S. has 1,000 more troops in Afghanistan than it disclosed. *New York Times*. https://www.nytimes.com/2021/03/14/world/asia/us-troops-afghanistan.html

Gigova, R. (2017, February 17). *US troops deploy to Bulgaria as part of NATO operation to support Eastern European allies*. CNN. https://edition.cnn.com/2017/02/17/us/us-troops-deploy-to-bulgaria-as-part-of-nato-operation-to-support-eastern-european-allies/index.html

Glaser, J. (2017, July 18). *Withdrawing from overseas bases: Why a forward-deployed military posture is unnecessary, outdated, and dangerous*. Cato Institute Policy Analysis 816. Retrieved February 4, 2021, from https://www.cato.org/sites/cato.org/files/pubs/pdf/pa_816.pdf

Gold, J. (2018, February 23). Trump administration: Troops can stay in Iraq, Syria indefinitely. *Defense News*. https://www.defensenews.com/congress/2018/02/23/trump-administration-troops-can-stay-in-iraq-syria-indefinitely/

Gronholt-Pedersen, J. (2019, August 27). *In spotlight after Trump offer, Greenland sees chance for an economic win*. Reuters. https://www.reuters.com/article/us-usa-trump-greenland/in-spotlight-after-trump-offer-greenland-sees-chance-for-an-economic-win-idUSKCN1VH1AB

Haltiwanger, J. (2017, November 21). Trump's secret war? U.S. military's presence in Middle East has grown 33 percent in past four months. *Newsweek*. https://www.newsweek.com/trumps-secret-war-us-militarys-presence-middle-east-has-grown-33-percent-past-718089

Harkins, G. (2020, May 6). *Australia OKs marine corps deployment during pandemic*. Military.com. https://www.military.com/daily-news/2020/05/06/australia-oks marine corps-deployment-during-pandemic.html

Henry, R. (2006). Transforming the U.S. global defense posture. *Naval War College Review, 59*(2), 12–28.

Herrera, R., & Cicchini, J. (2013). Some notes about the U.S. military bases and personnel abroad. *Journal of Innovation Economics Management, 2*(12), 127–149.

Holland, S. (2018, December 26). *Trump defends Syria pullout during surprise visit to Iraq*. Reuters. https://www.reuters.com/article/us-usa-trump-syria/trump-defends-syria-pullout-during-surprise-visit-to-iraq-idUSKCN1OP175

Jaffe, G., & Ryan, M. (2018, January 21). Up to 1,000 more U.S. troops could be headed to Afghanistan this spring. *Washington Post*. https://www.washingtonpost.com/world/national-security/up-to-1000-more-us-troops-could-be-headed-to-afghanistan-this-spring/2018/01/21/153930b6-fd1b-11e7-a46b-a3614530bd87_story.html?utm_term=.eb80d3a8374f

Kane, T. (2004). *Global U.S. troop deployment, 1950–2003*. Heritage Foundation. Retrieved March 8, 2020, from https://www.heritage.org/defense/report/global-us-troop-deployment-1950-2003

Kane, T. (2006). *Global U.S. troop deployment, 1950–2005*. Heritage Foundation. Retrieved March 26, 2021, from https://www.heritage.org/defense/report/global-us-troop-deployment-1950-2005

Kube, C., & Lee, C. E. (2020, October 14). *Trump weighs pre-election foreign policy speech that reveals plan to pull more troops from Afghanistan*. NBC News. https://www.nbcnews.com/politics/2020-election/trump-weighs-pre-election-foreign-policy-speech-reveals-plan-pull-n1243305

Lamothe, D., Dawsey, J., Ryan, M., & Sonne, P. (2018, December 21). Trump orders major military withdrawal from Afghanistan as Mattis departs. *Washington Post*. https://www.washingtonpost.com/world/national-security/trump-agitating-for-major-military-withdrawal-from-afghanistan-advisers-say/2018/12/20/0c35f874-04a3-11e9-b5df-5d3874f1ac36_story.html?utm_term=.3dc80a77fb57

Lawrence, J. P. (2020, March 13). US troops have left bases in Helmand, Laghman provinces, officials say. *Stars and Stripes*. https://www.stripes.com/news/middle-east/us-troops-have-left-bases-in-helmand-laghman-provinces-officials-say-1.622323

Lee, C. E., & Kibe, C. (2020, April 27). *Trump tells advisers U.S. should pull troops as Afghanistan COVID-19 outbreak looms*. NBC News. https://www.nbcnews.com/politics/national-security/trump-tells-advisers-u-s-should-pull-troops-afghanistan-covid-n1191761

Lekic, S. (2020, June 24). Kosovo talks to focus on economy, less likely to affect US military role. *Stars and Stripes*. https://www.stripes.com/news/europe/kosovo-talks-to-focus-on-economy-less-likely-to-affect-us-military-role-1.634984

Lindsay-Poland, J. (2009). U.S. military bases in Latin America and the Caribbean. In C. Lutz (Ed.), *The bases of empire: The global struggle against U.S. military posts* (pp. 71–95). New York University Press.

Martinez Machain, C., & Morgan, T. C. (2013). The effect of US troop deployment on host states' foreign policy. *Armed Forces & Society, 39*(1), 102–123.

Mashal, M. (2020, June 19). U.S. Troops in Afghanistan reduced to 8,600, General says. *New York Times*. https://www.nytimes.com/2020/06/19/world/asia/afghanistan-us-troop-withdrawal.html

McKay, R. (2020, January 7). *'We're going to war, bro': Fort Bragg's 82nd Airborne deploys to the Middle East*. Reuters. https://www.reuters.com/article/us-iraq-security-usa-fort-bragg/were-going-to-war-bro-fort-braggs-82nd-airborne-deploys-to-the-middle-east-idUSKBN1Z616W

Mehta, A., & Copp, T. (2017, August 21). *Trump Afghanistan strategy calls for more troops, regional pressure*. Military Times. https://www.militarytimes.com/news/2017/08/22/trump-afghanistan-strategy-calls-for-more-troops-regional-pressure/

Moore, J. (2017, November 17). Trump's Afghanistan troop surge is complete, raising total number of U.S. servicement to 14,000. *Newsweek*. https://www.newsweek.com/trumps-afghanistan-troop-surge-complete-raising-total-number-us-servicemen-714588

Morello, C. (2020, April 23). U.S. to give aid to Greenland, open consulate in bid to counter Russia and China. *Washington Post*. https://www.washingtonpost.com/national-security/us-to-give-aid-to-greenland-open-consulate-in-bid-to-counter-russia-and-china/2020/04/23/369bf2d4-8568-11ea-9728-c74380d9d410_story.html

Myers, M. (2019, August 6). *Report: ISIS has been rebuilding as US troops withdraw from Syria*. Military Times. https://www.militarytimes.com/news/your-military/2019/08/06/report-isis-has-been-rebuilding-as-us-troops-withdraw-from-syria/

O'Mahoney, A., Priebe, M., Frederick, B., Kavanagh, J., Lane, M., Johnston, T., Szayna, T. S., Hlavka, J. P., Watts, S., & Povlock, M. (2018). *U.S. presence and the incidence of conflict*. RAND Corporation.

Orenstein, M. A. (2019). *The lands in between: Russia vs. the West and the new politics of hybrid war*. Oxford University Press.

Paquette, D., & Warrick, J. (2020, February 22). Al-Qaeda and Islamic State groups are working together in West Africa to grab large swaths of territory. *Washington Post*. https://www.washingtonpost.com/world/africa/al-qaeda-islamic-state-sahel-west-africa/2020/02/21/7218bc50-536f-11ea-80ce-37a8d4266c09_story.html

Rapp-Hooper, M. (2019, September 23). Trump will send troops to Saudi Arabia. Here's why it matters that there's no formal defense alliance. *Washington Post*. https://www.washingtonpost.com/politics/2019/09/23/trump-will-send-troops-saudi-arabia-heres-why-it-matters-that-theres-no-formal-defense-alliance/

Reuters. (2019a, December 5). *U.S. military could add troops to Middle East, but struggles to explain plans*. https://www.reuters.com/article/us-usa-middleast-military/u-s-military-could-add-troops-to-middle-east-but-struggles-to-explain-plans-idUSKBN1Y92G2

Reuters. (2019b, December 14). *U.S. sent troops to Saudi Arabia on defensive basis: Mnuchin*. https://www.reuters.com/article/us-qatar-forum-usa/u-s-sent-troops-to-saudi-arabia-on-defensive-basis-mnuchin-idUSKBN1YI0BI

Reuters. (2019c, December 17). *Syria hands oil exploration contracts to two Russian firms*. https://www.reuters.com/article/us-syria-oil-russia/syria-hands-oil-exploration-contracts-to-two-russian-firms-idUSKBN1YL0VK

Reuters. (2020a, January 3). *Pentagon to deploy thousands of additional troops to Middle East*. https://www.reuters.com/article/us-iraq-security-usa-middleeast/pentagon-to-deploy-thousands-of-additional-troops-to-middle-east-idUSKBN1Z21U4

Reuters. (2020b, January 8). *Factbox: U.S. forces in Gulf region and Iraq*. https://www.reuters.com/article/us-iraq-security-usa-presence-factbox/factbox-u-s-forces-in-gulf-region-and-iraq-idUSKBN1Z72GF

Reuters. (2020c, April 10). *NATO mission in Kosovo reports first COVID-19 case*. https://www.reuters.com/article/instant-article/idUKKCN21S1P1

Reuters. (2020d, August 9). *Esper: U.S. will cut troop levels in Afghanistan to 'less than 5,000.'* https://www.reuters.com/article/us-usa-afghanistan-military/esper-u-s-will-cut-troop-levels-in-afghanistan-to-less-than-5000-idUSKCN255031

Reuters. (2020e, August 28). *U.S. expected to reduce troops in Iraq by a third to about 3,500: official*. https://www.reuters.com/article/us-usa-iraq-military/u-s-expected-to-reduce-troops-in-iraq-by-a-third-to-about-3500-official-idUSKBN25O2DI

Reuters. (2020f, September 9). *United States formally announces troop reduction in Iraq*. https://www.reuters.com/article/us-usa-iraq-troops/united-states-formally-announces-troop-reduction-in-iraq-idUSKBN2601RZ

Rosenberg, C. (2017, August 21). Trump's Pentagon wants to spend almost $500 million on Guantánamo construction. *Miami Herald*. https://www.miamiherald.com/news/nation-world/world/americas/guantanamo/article168273127.html

Salama, V., Ballhaus, R., Restuccia, A., & Bender, M. C. (2019, August 16). President Trump eyes a new real-estate purchase: Greenland. *Wall Street Journal*. https://www.wsj.com/articles/trump-eyes-a-new-real-estate-purchase-greenland-11565904223

Schmidt, S. (2020). *Armed guests: Territorial sovereignty and foreign military basing*. Oxford University Press.

Schmitt, E. (2019, March 29). U.S. troops leaving Syria, but some may stay longer than expected. *New York Times*. https://www.nytimes.com/2019/03/29/world/middleeast/us-troops-syria-isis.html

Schmitt, E. (2020, September 9). U.S. to reduce troop levels in Iraq to 3,000. *New York Times*. https://www.nytimes.com/2020/09/09/us/politics/iraq-troops-trump.html

Seck, H. H. (2020, December 21). *In first, NATO forces now outnumber US troops in Afghanistan*. Military.com. https://www.military.com/daily-news/2020/12/21/first-nato-forces-now-outnumber-us-troops-afghanistan.html

Sediqi, A. Q., & Torode, G. (2019, February 6). *Taliban reject Trump's suggestion of lingering U.S. counter-terrorism presence*. Reuters. https://www.reuters.com/article/us-usa-afghanistan-talks/taliban-reject-trumps-suggestion-of-lingering-u-s-counter-terrorism-presence-idUSKCN1PV16P

Seldin, J. (2019, May 31). *US troop levels in Syria 'quite a bit lower' with help on the way*. VOA News. https://www.voanews.com/middle-east/us-troop-levels-syria-quite-bit-lower-help-way

Shanker, T., & Landler, M. (2007, February 11). Putin says U.S. is undermining global stability. *New York Times*. https://www.nytimes.com/2007/02/11/world/europe/11munich.html

SIPRI. (2021). *SIPRI military expenditure database*. Stockholm International Peace Research Institute. Retrieved September 12, 2021, from https://www.sipri.org/databases/milex

Sly, L., & Morello, C. (2018, January 17). U.S. troops will stay in Syria to counter 'strategic' threat from Iran. *Washington Post*. https://www.washingtonpost.com/world/middle_east/us-troops-will-stay-in-syria-to-counter-strategic-threat-from-iran/2018/01/17/eeed9d16-fb8f-11e7-9b5d-bbf0da31214d_story.html?utm_term=.c472a5bc264a

South, T. (2019, August 12). Deep in the mountains of Honduras, few know what this US military task force does. *Army Times*. https://www.armytimes.com/news/your-army/2019/08/12/deep-in-the-mountains-of-honduras-few-know-what-this-us-military-task-force-does/

Stewart, P., & Houreld, K. (2020, November 17). *Trump may withdraw troops from Somalia as part of global pullback*. Reuters. https://www.reuters.com/article/us-usa-somalia-withdrawal-idUSKBN27X1YU

Stravers, A., & El Kurd, D. (2020). Strategic autocracy: American military forces and regime type. *Journal of Global Security Studies, 5*(3), 427–442.

Taylor, A. (2019, October 22). Why U.S. presidents find it so hard to withdraw troops from the Mideast. *Washington Post*. https://www.washingtonpost.com/world/2019/10/22/why-us-presidents-find-it-so-hard-withdraw-troops-mideast/

The Economist. (2020, January 9). *How Iran can respond to the killing of Qassem Suleimani*. https://www.economist.com/briefing/2020/01/09/how-iran-can-respond-to-the-killing-of-qassem-suleimani

The Economist. (2021, June 5). *The new geopolitics of global business: China and America dominate like never before*. https://www.economist.com/leaders/2021/06/05/the-new-geopolitics-of-global-business

Times of Israel. (2020, May 8). *US defense department said pushing to withdraw troops from Sinai; Israel opposed*. https://www.timesofisrael.com/us-defense-dept-reportedly-pushing-for-withdrawal-of-troops-from-sinai/

Turse, N. (2017, April 27). Secret US military documents reveal a constellation of American military bases across Africa. *The Nation*. https://www.thenation.com/article/archive/secret-us-military-documents-reveal-a-constellation-of-american-military-bases-across-africa/

Turse, N. (2018, December 1). *U.S. military says it has a "light footprint" in Africa: These documents show a vast network of bases*. The Intercept. https://theintercept.com/2018/12/01/u-s-military-says-it-has-a-light-footprint-in-africa-these-documents-show-a-vast-network-of-bases/

Turse, N. (2019, January 9). The Pentagon's shadow military bases. *The Nation*. https://www.thenation.com/article/syria-iraq-pentagon-overseas-military-bases/

U.S. Department of Defense. (2018). *Base structure report—Fiscal year 2018 baseline*. Retrieved March 15, 2021, from https://www.acq.osd.mil/eie/Downloads/BSI/Base%20Structure%20Report%20FY18.pdf

U.S. Department of Defense. (2021). *Military and civilian personnel by service/agency by state/country*. U.S. Defense Manpower Data Center. Retrieved November 23, 2021, from https://dwp.dmdc.osd.mil/dwp/app/dod-data-reports/workforce-reports

U.S. Department of State. (2019, October 11). *Special representative for Iran and senior advisor to the secretary Brian Hook: Special briefing*. Retrieved August 2, 2021, from https://www.state.gov/special-representative-for-iran-and-senior-advisor-to-the-secretary-brian-hook-2/

Vine, D. (2015). *Base nation: How U.S. military bases abroad harm America and the world*. Metropolitan Books.

Walt, S. M. (2018). *The hell of good intentions: America's foreign policy elite and the decline of U.S. primacy*. Farrar, Straus and Giroux.

Ward, A. (2020, September 21). Trump says the US is "out of Syria." But more US troops are heading there. *Vox*. https://www.vox.com/2020/9/21/21448845/syria-russia-usa-troops-trump

Williams, K. B. (2020, November 12). *Outgoing Syria envoy admits hiding US troop numbers; praises Trump's Mideast record*. Defense One. https://www.defenseone.com/threats/2020/11/outgoing-syria-envoy-admits-hiding-us-troop-numbers-praises-trumps-mideast-record/170012/

World Bank. (2021). *World Bank open data: GDP (constant 2015 US$)*. https://data.worldbank.org/indicator/NY.GDP.MKTP.KD

Zengerle, P. (2020, June 18). *U.S. lawmakers offer bill to block Trump from German troop withdrawal*. Reuters. https://www.reuters.com/article/us-usa-germany-military-congress/u-s-lawmakers-offer-bill-to-block-trump-from-german-troop-withdrawal-idUSKBN23P2LA

Zenko, M. (2018, October 18). *US military policy in the Middle East: An appraisal*. Chatham House. Retrieved February 14, 2020, from https://reader.chathamhouse.org/us-military-policy-middle-east-appraisal

CHAPTER 7

Regional Domino Narratives and the Geopolitics of Withdrawal

*

This chapter is organized by regions. I start with developments in Africa. Thereafter comes the Middle East, before I move on to Afghanistan, the most dramatic case of U.S. troop withdrawal in recent years. I end that tour with analyses of Europe, which is the "old" geopolitical center of the world, and Asia–Pacific, the new and rapidly rising one. Along the way, I review several cases of proposed or threatened withdrawal, most of which came to naught. These include Germany, South Korea, Afghanistan, Iraq, Syria, and Somalia. Supplementing these cases are the many instances where rationales are constructed that highlight the necessity of instead *increasing* the United States' military presence—in Central and Eastern Europe, in the Middle East, and, most notably, in the Asia–Pacific region.

The African Dominoes

Africa is the continent with the lowest number of U.S. troops, even if the real amount of U.S. military activity is severely understated by the official figures. The vast size of the total area of operation, and the often limited infrastructure, mean that the United States has only a single true main operation base on the mainland. This is Camp Lemonnier in Djibouti,

J. Jakobsen, *The Geopolitics of U.S. Overseas Troops and Withdrawal*, Palgrave Studies in International Relations,
https://doi.org/10.1007/978-3-030-94488-9_7

which serves as a hub from which a much larger network of smaller locations or outposts can be reached and supplied. From Camp Lemonnier, the United States can watch over the highly important Red Sea shipping route and the bottleneck at its entrance that is the Bab el-Mandab Strait. Djibouti is also close to Yemen and Somalia, both of which have long been central theaters in U.S. counterterrorism efforts. The overriding concern for Washington in Africa as a whole has been, and still is, international terrorism, which is reflected by its base network's "lily-pad" character. The military facilities are low in "permanent" troop numbers; many of the known locations hardly register any official U.S. troops at all. We know from other sources, however, that U.S. forward-operating sites or military outposts in Africa are quite prevalent, and they are key to America's global, and mostly secret, "shadow war" against several *jihadist* groups (Schmitt, 2018).

In Somalia, the U.S. military has had a significant presence for a few years. In 2017, at the request of authorities in Mogadishu, the Trump administration authorized the first official deployment of U.S. troops there since its forces were pulled out in 1994 following the infamous and much-publicized killings of 18 U.S. service members the year before (Gaffey, 2017; Morgan, 2017). In late 2020, however, it turned out that the Trump administration had had a change of mind: it was now considering leaving Somalia or at least abolishing that part of the mission where Somali forces were receiving training by U.S. troops. The fight against the al-Shabaab militant group would reportedly continue, but Somalia-based U.S. troops would be repositioned in neighboring states, primarily Djibouti and Kenya (Youssef & Phillips, 2020). Despite that these troops, which were classified as "non-permanent," were to be relocated within the region, and even if air strikes against the al-Shabaab would continue, this was widely perceived to be a case of U.S. retrenchment, against which sundry voices warned. Especially citing concerns about the "vacuums" that would be created and filled by Washington's local adversaries if the United States withdrew, Somalia experts and security analysts called the decision "shocking" and "disastrous" for the country's stability and security (Anna, 2020).

Domino warnings abounded. The decision, said analysts, would likely work to hand the al-Shabaab militants a considerable edge in Somalia's armed conflict, and it would significantly heighten the risk of terrorist attacks in the region, including against Western and U.S. targets (Schmitt, 2020a; Yusuf, 2021). Some pointed to the risk that Somali elite forces, which had been trained and supported by the United States, would soon disintegrate. Others specifically feared the damage that such a move would inflict on U.S. credibility, and on the perceptions others held of the United States' ability to lead. The al-Shabaab would, purportedly, frame the withdrawal as a victory, and as a corresponding U.S. loss. This would boost the group's propaganda efforts, and it would send an ominous signal to other states in the region. In short and more broadly, as lamented by Major General J. Marcus Hicks, who used to command U.S. Special Operations forces in Africa, any rapid withdrawal from Somalia or other trouble spots would "undermine our security, damage our ability to lead internationally, and strengthen our adversaries"—including *state* adversaries, such as Russia, China, and Iran (Dahir & Schmitt, 2020).

Despite these warnings, the withdrawal commenced, and it was ostensibly completed only a few days before President Trump left office (Anna, 2021). This represented the *only* notable withdrawal of U.S. troops from an African country during Trump's tenure (with Niger possibly constituting a second such case). And the troops didn't actually leave the continent; instead, they were relocated to more secure bases in Djibouti and Kenya, whence the fight against the al-Shabaab continued. There is, therefore, little to indicate that the fairly meager numbers of U.S. troops in Africa have shrunk to an even lower level during the Trump presidency. This is so even if discussions within the Trump administration soon started focusing on the possibility of downscaling or ending some of the missions of the U.S. Africa Command (AFRICOM), to aid a sought-after repositioning of forces into regions more clearly characterized by growing great-power rivalry (Cooper et al., 2019). Proposals to this effect were made at least as early as 2018 (Ali, 2018). This followed the December 2017 publication of the Trump administration's first National Security Strategy, which was more heavily focused on state-based security threats (The White House, 2017). Among the decisions over which authorities now mulled was commencing on a significant pullback from Central and West Africa, which could include leaving a newly built drone base in Niger and ending cooperation with the French forces that battle extremist groups in the Sahel (Cooper et al., 2019, Phillips, 2020).

Any such retrenchment would mesh well with the argument that defeating *jihadists* in these vast areas of operations is a near-impossible task. But defeating and keeping in check, solving and containing, are not the same. Rather unsurprisingly, therefore, the bulk of reactions to these retrenchment discussions and plans have tended to follow a familiar pattern, encompassing arguments that lean heavily on geopolitical chain-reaction—or domino—thinking. Some argue that the partial pullback of (the secretly deployed) U.S. service members from Niger following an ambush in October 2017, in which four U.S. Army soldiers on patrol were killed by the Islamic State (BBC, 2017b), helped invigorate *jihadist* groups in the area; this, therefore, illustrates the unintended effects of a premature withdrawal of U.S. troops from an area of operations where the militants had not yet been sufficiently subdued (Schmitt, 2019). According to Herman Cohen (2020), formerly a U.S. Ambassador and Assistant U.S. Secretary of African Affairs, a pullback of the already limited U.S. forces deployed in the broader region *might* force the United States into "a broader and deeper foreign entanglement, by allowing these groups to grow and attack U.S. bases or even the mainland." This, he warns, could ultimately result in yet "another endless war" for America.

The seeming momentum gained by the withdrawal plans also sparked critical newspaper editorials (*Washington Post*, 2020), and it incited protests and warnings from numerous members of the U.S. Congress (Mitchell, 2020). Furthermore, some of the concerns that were raised came from Washington's regional allies, whose counterterrorism success is often heavily dependent on collaboration with U.S. forces (Cooper, 2020). Warnings also came from high-ranking U.S. military officials, one of whom stated, with reference to the operations in West Africa, that they amount to a low-cost "insurance policy" that the nation would be well-advised to continue financing (Schmitt, 2018). Such a view is shared by the head of AFRICOM, General Stephen Townsend. In a February 2020 interview with *CNN*, Townsend emphasized that China and Russia stood ready to "fill the void" created by any U.S. withdrawal; regional terrorist groups would "grow unchecked" and also "ultimately threaten the homeland"; while the United States would also lose out on the considerable economic opportunities that the African continent offered (Browne, 2020a).

Enduring Dominoes in the Middle East

As we now know, the withdrawal of U.S. troops from Africa never really materialized on Donald Trump's watch. The matter was by and large deferred to the incoming administration, which would soon start working on a Global Posture Review, whose mission was to assess the global role of America and its overseas bases and troops (The White House, 2021). That review would surely focus much energy on the size, role, and efficacy of U.S. troop contingents in the Middle East, which have constituted a controversial issue in U.S. politics ever since the United States invaded Iraq in 2003. Here as well, though, the concerns of the *Stayers* have dwarfed the arguments of the *Leavers*. As is the case regarding U.S. troops in Africa, but here at a much larger scale, the *Stayers* worry about perceived threats from global and regional major powers *and* from terrorist organizations. The immediate fear is that if the United States fails to engage militarily in that region, by fighting or by deterring adversaries, this will play into the geostrategic hands of one major regional power (Iran), one external major power with extra-regional ambitions (Russia), and several non-state actors, in particular the Islamic State of Iraq and Syria (Islamic State, or ISIS, for short). Inclinations to withdraw troops consistently run up against worst-case scenarios: the specter of more failed states, an expansion (or re-emergence) of the "caliphate," a boost to an increasingly assertive Russia, and Iranian regional hegemony or quasi-hegemony.

On the one hand, the Middle East is *the* region—at least if we count Afghanistan as belonging to the "Greater" Middle East—toward which President Trump had directed the better part of his withdrawal statements. Becoming involved here was, according to an October 2019 Tweet of his, "the worst decision ever made in the history of our country." Consequently, the United States would now "slowly & carefully [bring] our great soldiers & military home" (France24, 2019).

Troop Withdrawal and Domino Warnings, the First Syrian Version

The backdrop of these comments were developments in the war in Syria. Yet, the United States actually *increased* its presence there very soon after the inauguration of the new administration. In fact, troop numbers were raised sharply around the entire Gulf region from the second to the third quarter of 2017. The president, it seemed, chose to continue on the path taken by his predecessor in the White House. He even went one step further, resolving to deal with ISIS by way of more bombs and more soldiers. This was apparently a president who had ultimately decided to place his faith in the Pentagon and his sundry other military advisors (Gaouette & Browne, 2017; Haltiwanger, 2017). These were military people who strongly believed that the terrorist group had to be militarily beaten, or at least militarily contained, for the threat to be reduced to acceptable levels. And only the United States had the wherewithal to do that job.

In fall 2017, thus, the number of U.S. troops stationed at bases around the Persian Gulf was quite massive, although numbers neither were nor had ever been massive in Syria itself. Notwithstanding that this was an internationalized civil war the U.S. influence over had always been quite limited, the increased military efforts did yield results with respect to Washington's main target. The Islamic State suffered huge losses of territory in 2017 and 2018, which soon made the president revert to his earlier retrenchment mutterings. He was now adamant that the around 2000 troops stationed there had to leave, preferably quite rapidly, an objective that he had "been campaigning on (…) for years" (DeYoung, 2018). As the United States, in spring 2018, and in the president's words, was "knocking the hell out of ISIS," they would "be coming out of Syria, like, very soon," going "back to our country, where we belong" (Jones, 2018).

Even if this stated intention of his had been repeated over and over again, it still amounted to something of a shock when Trump, in December 2018, announced the immediate withdrawal of all U.S. troops from Syria (Landler et al., 2018). Yet, this very announcement also illustrates the counterforces that almost inevitably come into play in such circumstances. In this case, Trump's decision to withdraw from Syria (and from Afghanistan, more about which later) stunned U.S. allies. It also ran squarely against the advice of most of his advisors, especially given that it would undermine the overall U.S. strategy in the region, which prompted the resignation of Trump's defense secretary James Mattis (DeYoung, 2018). Warnings about negative consequences were plentiful and forceful. They centered on future prospects for Syria, for the region, and for U.S. security; they centered on hypotheticals and geopolitical chain reactions that *could—possibly*—be set off by the spark provided by a withdrawal of what was in reality quite a limited number of U.S. troops.

It is true that U.S. troop numbers in Syria did halve in the space of the first few months of 2019, even as alarm bells sounded about the dominoes that would *likely* fall from the decision. Quite a few notable voices had much earlier pointed to the almost certainty that a U.S. withdrawal would significantly strengthen the relative power of Washington's state adversaries. For example, the president's then-national security advisor

John Bolton had been insistent that U.S. troops would not leave Syria "as long as Iranian troops are outside Iranian borders" (Associated Press, 2018). Still, now the main focus was on ISIS and other *jihadists*, notably those affiliated with Al-Qaeda. Even if the Islamic State had been thoroughly beaten territorially at the time of Trump's decision, the specter of its resurgence loomed especially heavily over Washington—and also over Arlington, Virginia: The Pentagon's own assessment highlighted that, in the absence of sufficient pressure, ISIS could stage a comeback within a year (Stewart & Ali, 2019). The general overseeing U.S. military operations in the region, Frank McKenzie, weighed in. He asserted that, although the *jihadists* in Syria had been "severely hobbled," they had not been eliminated. But what was vital to recognize, according to the general, was that it was the sheer presence of the U.S. military in the region that greatly limited the risk these groups would pose to the United States (Burns, 2020). At Capitol Hill, meanwhile, a bipartisan agreement seemed to coalesce around the view that the United States needed to invest heavily in the stability of Syria, to ensure that they did not leave behind "fertile recruiting grounds for the next generation of ISIS" (Detsch, 2020). Furthermore, the Republican-dominated Senate passed a (non-binding) resolution in February 2019 in opposition to any rapid withdrawal of troops; they argued that "a precipitous withdrawal" could create a vacuum that not only ISIS—but also Iran and Russia—stood ready to fill (Zengerle, 2019a).

Troop Withdrawal and Domino Warnings, the Second Syrian Version

With respect to these issues, the next dramatic phase of the Syrian War took place in early October 2019. Facing a Turkey growing increasingly eager to establish, through an invasion, a buffer zone in north-eastern Syria, President Trump abruptly and extremely controversially decided to order U.S. troops out of Ankara's way (Schmitt et al., 2019). The announcement sowed much confusion in Washington and beyond. Yet again, the criticism and the warnings centered on the chain reactions that were ostensibly bound to follow. And yet again, the president faced admonitions from within his own party: The Senate Majority Leader Mitch McConnell (R-KY) urged Trump to "exercise American leadership," warning that the decision would benefit Iran, Russia, the Syrian regime, and the Islamic State; Senator Lindsay Graham (R-SC), normally a staunch ally of the president, described the move as "shortsighted and irresponsible" and a major victory for both Iran and Syria's authoritarian president, adding that the idea that the Islamic State had been defeated was "the biggest lie being told by the administration"; while other Republicans in Congress labeled it "a catastrophic mistake," one that would have repercussions "far beyond Syria" (Baker & Jakes, 2019).

The latter quote alluded to the signal effect being created by what was essentially a decision to end the U.S.'s alliance with the Syrian Democratic Forces (SDF). The core of the SDF is formed by the Kurdish People's Protection Units (YPG), deemed terrorists by Turkey, which would now have to face Turkish forces more or less on their own. Overall U.S. credibility—its reputation and position in the world in the eyes of both allies and adversaries—was allegedly at stake here, according to the U.S. foreign-policy community at large (Barkey, 2019). Republican Senator Mitt Romney (R-UT) stated that the decision was "a betrayal" (Baker & Jakes, 2019); whereas Nikki Haley (2019), President Trump's former Ambassador to the United Nations, tweeted that "[t]he Kurds were instrumental in our successful fight against ISIS in Syria. Leaving them to die is a big mistake." According to some, such as Elizabeth Dent and retired General Joseph Votel, the latter of whom had just stepped down as leader of the U.S. Central Command, this would not only threaten to reverse the gains made from over half a decade of battling ISIS; it would also "severely damage American credibility and reliability in any future fights where we need strong allies" (Votel & Dent, 2019).

The plethora of dire warnings about adverse geopolitical chain reactions based on worst-case scenarios—a rejuvenation of ISIS; an opening for Russia and Iran to increase their influence in Syria and beyond; a de facto green light for the Assad regime to expand their war-fighting into new areas of Syria; a signal to Turkey that its envisioned buffer zone could be extended at little cost; and a severe dent in U.S. credibility vis-à-vis allies and adversaries worldwide (ostensibly forever!)—did seem to have some effect on the president and his administration's decision-making. As it turned out, the oft-repeated decision to leave Syria was only half-heartedly effectuated: a mere couple of weeks after the start of that new phase of the Syrian debacle, President Trump changed course, indicating that at least some troops would stay in eastern Syria; this was allegedly to ensure that Syrian oil fields did not end up in the hands of ISIS (DeYoung et al., 2019; Schmitt & Haberman, 2019)—or of the Syrian government and their Russian allies (*Middle East Monitor*, 2020a). By early December, the force levels seemed to stabilize, with the United States keeping either

600 or 900 troops in the country, depending on the estimate chosen (Cloud, 2021; Stewart, 2019b; Williams, 2020). Whatever is the correct number, Donald Trump's earlier decision to withdraw completely from Syria simply did not manifest. In fact, a new base was built in spring 2020 in the oil-rich Deir Ez-Zor region, while an existing base at the Al-Omar Oil Field was expanded (*Middle East Monitor*, 2020b).

The year 2020 also saw numerous near-clashes, sometimes almost on a daily basis, between U.S. and Russian troops in Syria, in what was largely a "forgotten war" (O'Connor & Jamali, 2020). These brushes or militarized interactions between two nuclear-armed great-power adversaries were ultimately judged to constitute a price worth paying to increase the likelihood that the uneasy status quo was upheld in a country where it was not at all certain that U.S. *core* interests were really at stake. Except that they were; for it is usually so that when a superpower has already decided to make investments—and sink some costs—in some place of primary or secondary importance, these very commitments and sunk costs, ***in and by themselves***, are deemed vital to preserve, if for nothing else than to uphold the reputation of the superpower. And preserving them presumably also worked to ***minimize the risk of*** other large and small dominoes falling. The United States actually beefed up their troop presence in Syria in fall 2020, adding both military equipment and a hundred soldiers to their existing contingent (BBC, 2020). The U.S. envoy to Syria, Ambassador James Jeffrey, gave a fitting summary of the status of the U.S. presence in Syria in late 2020, following months and years of withdrawal threats and attempts:

> What Syria withdrawal? There was never a Syria withdrawal. (...) When the situation in northeast Syria had been fairly stable after we defeated ISIS, [Trump] was inclined to pull out. In each case, we then decided to come up with five better arguments for why we needed to stay. And we succeeded both times. That's the story. (Williams, 2020)

Washington's Syria experience does, on the one hand, show that it *is* possible to fulfill pledges to withdraw within some loosely specified time frame. On the other hand, this also represents an illustrative case example of the considerable constraints that the U.S. president faces when attempting to act on such a decision. These constraints are predominantly personified by members of the broader U.S. foreign-policy community. Yet, this group of people consistently, and almost exclusively, concoct or contrive their warnings and their narratives based on interpretations and perceptions of what the international system allows and recommends—or on what it does not allow and recommend. Once committed to Syria through a material presence, any withdrawal of U.S. troops was seen as *potentially* calamitous. This commitment had created a status quo any alternatives to which were rapidly deemed inferior. A pullout of troops, for its part, would eventually create another status quo, but one to which a quick and frictionless "renormalization" seemed a very unlikely prospect indeed. Moves to create a new status quo or reality in a Syria devoid of U.S. troops would not lead to much if any gain for the United States, was the near-univocal message from the U.S. foreign-policy community. Rather, such moves would immediately affect the balance of power in the region and beyond, weakening the United States and strengthening Russia and Iran, leaving both undeterred; it would change the overall order in the Middle East for the worse; it would create a vacuum into which these states (and Turkey) and *jihadists* would eagerly jump; it would leave the multinational fight against ISIS bereft of U.S. leadership, and therefore of effective leadership at all; it would thus vastly increase the likelihood that terrorists would once again target the U.S. homeland, for a secure haven for *jihadists* in a highly interdependent world means that not even states on the other side of the globe are secure; and it would damage U.S. regional and worldwide credibility, perhaps irreparably.

It also took a major *system-level* disruptive event (i.e., Turkey's warnings about an immediate invasion) to set in motion the initial moves that at least led to a repositioning and partial reduction of U.S. troops. But that event was insufficient to make the United States fully withdraw. When Donald Trump left the White House in early 2021, the United States had more troops in Syria than it did when he moved in.

Push Factors and a Partial Withdrawal from Iraq

The October 2019 Syrian debacle also affected neighboring Iraq, which was set to host some of the troops that were withdrawn from Syria (Ali, 2019a). However, the question of the legitimacy of the presence of U.S. troops in Iraq itself took center stage in December 2019 and January 2020 following a series of dramatic events. On December 29, 2019, the United States launched airstrikes against Kata'ib Hezbollah, an Iraqi-based Shia militia supported by Iran. This act was retribution for the militia's rocket attack against an Iraqi military base two days earlier, which killed an American contractor (Barnes, 2019). The U.S. attack drew strong criticism from Iraqi authorities, which said it would review its relationship with Washington considering that such an action severely infringed on Iraq's sovereignty (Salim et al., 2019). On December 31, supporters of Kata'ib Hezbollah stormed the U.S. Embassy in Baghdad, further inflaming the situation and leading the U.S. administration to start preparing to send additional troops to the region (Rasheed & Ali, 2019). Then, on January 3, the United States carried out an airstrike by Baghdad International Airport that killed, among others, Abu Mahdi al-Muhandis, a leader of Kata'ib Hezbollah, and General Qassem Soleimani, head of Iran's Islamic Revolutionary Guard Corps' Quds Force (Loveluck, 2020).

These events had three notable effects of relevance here. First, they led to a further escalation of the conflict, particularly between the United States and Iran (and their Iraq-based proxies). This was most clearly symbolized by Iran's January 8 missile attack on bases in Iraq hosting U.S. troops, which killed no one but was later reported to have left over one hundred personnel with traumatic brain injuries (Lamothe, 2021; Safi et al., 2020). Second, the tensions goaded the United States into increasing its presence in the region quite significantly, as it responded to these events by sending thousands of additional troops to other countries in the region (Reuters, 2020a). Third, they prompted the Iraqi parliament to demand the withdrawal of U.S. troops from the country. This, in turn, caused President Trump to threaten to levy economic sanctions on Iraq. The president added that if the troops were ordered to leave, Baghdad would have to pay the costs of the Americans' "very extraordinarily expensive" air base (Rasheed et al., 2020).

The initial signals coming from Washington was that the U.S. troops would stay irrespective of any host-country opposition, not least as they were still key in the fight against the Islamic State (Wong & Specia, 2020). Yet, the tit-for-tat moves by the United States and Iranian-backed groups in Iraq continued with some regularity: threats, counterthreats, attacks, and counter-attacks were the order of the day in spring 2020 (Ali & Stewart, 2020a; Cunningham, 2020; Loveluck & Ryan, 2020; O'Connor, 2020), through the summer (Reuters, 2020d) and fall (Tawfeeq & Starr, 2020), and all the way through to the end of the Trump presidency (Rasheed & Ismail, 2020; Schmitt, 2020b). The spring attacks presumably helped convince the U.S. administration, in March 2020, to start relocating troops from three of its eight Iraqi bases (Kohnavard, 2020; Reuters, 2020c). Further base transfers were soon to follow (Nazeh & Al-Sudani, 2020; Snow, 2020); these were evidently caused, or at least the process was accelerated, by the "push" forces associated with Iran's ambitions to pressure the United States to leave the country by encouraging Iranian-linked proxy groups to continue attacking U.S. and coalition troops in Iraq (Brennan, 2020). "We will make them leave if they don't want to leave," was the message coming from Kata'ib Hezbollah and other militias (Kohnavard, 2020); whereas CENTCOM's leader, General Frank Mckenzie, appeared adamant that the United States was never going to withdraw from the Middle East "in response to Iranian pressure" (South, 2020).

Iran and the militias seemed to be successful, however. The United States did move to withdraw a significant share of its troops in the fourth quarter of 2020. This happened amid the usual wave of stark warnings from the foreign-policy community about the consequences that retrenchment, in particular a *complete* withdrawal of U.S. troops, would have on the seemingly eternal fight against Washington's enemies. A risk assessment by the RAND Corporation in May 2020 concluded that a "full withdrawal and disengagement (…) would have potentially catastrophic consequences for the ongoing counter-Islamic State campaign (Connable et al., 2020, p. 10); Iraqi officials, and U.S. diplomats, feared that Iraq would be turned into a "battle zone" (Davison, 2020); a common assessment of analysts was that Iran, in addition to ISIS, would benefit enormously from such a decision (Falk, 2020), and that Iraq and the United States were still highly dependent upon one another (O'Hanlon & Allawi, 2020); while retired U.S. Navy Admiral and former Supreme Allied Commander of NATO, James Stavridis (2020), lamented in an opinion piece for *Bloomberg* that the withdrawal process entailed a "gift to Iran" and a victory for ISIS, a group that was admittedly decimated but whose "remaining embers could reignite."

In the end, though, a complete withdrawal was not realized. When Donald Trump left office in early 2021, troop numbers were down to 2500 (Al Jazeera, 2021), which constituted a reduction of around 3000 troops since the start of his presidency. Only four weeks into the new administration of Joe Biden, however, NATO member states agreed to increase the number of troops for its *multinational* mission in Iraq by 3500, thereby actually increasing the total number of foreign military-service personnel in the country (Garamone, 2021b).

The Persistence of U.S. Military Bases in the Middle East

The Trump era saw rumblings about a cutback of U.S. troops from other states in the region as well. As was the case in Iraq, Turkey has also sometimes indicated its apparent displeasure with the U.S. presence. Prominent voices there, including President Recep Tayyip Erdogan, have intermittently argued for a shutdown of the key Incirlik Air Base over U.S. support of Kurdish fighters in Syria (deGrandpre & Snow, 2017), and over general dissatisfaction with U.S. policies toward Turkey (Ali, 2019b). In other instances, withdrawal discussions have instead originated in Washington. This was so with respect to the massive Al-Udeid Air Base in Qatar in the summer and fall of 2017 (Al Ansari, 2017). At the start of the diplomatic and economic conflict between Qatar, on the one hand, and Saudi Arabia, Bahrain, the United Arab Emirates, and Egypt on the other, the U.S. president seemed to side with the latter group, rather ambiguously hinting that leaving Qatar was an option he seriously considered (Karam, 2017).

However, such suggestions tend to die down quite quickly as they are supplanted by concerns about the negative consequences for the security of the United States—and normally also for the host itself—of any withdrawal. With respect to Turkey, it is commonplace for the United States to disregard such threats considering the strategic and political benefits that ultimately flow in Ankara's direction from its relationship with Washington. The potential and often probable fallout from a withdrawal tend to weigh heavily on any such discussion. This is the case for Bahrain as well. When voices appeared recommending that the United States leave the tiny island state, which hosts the U.S. Fifth Fleet, over human-rights concerns, one analyst remarked that this would "send shockwaves across the region" and severely damage the reputation of the United States in the eyes of regional allies and partners (Saab, 2018). The United States does not seem to have any "Plan B" if they ever choose to leave Bahrain, which arguably hosts many of the most important U.S. military facilities in the whole region (McDaniel, 2013). It is clear that the current key goals of the United States in the Middle East—first and foremost deterring and containing Iran and blocking any resurgence of the Islamic State—would be significantly hampered if the U.S. regional, highly interdependent network of bases started to unravel, which an abandonment of Bahrain could well help effectuate.

Yet, several other states in the region are also center stage in the geostrategy of Washington. The Al Dhafra Air Base in the United Arab Emirates, whose use by the U.S. Air Force was kept secret for over a decade until the summer of 2017 (Pawlyk, 2017), has been much-employed as a launchpad for U.S. attacks on targets in Syria and Iraq (*Middle East Monitor*, 2017), as a staging post for drones (Reuters, 2019b), and as a military-logistical hub in the containment of Iranian proxies in the region (Stewart, 2017). The Incirlik Air Base in Turkey proved vital in the fight against ISIS (deGrandpre & Snow, 2017). Oman signed a deal with Washington in 1980 giving the United States access to military sites in the Sultanate, including several military airfields (Coffey, 2019); while a recent deal hands the United States improved access to facilities and ports in Oman, which reduces Washington's need to sail through the bottleneck that is the Strait of Hormuz (Stewart, 2019a). In the middle of the Indian Ocean, moreover, we find the "unsinkable aircraft carrier" of Diego Garcia, which is a highly disputed part of British Indian Ocean Territory whence the indigenous Chagossians were evicted by London in the 1960s. There, the United States has recently extended its lease for a further twenty years of a military base that is perfectly shielded and perfectly located for power projection in the whole of the Greater Middle East and beyond (Black, 2019; Mance, 2016). The United States also keeps thousands of troops at several bases in Kuwait, and it has numerous bases in Saudi Arabia (Hussein & Haddad, 2021). It was to these two countries that the United States sent thousands of *additional* troops (classified as "non-permanent") in late 2019 and early 2020, in order to bolster its deterrence against Iran.

Furthermore, Al-Udeid in Qatar is, along with the NSA Bahrain Navy Base, surely the lynchpin of the U.S. military-base network around the Persian Gulf, and it is the largest U.S. air base in the Middle East (McCarthy, 2017). Irrespective of President Trump's apparent dissatisfaction with the regime in Doha in the early months of his presidency, and despite his rather thinly veiled threats in summer 2017 that a U.S. withdrawal could be in the cards, most observers would claim that this was and is a highly unrealistic prospect for several reasons (Karlin & Dalton, 2017). Andrew Stravers and Dana El Kurd, for example, note that previous U.S. administrations have also contemplated reducing the country's dependence on Qatar; but "with every attempt at decreasing dependence, the US military has discouraged administration officials from pursuing the option" (Stravers & El Kurd, 2020, p. 433). All potential alternatives to Al-Udeid are ultimately deemed inadequate, not least by U.S. military officials, one of whom curtly stresses that the strategic location of the base is simply unique, situated as it is "right in the center of everything" (Taylor, 2019b).

The prevailing status quo, characterized by ample redundancy and the dispersal of U.S. troops and military assets among several regional allies and partners, is credibly believed to bolster both deterrence and reassurance. Considerable costs are sunk in logistics and infrastructure and political-strategic relationships, which also adds greatly to the barriers against withdrawal. As a further testament to the allure of (bolstering) the status quo, indications are that the United States, and their hosts, are in the process of *expanding* some of the key U.S. bases in the region, such as in Jordan (Trevithick, 2019) and Qatar (Reuters, 2018; Vandiver, 2018a). In September 2017, moreover, the United States officially opened its first-ever permanent military base in Israel (Gross, 2017).

Why the Stayers Prevailed

Under President Trump, the United States did eventually downsize its troop contingents in Iraq, and also in Afghanistan (more on which in the next section); that is to say, the United States partially withdrew from the two countries that it invaded and occupied two decades ago. Yet, the overall base pattern and troop numbers in other Middle Eastern states have overwhelmingly remained the same. One estimation, by the Arab Center Washington DC, held that about 65,000 U.S. troops were still located in the region near the end of 2020, if one includes naval and air forces on regular rotation (a few thousand soldiers that left Iraq and Afghanistan in the very last weeks of the Trump administration are presumably included in this estimation) (Brumberg, 2021).

That the United States' core military-base infrastructure, and its troops, remain firmly in place runs counter to the very widespread notion that the United States, "after a wildly erratic spree of misadventures, is backing out of the region" (Kramer, 2016, p. 53). F. Gregory Gause III (2019, pp. 10–12), a professor at Texas A&M University, finds it "remarkable how many observers, both in the United States and in the Middle East, think that the United States is *already* in the midst of withdrawing from the region," even when the "empirical reality is not in line with the perception of withdrawal." If 2005 is used as a base year for comparison, he continues, the United States *has* indeed retrenched somewhat. Back then, however, the United States had 200,000 soldiers actively engaged in theaters of war in the "Greater" Middle East. Wars are won, or lost, or they end in any case, and troops are eventually withdrawn from these wars. But bases and troops—*U.S.* bases and troops—remain in high numbers right outside of the war zones. In fact, they remain in *much higher numbers* than they did before 9/11, and the pre-2001 period provides a more natural basis for comparison than do the years of outright war. Though many have been eagerly awaiting a downscaling of U.S. operations in the region's war zones, most members of the U.S. foreign-policy establishment have long since "renormalized" to a status quo in the Middle East that involves far greater troop numbers than what obtained in the 1990s, the unquestionable height of unipolarity.

Gause III (2019, p. 13) asserts that the widespread perceptions that a withdrawal is in the process are only correct if one also factors in intentions, as it is certainly the case that "the last three American presidential elections were won by the candidate who was more dovish on the Middle East." And none were more so than Donald Trump. So why didn't retrenchment manifest in the Middle East outside the zones of (abandoned) war? The answer certainly rests not with U.S. public opinion; it is as ambivalent about U.S. troops in the Middle East in general as it is about U.S. troops elsewhere and everywhere (Zenko, 2018). In fact, opinion polls in summer 2020 showed that a considerable majority of the American public wanted to pull out the troops deployed in Afghanistan and Iraq (Kheel, 2020b). And, if we consider the most publicized case of them all, which is Syria, there is also nothing to indicate that vested economic interests played any role when the president in the end chose to reverse his withdrawal order, which he ostensibly did to ensure that the United States was "keeping the oil" (Ali, 2019b). Syria's oil resources are fairly meager anyway. Moreover, the country has been and is heavily sanctioned by Washington, and U.S. companies are generally not allowed to do business there—unless they get a sanctions waiver, that is, which happened only to the small (albeit politically connected) oil firm Delta Crescent (Seligman & Lefebre, 2020). It is far more plausible that some of the president's advisers came to believe that propounding the objective of "securing the oil" would convince their boss to keep most of the troops in north-eastern Syria. This pretext came as a relief to most U.S. military commanders, who saw a continuation of the U.S. military presence as absolutely necessary to contain ISIS and to check the power of both Iran and Russia (Rosen, 2021). It also helped the U.S. envoy to Syria, Ambassador James Jeffrey, who reported that he and his team had obscured the real troop tally in order to keep it at levels deemed high enough to support most of the wobbling Syrian dominoes (Williams, 2020).

It is also plausible, however, that when President Trump bemoaned the U.S. presence in the Middle East altogether, this was in reality a far cry from *actually* ordering a U.S. withdrawal from Qatar, from Saudi Arabia, from Turkey, and from most of the rest of the U.S. base hosts in the region. But threats to this effect were indeed made; they did prove highly credible with respect to Syria, Iraq, and Afghanistan; and they did fit with what was a consistent narrative long since constructed and thereafter continuously repeated by Donald Trump. Institutional and physical constraints certainly play a part here as well: dismantling such a vast *network* of bases does face quite natural, often unspoken resistance from sheer on-the-ground material and bureaucratic realities, and from the considerable costs that are sunk into this network. Yet, this on-the-ground reality, in turn, rests on a different dimension of reality that is usually only tacitly revealed—although sometimes it becomes eerily explicit. When the United States started pulling its forces out of some of its Iraqi bases in March 2020, this sparked an increase in aggressive acts directed against U.S. and coalition forces by Iranian-linked militias; it augmented Iran's power in Iraq; and it correlated with a significant uptick in ISIS attacks in the country (Cornish & al-Omar, 2020).

To a certain extent, Iraq illustrates what *can* happen if troops are pulled out, or if they, as in this case, are *in the process* of being pulled out. When the U.S. foreign-policy community warn against troop withdrawal, they inexorably point to these allegedly imminent chain reactions. The U.S. military establishment, the U.S. Congress, analysts and commentators, think tanks, and the media (and also the host countries, in most instances) may be domestic-level actors. But their arguments, and their fears, and the logics that they manufacture, are arguments, fears, and logics centering on *system-level developments* that will, or that *may*, damage U.S. national security. Mere proposals suggesting a large-scale U.S. withdrawal from the Middle East conjure up memories of 9/11, of new uncontrollable vacuums, and of a region dominated by Iran and Russia. The objective of leaving the region is really incompatible with the objective of upholding U.S. regional power and influence—and it is incompatible with promoting regional security; and the former objective almost always loses out in the battle of arguments. This is all the more so as the known is known and has the scent of the comforting and familiar; while the unknown is unknown and (perceived to be) far less manageable, and it easily evokes disturbing images of rapid and detrimental change.

The *Leavers* of the scholarly world argue forcefully that the United States is better off with a "hands-off" policy toward states in the Middle East whose problems cannot be solved by the United States anyway (Mearsheimer, 2014). Whether or not this is a sensible recommendation, it first and foremost represents a ***normative*** position. The descriptive ***reality*** is that "reality" tends to fly in the face of presidents. And this "reality," Middle East expert Marc Lynch (2016, p. 128) predicted some months before Donald Trump's assumed office, "is that Trump is likely to experience the same difficulties in transforming America's role in the region as his predecessors, whose transformational ambitions have also been defeated by the structural realities of the United States' position in the Middle East." No matter his intentions, Lynch went on, "an inexperienced president will face constant pressure from U.S. allies, the Republican foreign policy establishment, and the national-security bureaucracy to adopt more conventional positions." In hindsight, it was pretty much just this combination of structural or systemic pressures and the agency, arguments, and preferences of the *Stayers* of the Blob that ensured that the United States, for all intents and purposes, eschewed the option of leaving the Middle East—even with a self-avowed *Leaver* as president.

The Impending Fall of the Afghan Dominoes

Afghanistan is in South Asia. It was for political-strategic reasons that it was "transferred" by decree, so to speak, to another region by the "Greater Middle East Initiative," a plan presented in 2004 by the administration of George W. Bush which was part of a broader scheme to reform Muslim-majority countries (Al Jazeera, 2004). The conception of Afghanistan as part of the Middle East has since stuck to a surprising extent, even if reforms in the (Greater) Middle East have not. By the time of writing, the United States is no longer present in Afghanistan, which is again ruled by the fundamentalist Taliban, as it was before 9/11. Although the withdrawal process was set in motion in the last year of the Trump presidency, the pullout of the last of the U.S. troops took place under the Biden administration.

The Afghanistan debacle has brought numerous lessons. Since January 2017, and with respect to the issues covered in this book, we have learned, first, that the route to withdrawal is very long, winding, and arduous, but it is still one that is possible to complete. Second, we have learned that the counterforces at work are plentiful and vigorous and immensely focused on the negative consequences a withdrawal *might* have. Third, we have learned that in the latter stages of the withdrawal process—and after—many of the adverse geopolitical chain reactions that the *Stayers* worry about *can* indeed manifest. The Afghan case, though rife with tragedy, also tells a vital story about the prescience, or lack thereof, of the pre-withdrawal domino warnings that originally mostly belonged in the realm of hypotheticals or possibilities. Afghanistan is thus also a case that illustrates the *manifestation* of adverse geopolitical chain reactions, even if these are still early days.

A Leaver and His Troop Surge

The latter events belong mostly to the Biden era, which I will return to in the concluding chapter. Here, I focus on the four years of Trump. From the very start and to the very end of his presidency, the tension between an unambiguous wish to pull out troops and the specter of adverse chain reactions that *might* result from a withdrawal came to define much of his administration's Afghanistan policies. In a November 2013 Tweet that was fittingly representative of his retrenchment inclinations, Trump, at that time still a private citizen, had stated that the United States had "wasted an enormous amount of blood and treasure in Afghanistan. Their government has zero appreciation. Let's get out!" (Pramuk, 2017b). In contrast, the domino dynamics, or the *potential* domino dynamics, were well captured around the same time in *The Revenge of Geography* by Robert Kaplan (2012, pp. 326–327), who wrote rhetorically about the issue of leaving Afghanistan:

> Would the intelligence that has led to successful drone attacks on al Qaeda in Waziristan dry up? Would Ayman al-Zawahiri and other surviving luminaries of al Qaeda make triumphal entries in front of al Jazeera television cameras into Jalalabad? Would Afghanistan become a radicalized Taliban state under the tutelage of Pakistani intelligence? Would India, the global

> pivot state of the twenty-first century, lose respect for the United States as a consequence? Would Iran informally annex western Afghanistan? (...) Truly, withdrawing precipitously from Iraq or Afghanistan would be irresponsible because – like it or not – merely by invading these places and staying there so long, we have acquired substantial stakes in the outcomes.

In other words, the withdrawal objective ran up against the *possible* detrimental consequences of withdrawal; and it ran up against the logic of, and the widespread human tendency to overvalue, sunk costs and the status quo, even when the status quo was eminently problematic, as it most certainly was in Afghanistan. Initially, after some intense debate within the Trump administration (Landler & Schmitt, 2017), the decision about troop levels was delegated by the president to Secretary of Defense James Mattis. Seeking to break the stalemate in the battle against the Taliban insurgency, Mattis recommended sending some 4000 additional troops to the country, with NATO allies adding a further 2000 soldiers (*The Guardian*, 2017). This modest "troop surge" was completed by November 2017, with the total number of U.S. troops in Afghanistan by then standing at around 14,000 (Moore, 2017).

This decision obviously risked spawning accusations that the president was "flip-flopping" on the issue. Likely even more problematic for Trump personally, it signaled continuity from the Obama era. In any event, the president admitted that his advisers, notably Defense Secretary Mattis and National Security Adviser H. R. McMaster, had succeeded in moderating his impulses to withdraw (BBC, 2017a; MacDonald, 2018, pp. 427–428). Wowing to win the war against the Taliban, Trump acknowledged that even though his "original instinct was to pull out, and historically, I like following my instincts," he had all his life "heard that decisions are much different when you sit behind the desk in the Oval Office" (Mehta & Copp, 2017). Now was apparently not the right time to pull out of Afghanistan after all; "[c]onditions on the ground, not arbitrary timetables, will guide our strategy from now on," was the new message coming from Trump (Garamone, 2017). Thus it was that U.S. troop numbers in Afghanistan would *rise*, rather than fall, from the Obama administration to its successor. Even so, the ultimate objective was still withdrawal—or at the very least a major reduction of troop numbers.

The Withdrawal Begins

In late 2018, the topic again rose to the top of the agenda, with the president directing the Pentagon to pull out half of the 14,000 troops (Lamothe et al., 2018). This announcement immediately sparked fears among U.S. and foreign officials alike (Witte & Stanley-Becker, 2018). James Mattis resigned as defense secretary over the issue, which also involved a simultaneous order to withdraw troops from Syria (Sonne et al., 2018). Republican Senator and ardent Trump ally Lindsay Graham (R-SC) called the troop withdrawals "a high risk strategy" that *could* in the end lead to "a second 9/11" (Ali & Stewart, 2018). Other U.S. officials argued that such a drawdown would seriously hamper U.S. credibility vis-à-vis other allies as well as create a vacuum the filling of which would likely be done by Al-Qaeda and the Islamic State (Ali & Landay, 2018).

Such warnings seemed to have a moderating effect on the president, for the order was put on hold even if the end goal presumably remained. The president himself seemed to recognize the risks involved if the United States pulled out from what he called "the Harvard of terrorists"; again, he alluded to the advice he had been given by military officials, who argued that it would be far better to fight terrorists in Afghanistan than in the United States itself (Reuters, 2019c). One of these officials, Chairman of the Joint Chiefs of Staff General Joseph Dunford, envisaged a U.S. counterterrorism presence that would not end until the Afghan insurgency was over (Reuters, 2019a). All the while the president himself continued to signal his univocal ambition to withdraw, at the very least before the 2020 presidential election (Wroughton & Jain, 2019).

These plans appeared to gain renewed momentum in early September 2019, when it was announced that Washington had reached a deal "in principle" with the Taliban that would include the withdrawal of 5400 U.S. troops (BBC, 2019). What was apparently agreed was a reduction rather than a full pullout, with the president himself stating that "you have to keep a presence (...) we're going to always have a presence" (Reuters, 2019e). Even so, the proposed deal with the Taliban again led to warnings about the consequences that would *probably* ensue from a premature pullout, with some foreseeing a full-blown civil war (Landay, 2019).

As it happened, during the first three quarters of 2019, the United States did undertake what the administration referred to as a "quiet" withdrawal of around 2000 troops from Afghanistan (Gibbons-Neff & Mashal, 2019). Of greater import, in late February 2020, the United

States reached another, seemingly more concrete, peace deal with the Taliban (Sediqi & Cornwell, 2020). Under its terms, the United States would reduce its forces in Afghanistan by one-third within 135 days, which was to be followed by a full withdrawal of the remaining troops within a year (DeYoung, 2020). Almost immediately after this announcement, the pullout commenced (BBC, 2020). As accounted for in Chapter 6, the withdrawal thereafter continued pretty much according to plan. By January 15, 2021, the Pentagon declared that only 2500 U.S. troops remained in Afghanistan (Garamone, 2021a).

Some sources dispute that number (Gibbons-Neff et al., 2021). Yet, whether the actual number was 2500 or 3500 is of less importance than the observation that the United States, after nearly two decades of war, was now quite close to pulling the final plug on an Afghanistan mission that, for its part, had hardly gone according to plan. In the last year of his term, Trump seemed insistent that his move was the right one. Responding to a May 2020 editorial in *The Wall Street Journal* (2020) criticizing the pullout, the president retorted that his country had been there for almost two decades but had "never really fought to win" (Ryan, 2020). A few days later, he laconically added that, although two decades years was more than enough, the United States could "always go back if we want to" (Al Jazeera, 2020).

The Domino Warnings That Came to Naught

Trump did find support for his stance. Exasperation with the Afghan War had surely been rife for the better part of these 19 years. The *Leavers* tended to highlight that U.S. troop levels had fluctuated quite massively over this period; but no one specific number of soldiers could really fix what was a multifaceted problem "in need of a local, political solution developed by the people who must live with its results" (Kristian, 2020). Still, the *Leavers* among the foreign-policy community were solidly dwarfed by the *Stayers*, even if the narratives of the latter—this time—failed to convince the president that the withdrawal should be reversed.

As usual, the arguments of the *Leavers* spanned the whole gamut of possible sources of adverse geopolitical chain reactions. In general, though, they centered on the presumed critical role that U.S. troops—even a limited number of U.S. troops—had in supporting wobbling dominoes. Afghanistan was more about *not losing* than it was about winning; it was about providing insurance against even worse things, like a repeat of 9/11. Such a notion was often only implicitly expressed. Sometimes, however, arguments to this effect were more explicitly voiced. In an analysis published in late November 2020 in *The Atlantic*, Jim Golby (2020), who had earlier served as a special adviser to Vice Presidents Joe Biden and Mike Pence, bluntly argued that pursuing "victory" was not at all the main point of staying. Even the 14,000 U.S. troops that were in Afghanistan prior to the withdrawal decision were insufficient to *beat* the Taliban.

> As unpalatable as it sometimes may seem, putting U.S. troops in harm's way isn't always about winning or solving problems. Sometimes it's about managing risk – keeping minor problems from turning into major crises. For years, America's de facto policy in Afghanistan has been to maintain enough troops to prevent the Taliban from overrunning the U.S.-backed government in Kabul and allowing terrorist groups to use the country as a base of operations again. (...) [A] complete withdrawal would likely have devastating results.

U.S. troops, a report by the RAND Corporation concurred, were not there to win, for "[w]inning may not be an available option, but losing certainly is"; and a "precipitous departure, no matter how rationalized, will mean choosing to lose" (Dobbins et al., 2019). Losing, which in this case equaled withdrawing, *could* have deleterious effects. Most observers drew their worries from the system level, even if often only tacitly. The United States is the leading state in the international system, one Afghan-based analyst stated, and its presence in Afghanistan, even if only symbolic, was of immense value to a very fragile country whose neighbors were armed with nuclear weapons (Gibbons-Neff, 2020). In an opinion piece published in *The Washington Post*, and despite that he served as leader of the very country whose intelligence services are widely believed to have abetted the Taliban for decades, Pakistani Prime Minister Imran Khan (2020) expressed worries about "regional spoilers" that did not seek stability in Afghanistan but rather saw its instability as a boon to their own geopolitical objectives. Virtually all *Stayers* pointed to the risk of governance vacuums being created by a withdrawal. These *could*, or they would, be filled by various combinations of the Taliban, local warlords, and terrorist groups. NATO General Secretary Jens Stoltenberg, who rarely if ever criticized the President of the United States, feared that Afghanistan would again become base from which transnational terrorists could "plan and organize attacks on our homelands," and he emphasized the *possibility* that ISIS would rebuild there "the terror caliphate" it had earlier established, and then had been forced to abandon, in Iraq and Syria (Taylor & Birnbaum, 2020).

Regional powers and terrorists aside, the broader consequences for the United States' position in the hierarchy of states would also, *possibly*, be threatened by a hasty withdrawal. This was yet another case of U.S. worldwide credibility "taking a sledgehammer" blow, lamented Laurel Miller, a former U.S. special representative for Afghanistan and Pakistan (Ali & Stewart, 2020b). Any "precipitous withdrawal," predicted Earl Anthony Way, another former U.S. diplomat with ample experience from Afghanistan, would undermine U.S. interests and others' perceptions of America's resolve, and it would make the country pay a huge price "in the short, medium and long term" (Cloud & Wilkinson, 2020). Purported lessons from history supplemented these system-based worries: Many saw in Afghanistan echoes of Vietnam—and of Iraq. Colin Jackson, a former senior Pentagon official who used to oversee President Trump's Afghanistan policy, coarsely stated that there didn't exist any examples "where pulling the plug has gone well—Vietnam, Iraq. Not one" (Lamothe et al., 2020).

Adding to these gloomy predictions was a rapidly deteriorating security situation in Afghanistan once the withdrawal process had begun, which mirrored similar and simultaneous dynamics in Iraq. The Taliban stepped up their attacks quickly and significantly, targeting Afghan government forces and civilians and establishing control over an increasing slice of Afghan territory (George & Tassal, 2020; Shalizi et al., 2020). This time, however, the president and his team didn't budge; the drawdown of troops continued. And it continued even as the U.S. Congress twice—in the summer of 2020 and at the year's end—voted to halt the withdrawal under amendments to the National Defense Authorization Act. These moves were aimed at blocking the Pentagon's access to funds necessary to proceed with the drawdown (Al Jazeera, 2021; Kheel, 2020a). They did not succeed, however. Under President Trump, America now seemed—at last—fully committed to withdrawing as many troops as logistically possible. The U.S. mission in Afghanistan, after almost two decades, was about to end. It hadn't *fully* ended, however; the drama of 2020 would continue into 2021. The new president, Joe Biden, was eventually tasked with making the ultimate decision on whether the United States' Afghan War would *really* come to its conclusion.

The Geopolitics of U.S. Troops and Withdrawal in Europe

Russia's 2014 invasion and annexation of the Crimean Peninsula, and its subsequent intervention in Ukraine's civil war, broke the relative tranquility of Europe. Notwithstanding the altered threat perceptions among the United States' European allies, Donald Trump had, during his election campaign, focused far less on the possible threat posed by Russia than he did on the purported "free-riding" of the United States' European allies. Protecting European members of the North Atlantic Treaty Organization (NATO), he deplored, "is costing us a fortune" (Rucker & Costa, 2016). This was a message widely perceived to entail a veiled threat that the United States' commitment to its allies' defense was not any longer unequivocal. The message was also repeated with consistency during the election campaign: U.S. allies were "not paying their fair share" for the long-standing protection they received from the United States; if this situation continued, America "must be prepared to let these countries defend themselves" (*New York Times*, 2016).

These admonitions continued after Trump entered office, and U.S. allies were soon even more alarmed by the rhetoric of the new president. In a meeting with NATO allies in Brussels, Belgium, in May 2017, Trump startlingly refused to endorse the alliance's critical Article 5, whose essence is that an attack on one member state is to be regarded as an attack on all (*New York Times*, 2017). This was an ominous sign for the allies, and one that was repeated at the NATO Summit a year later (Birnbaum & Rucker, 2018). The NATO alliance, in effect since 1949, was thus perceived to be under serious threat; and with it, there reigned uncertainty about the U.S. military-base network in Europe. After over seven decades of heavy U.S. troop presence in Europe, would the Trump administration now *really* start dismantling the framework that had so long helped provide assurance toward allies, deterrence against adversaries, geopolitical stability, and U.S. influence in and over Europe?

A Network of U.S. Bases and Troops in Europe

Numbers suggest that it wouldn't. Or, at least, it didn't. As accounted for in Chapter 6, troop numbers stayed almost exactly the same: there were 62,101 permanently assigned active-duty U.S. troops stationed in Europe when Trump entered office, and 62,248 when he left. The largest host, Germany, saw only a minuscule reduction, despite the unambiguous signals by the president that troop numbers there would be drastically cut (more about this later). Troop levels also stayed the same in Italy, the second biggest host. The United Kingdom, for its part, witnessed an *increase* of 11.1%, hosting 9424 U.S. troops by the end of 2020. The general picture was one of continuity: Spain saw a small decrease, as did Greece; while the United States added a few troops to its contingents in Belgium and the Netherlands.

In Germany, the United States has over one hundred military sites or bases where, among others, around 20,000 members of the U.S. Army and 13,000 U.S. Air Force personnel are stationed, often along with their family members. These facilities form much of the core of the U.S. base network in Europe. They include, inter alia, Ramstein Air Base in Rhineland, which also serves as the U.S. Air Force's headquarters in Europe; Spangdahlem Air Base, near Luxembourg, which for half a century has been a vital cog in the U.S. overseas military machinery (Militarybases.com, 2021); and the U.S. European Command (EUCOM) headquarters in Stuttgart, which coordinates all U.S. forces stationed in Europe (Knight, 2020). In fact, the bases in Germany have been central to U.S. operations beyond Europe as well, not least in the Middle East and North Africa (MENA), in countries such as Iraq and Libya (Vandiver, 2017).

Military bases in other European states have also served as staging posts in the battle against and deterrence of non-state actors, such as the Islamic State, and aspiring major powers in neighboring regions, such as Iran and Iraq (Yost, 1995, p. 73). This is the case for the many air bases in the United Kingdom that are run and operated by the U.S. Air Force. Moreover, fighter planes, drones, and military vessels stationed at major bases in Italy—for example, Aviano Air Base at the foot of the Dolomites; Naval Air Station Sigonella on the island of Sicily; and the U.S. Naval Support Activity Naples, home to the United States Sixth Fleet—have played a key role in several of Washington's military missions in MENA in the entire post-Cold War period (Ottens, 2018; Politi, 2016; Vandiver, 2018b).

Along with Germany, Belgium, the Netherlands, and Turkey, Italy also hosts some of the around 150 tactical nuclear weapons that the United States still keeps in Europe (Taylor, 2019a).

Italy's strategic location is in this respect obvious. This is even more clearly the case for Greece, however. In particular, the importance to U.S. operations in MENA of the major U.S. naval base in Souda Bay, Crete, seems to have risen markedly in recent years (Ekathimerini, 2018; Ioannou, 2017). This is so even if Souda Bay has proven its usefulness for decades. The base has been central to a number of U.S. post-Cold War operations beyond Europe, including Operations Desert Shield and Desert Storm (in Iraq), Operation Enduring Freedom (post-9/11 counterterrorism), and Operation Iraqi Freedom (from 2003) (Goure, 2016). More broadly, U.S. activities in Greece are a testament to the many security issues in and around Europe that are still a concern to the U.S. foreign-policy community. Plans to expand the U.S. military presence there were revealed in September 2018, amid concerns about Russia's geostrategic ambitions, uncertainties over Turkey's more assertive foreign policies, and the non-abating instability in the nearby Middle East (Maza, 2018; Youssef, 2018). Considering these challenges, and also considering Greece's highly strategic location, an *expansion* of bases there seems more logical than a contraction; as long, that is, as the United States remains committed to deterring adversaries and intervening in crises in key regions of the world for the purpose of hindering such geopolitical issues snowballing into less controllable geopolitical chain reactions. A little more than a year later, in October 2019, the United States and Greece signed a revised mutual defense cooperation deal that would heighten the role of Athens in Washington's strategic portfolio and lead to significant new investments by the United States in Greek military facilities (Psaropoulus, 2019).

U.S. Troops for Deterrence of a Russian Bear

Whatever the importance of the Middle East and North Africa, the non-withdrawal of U.S. troops from Europe over the period under study is, in the main, caused by concerns about other great powers—by those inhabiting the "Heartland" of Eurasia. China, though, is, for the time being at least, not the rival of key concern here. This is so even if members of the U.S. foreign-policy community have been insistent that the United States must energetically guard its long-lasting foothold on Portugal's Azores

Islands, on which Beijing has reportedly set its sights (Ames, 2016; Kates, 2017). A more immediate worry is Russia. Thus, the observed continuity in troop numbers and base locations rests on the notion that U.S. disengagement would only work to boost Moscow's confidence that it can, with little risk to itself, increasingly exert its influence abroad, by using "hybrid" or even outright military means.

Indeed, even prevailing U.S. and NATO force levels were by many deemed inadequate. A widely publicized 2016 war-game study by David Shlapak and Michael Johnson of the RAND Corporation suggested that the capitals of Estonia and Latvia *could* be besieged by Russian troops within 60 hours at most—and there was nothing allied forces could do to stop them (Shlapak & Johnson, 2016). Prospects and options were univocally bad (even if they rested on the not obviously plausible scenario in which Moscow would ever really *want* to invade its small neighbors again): The United States and its NATO allies could choose to stage "a bloody counteroffensive, fraught with escalatory risk, to liberate the Baltics; to escalate itself (...); or to concede at least temporary defeat, with uncertain but predictably disastrous consequences for the Alliance and, not incidentally, the people of the Baltics" (Shlapak & Johnson, 2016, p. 1).

The message was thus that, unless the Allies endeavored to strengthen its presence and capabilities, especially along Europe's eastern and northern rims, Russia *could* be tempted to push over two Baltic dominoes, which *could* then cause numerous other dominoes to tip over. Such concerns as expressed by that study, and by others (Fabian et al., 2019), account for much of the reason why several states in Europe, especially but not exclusively in Central and Eastern Europe, have been included in the Pentagon's plans for a further expansion of U.S. military presence in this region. The 2014 European Reassurance Initiative, which was later relabeled the European Deterrence Initiative (EDI), was initiated by the White House to deter any further Russian expansion into its Western neighborhood. Increased funding was allocated within this scheme for major investments in bases and military facilities, and the stationing of additional U.S. troops (Shinkman, 2017). Although there were at first doubts as to whether or not President Trump would support the EDI, considering his earlier campaign rhetoric, the initiative continued under his watch as well (Posen, 2018, p. 22; Schreer, 2019, p. 13). In fact, budget allocations to the EDI soon increased, from \$4.8 billion in 2018 to \$6.5 billion the year after, with much of the money going to the

U.S. Air Force in Europe (Gould & Insinna, 2018). To wit, far from reducing the U.S. military footprint in Europe, a general belief among the broader U.S. foreign-policy establishment—namely, that the benefits of deterrence far outweighed the risk of augmenting security-dilemma dynamics in Europe—instead effectuated the opposite.

Hence, significant investments were made in the Trump era at U.S. (and NATO) military facilities in Romania and Bulgaria (Euronews, 2017; Luca, 2018); in Hungary (Gould & Insinna, 2018); and in Estonia, Latvia, and Slovakia (Shinkman, 2017). Further north, hundreds of U.S. Marines were shipped to Norway near the end of the Obama presidency to conduct year-round training with allies and to reinforce deterrence against Russia. This mission continued—indeed, in terms of troop numbers, it was significantly expanded—under Trump (though it was downscaled near the end of his term) (Barnes & Jervell, 2017; Browne, 2020b). Moreover, Iceland saw the return of some of its Cold War status as an island state of substantial geopolitical importance, on account of the perceived need to keep tabs on Russian submarine activities in the North Atlantic (Cotovio & Pleitgen, 2018).

Most of these new investments, however, were conducted at bases where U.S. troops are serving on an explicit *rotational* basis, which means that Pentagon tallies of U.S. troops are understated in parts of Europe. This is the case for Poland, where a reported 4500 (rotational) U.S. troops have normally been stationed (*The Economist*, 2019), which contrasts to the mere 162 quoted by the Pentagon. In June 2019, moreover, Poland, which borders Russia's Kaliningrad exclave, signed an agreement with Washington to increase the number of (non-permanent) U.S. troops by around 1000; these were presumably to be stationed, according to the U.S. president, at a new, "world-class" base tentatively christened "Fort Trump" (Gearan & Sonne, 2019). Notwithstanding that the subsequent year indicated that the plans seemed to have been shelved amid disagreement over funding and location (Plucinska & Ali, 2020), the United States did send an additional 200 U.S. Army personnel to Poland in August 2020 (Youssef, 2020). Furthermore, in November 2020, Warsaw ratified the Enhanced Defense Cooperation Agreement, which the two countries had signed the previous year. At the same time, it formally opened a forward-based U.S. Army Headquarters in Poznań in the western part of the country (Żaryn, 2020). Thenceforth, the United States would have 5500 (rotational) troops stationed in Poland.

An Order to Leave Germany

In other words, and contrary to the message consistently relayed by Donald Trump both as presidential candidate and president, the administration of the self-declared *Leaver* actually *increased* the United States' military footprint in Europe. The first forty months of Trump's presidency saw practically no serious attempt at all to reduce the U.S. military presence anywhere in Europe, despite the president's repeated lambasting of Washington's NATO allies. Things only started to change in early June 2020. That was when *The Wall Street Journal* reported that Trump had directed the Pentagon to withdraw 9500 U.S. troops from Germany (Gordon & Lubold, 2020a). Later in the summer, that number was adjusted upward: The United States would reportedly pull out 12,000 troops from Germany, moving some of them to Italy, some to Belgium, and some back home to the United States (Ryan et al., 2020). Key command and control centers would ostensibly also be moved (Lendon, 2020), along with AFRICOM's headquarters in Stuttgart (Seldin, 2020).

What was seemingly behind this decision was the president's patent dissatisfaction with Berlin's overall foreign and security policies, including its trade and energy policies, and particularly its reluctance to raise military spending. In his usual forthright manner, Trump described Germany as a "delinquent" country that "treat us very badly on trade" (Mason & Mohammed, 2020). The United States gave Germany protection against Russia, Trump claimed, but in return the Germans were "paying Russia billions of dollars for energy coming from a pipeline, brand new pipeline" (Parker & Wingrove, 2020), a comment referring to the controversial Nord Stream 2 project. The United States would not "want to be the suckers any more" and would hence start pulling out troops, the president told reporters (Stewart & Ali, 2020). He then added that his country had "been taken advantage of for 25 years, both on trade and on the military" (Gordon & Youssef, 2020).

Donald Trump wanted to see the United States out—or at least halfway out—of Germany. In this, he did receive some support from parts of the U.S. foreign-policy community. Henry Olsen (2020), a columnist for *The Washington Post*, took his cue from apparent changes in the balance of power, arguing that Trump's move was the right one considering the vast distance that existed between the United States' "global commitments and its global power." Others weighed in, pointing to the relative lack of any real security threat in Europe. Even if the United States withdrew every single soldier from Europe, retired Lieutenant Colonel Daniel Davis remarked, it was not going to alter deterrence against Russia one bit considering the overwhelming conventional and nuclear capabilities of NATO (Borger, 2020). Besides, as another *Leaver* opined, Russia's military and economic weakness means that no plausible "hegemon is ascendant in northwest Eurasia"; hence, the United States have no need for stationing tens of thousands of troops on the continent (Cookson, 2020).

The German public didn't seem to be overly eager to keep all U.S. troops in their country either. A YouGov survey in midsummer 2020 revealed that close to half of Germans favored downsizing the U.S. troop contingent; only 4% preferred an increase in troop numbers; while 28% wanted numbers to remain as they were (around 21% declined to answer) (VanOpdorp, 2020). In other words, Germans at large appeared little bothered by the withdrawal order. Even weightier, the U.S. President and his administration had seemingly made up their mind. The withdrawal still failed to materialize. This could not be due to the preferences of *U.S.* public opinion, which is quite divided on the issue of U.S. military forces in Germany, according to survey evidence (Kafura, 2019). Neither is it plausible that economic concerns in those German towns and municipalities that benefit from the U.S. military presence have much if any bearing on national-security decisions in Washington. And, while it is more plausible that the "transactional" diplomacy of the U.S. President stressed the objective of burden-sharing—"operationalized" as increased defense spending by allies—more than it did that of troop withdrawal from Europe per se, this does not seem to be the case regarding Germany. On the contrary, by summer 2020, plans to close bases and withdraw troops from Germany were in full motion. There was little or nothing to indicate that President Trump would ever backtrack on his decision even should he ultimately manage to obtain a "better deal" from Berlin.

The Stayers Push Back

The troop withdrawal from Germany never materialized primarily because it met, from the very beginning, fierce resistance from a U.S. foreign-policy community that mobilized quickly and in force to have it stopped. Formally, it was the U.S. Congress that blocked the move. Starting in June 2020, Democratic lawmakers introduced legislation seeking to prohibit funding for any pullout of troops from Europe unless such a move was approved by Congress (Zengerle, 2020a). Republican members soon joined in, supporting a bipartisan proposal to freeze troop numbers in Germany (Gould, 2020). In December, the National Defense Authorization Act—the annual defense-policy bill—was unveiled by Congress; it formalized the U.S. Congress's support for upholding prevailing troop numbers in Germany, unless a reduction was justified by way of a thorough assessment of its consequences (Zengerle, 2020c). For all practical purposes, the issue had now been deferred to the incoming Biden administration (Gordon & Wise, 2020).

These acts by Congress were, fundamentally, motivated neither by self-serving objectives nor by party politics. Rather, they were bipartisan, and they built on rationales that drew heavily on notions about the negative impact that a troop withdrawal *could* have on the United States' position as the world's leading democracy; on allies' perceptions of U.S. power; on U.S. influence in Europe and beyond; on deterrence of Russia and of Russia's—and China's—perceptions of U.S. resolve; on U.S. credibility more broadly; and on the efficacy of the whole network of U.S. military bases—again in Europe and beyond. These rationales were tightly connected to the *system level of analysis*: they rested on assessments of the likelihood of adverse geopolitical chain reactions being set in motion by any troop withdrawal. Lawmakers from both sides of the aisle berated the withdrawal order. Not only was it a "petty and preposterous" move (Holland, 2020); it would also "significantly damage U.S. national security" while bolstering Russian power, self-confidence, and aggression (O'Brien, 2020); it would make champagne flow "freely (…) at the Kremlin" (Borger, 2020); and it was ignorant of the fact that the U.S. military forces in Germany supported America's geostrategic ambitions "beyond Europe and into the Middle East and Africa," where both China and Russia stood ready to fill any vacuum that emerged (Zengerle, 2020b).

The broader U.S. foreign-policy community concurred; their arguments, in turn, fit those of most European-based commentators. The narratives of the *Stayers* rested heavily upon system-level reasoning. References to the benefits that a withdrawal would have for Moscow abounded: for the Russians, one could be sure, "are quite happy" (Emmott & Chalmers, 2020) with this "gift to Putin" (McLeary, 2020). It was a gift entailing a *possible* reduction of the readiness of NATO; this would open up space for Russia to exert its influence abroad even more forcefully, which would aid Russian leaders in their glaringly evident goal of spawning discord within NATO (Rapp-Hooper, 2020). U.S. overall credibility would also suffer, was the near-unanimous verdict, as the pullout would "further erode allies' fate in NATO and U.S. defense guarantees" (Mason & Mohammed, 2020). The withdrawal order was also oblivious of the *interdependence* of this coherent network of bases: Germany, wrote *The Economist* (2020), "is a crucial cog in America's global military machine," adding that Trump's move was truly a "baffling" one that made "little sense and may never" manifest.

And, as it turned out, it didn't. Europe survived the Trump years—quite shaken and quite stirred, but it survived. Or, more precisely, the *status quo* of Europe's defense and security reality survived. NATO was to live on, and the U.S. troops were to remain where they were. In a sense, this should not have come as a surprise. Even with a *Leaver* at the helm of the country, the countercurrent created by the *Stayers* of the Blob is hard to overcome, and it drastically skews the odds in favor of the status quo. In another sense, however, the four years of the *Leaver* in the White House represented a near-perfect opportunity to get the long-promised "pivot" or "rebalancing" to Asia moving for real. For it was Asia, or Asia–Pacific, that, according to the administration of Trump's predecessor in the White House, represented the current and future fulcrum of geopolitics and international trade and investment. Europe was the old geopolitical center of the world. But the United States kept its massive military presence there under both Barack Obama and Donald Trump. For all intents and purposes, the rebalance to Asia would nevertheless continue, as the next section will show; but it did not, as was originally conceived, entail a rebalance *from* any other region.

U.S. Troops in a Rising Asia–Pacific

In Asia and the Pacific, as in Europe, the bulk of forward-deployed U.S. troops are a legacy of invasions, in this case those that took place during World War II and the Korean War. Unlike Europe, however, Asia never witnessed the formation of an overarching, multilateral, NATO-like alliance; what was instead constructed was, as Victor Cha (2009–2010) puts it, a "hub-and-spokes" system encompassing a few bilateral alliances between the United States and key regional states—notably, Japan, South Korea, and the Philippines. In addition, the United States did retain de facto control over most of the islands in the Pacific Ocean that it had acquired by force from Japan during the latter stages of World War II. The hub-and-spokes alliances, and the stepping stones of the Pacific, still form the backbone of U.S. military-force projection in this region.

There is also nothing to suggest that retrenchment from Asia–Pacific is in the cards. If a withdrawal of troops from this region was at all the objective of Donald Trump, it was one that was most certainly not pursued in any wholehearted, unambiguous, and very serious manner. From the beginning to the end of the Trump presidency, Asia–Pacific has seen U.S. active-duty troop numbers *rise* from from 67,626 to 89,173, an increase of 31.9%. Most of the U.S. troops in this region, as per December 31, 2020, are stationed in Japan (56,474, an increase of 43.5%), South Korea (26,184, +8.4%), and Guam (6183, +6.1%, plus 2606 National Guards and reserves). Moreover, in its last year, the Trump administration signaled that it would commence on a process of reviewing U.S. global troop commitments, with a view of increasing its presence in Asia even further, to balance the rise of China (Stewart, 2020). Indications in summer 2020 were that some of the troops that were based in Germany could be relocated to Asia–Pacific (Nagasawa & Miyasaka, 2020). The troop review never really gained any traction, though, and neither did the pullout from Germany. The incoming Biden administration soon announced similar plans to assess the United States' global military footprint (The White House, 2021), while it simultaneously pulled the definite plug on the withdrawal from Germany (Baldor, 2021). In other words, four years of Donald Trump as president had not witnessed a single U.S. soldier leaving Europe (and hardly any leaving the Middle East either). Even so, his administration had markedly *strengthened* U.S. military presence in Asia–Pacific, in large part to bolster the prevailing geopolitical status quo. This, of course, came on the back of similar objectives being pursued by President Trump's predecessor. The Obama administration's 2015 National Security Strategy, for example, pledged to "advance our rebalance to Asia and the Pacific" (The White House, 2015), even as many claimed that the U.S. reorientation toward this region had in fact started much earlier, in a "pivot before the pivot" (Silove, 2016; see also Christensen, 2016, p. 248).

Dominoes Big and Small

This recent troop increase must, therefore, be seen in the context of major-power rivalry and the fear in Washington that a failure to project sufficient power in the region will lead to accelerated Chinese expansion—economically, politically, and even territorially. Recent years have seen growing competition between the incumbent "hegemon" (i.e., the United States), in alliance with several secondary states in Asia–Pacific, and what undoubtedly appears to be the most serious challenger to U.S. regional and global leadership since the height of the Cold War, namely China. The scholarly literature is abundant with analyses and predictions of how this emerging great-power rivalry may play out in the near- and long-term future (e.g., Allison, 2017; Christensen, 2016; Coker, 2015; Medcalf, 2020; Shambaugh, 2021). However, this is surely a contest also taking place in the present. Under the administration of Donald Trump, the United States seemed to abandon much of its earlier part-strategy of engagement with China, instead pursuing a policy of quite intense strategic competition that bordered on containment. Yet, such a policy has ample backing within the corridors in Washington: there is now broad bipartisan agreement in the U.S. Congress that China poses a considerable threat to the United States, and indeed to the world.

The *risk* of dominoes being pushed over by a regional great power whose material resources and ambitions are rapidly growing is, furthermore, supplemented by the specter of adverse chain reactions set in motion by a lesser regional power, namely North Korea. What gave a vivid illustration of this was the dramatic 2017 nuclear crisis. This involved North Korean nuclear and ballistic-missile tests; threats and boisterous rhetoric from leaders in Pyongyang directed at the United States and South Korea (O'Connor, 2017); and warnings from President Trump that Pyongyang would be "met with fire and fury like the world has never seen" if it made any more threats against the United States (Pramuk, 2017a). The crisis also put a spotlight on the sprawling network of U.S. military bases in the region. This network is widely believed in Washington (and in Seoul) to constitute the ultimate deterrent against a North Korean attack on, or invasion of, the South; the standard reference employed is Pyongyang's onslaught on South Korea in 1950, whose permissive cause was the lack of any credible U.S. deterrent. To wit, the status quo may be unpleasant for most regional states, and for the United States; but the quite common argument is that even such "a dangerous situation" as that which prevailed in 2017, and which still prevails, "could get much, much worse absent American engagement" (Brands, 2018).

It was from the United States' regional network of military bases that the U.S. Commander in Chief was allegedly ready to launch his nation's forces into battle if the war-of-words in 2017 eventually turned into a kinetic fight. Key to such a scenario were the tens of thousands of U.S. troops operating from the air and naval facilities in Japan and, of course, South Korea. But the crisis also placed the U.S. Pacific island of Guam—another "permanent aircraft carrier"—at center stage: Any war over Korea would certainly also involve thousands of troops along with fast-attack submarines and a variety of bombers stationed at Anderson Air Force Base and Naval Base Guam (Burrows, 2017; Holmes, 2017). Moreover, nearly 6000 kilometers away from Pyongyang—in Darwin, Australia—over a thousand U.S. Marines reportedly stood "ready to fight" by April 2017 (Westbrook, 2017).

The Geopolitics of U.S. Bases and Troops in Asia–Pacific

While the war never materialized, and while 2018 witnessed a thaw in the relationship between the United States and North Korea, the nuclear crisis undoubtedly put into sharp relief one key part of the mission of U.S. overseas bases and troops in Asia and the Pacific. Still, except for the bases in South Korea, whose main purpose has been to bolster the credibility of U.S. commitments to Seoul in case of a North Korean attack, the primary long-term task of the forward-deployed U.S. forces is surely to act as a hedge against China. The bases in Japan are especially vital in this endeavor. "[E]ssential in almost any East Asian military contingency" (Christensen, 2016, p. 187), this network consists of some 120 different military sites, including the giant Misawa, Yokoto, and Kadena Air Bases (the latter of which is located in the middle of the disputed East China Sea), and the naval base at Yokosuka, home to the U.S. Seventh Fleet (Buckton, 2020; U.S. Department of Defense, 2018).

Concerns about Beijing's power and future intentions are the driving force behind U.S. military-base and troop developments among lesser hosts as well. This is perfectly in line with the analyses and policy recommendations offered in the first half of the twentieth century by two renowned classical geopoliticians, Halford Mackinder (1904, 1943) and Nicholas Spykman (2017 [1942]). This is so even if their works centered on the threat posed by two other Eurasian powers, Germany and (Soviet)

Russia. These, Mackinder and Spykman vehemently argued, had to be contained by the predominantly sea-based powers of the world. The United Kingdom and, later, the United States needed alliances, bridgeheads, and military bases along Mackinder's inner or marginal crescent, which Spykman relabeled the Rimland of Eurasia, whence sufficient pressure could be put on the threatening powers so that their otherwise almost inevitable expansion would come to an early halt. Rimland control, in turn, rested on the creation of a solid oceanic infrastructure, as highlighted even earlier by Admiral Alfred Thayer Mahan (1890); it rested on island nodes: resting places for naval vessels, bases on islands, and points on the map that gave the sea powers control over vital maritime choke points.

China has long since overtaken the role of Germany and Russia as Washington's main bogeyman. China represents, in one important sense, a different challenge: it is less of a continental state and more a state whose *maritime* ambitions seem at the forefront of its geostrategy and expansionist ventures. But U.S. policy should not be expected to deviate greatly from that of previous generations when, according to the contemporary geopolitical analyst Jakub Grygiel (2006, p. 168), the United States had focused on containing the expansion of land powers. It is true, Grygiel (2006, p. 170) goes on, that the United States now needs fewer land forces than it did in earlier great-power contests, for now its objective is to stop China from expanding at sea. This means that the United States will also have to draw on its maritime assets "to maintain its position of supremacy in East Asia, and for this purpose, unlike in the European theater, it can use only a handful of bases." But the point remains the same, as do the challenges: Washington must "implement a strategy similar to that pursued against the Soviet Union," by which Grygiel evidently means "encircling" China by way of a constructing and upholding a network of alliances and military bases in the region.

Worries in Washington are that, without the existence of this network, numerous developments detrimental to regional and U.S. security *could* follow. For example, a withdrawal from South Korea, Victor Cha (2019, p. 174) argues, would almost certainly create a huge "crisis of confidence among U.S. allies and partners, including Japan, Australia, and Taiwan, which would feed self-help inclinations in all capitals." This, furthermore, "would give confidence to China that it has achieved a watershed success in its long-term strategic objective to remove the United States from Asia." A failure to oppose China's ongoing expansion in the South China

Sea, moreover, *could* ultimately hand Beijing control over huge maritime areas, from which it *could* then project its power even further. *Possible* adverse geopolitical chain reactions abound, but the basic issue is really a very classic one: China is a potential regional hegemon—the ultimate "domino pusher," as seen from the status quo-inclined perspective of Washington—, a behemoth following in the footsteps of the Germany and the Soviet Union of bygone eras. China is *the* country that *can conceivably* soon rise to become a true "peer competitor" to the United States. Even offshore balancers, who are normally steadfast *Leavers*, make a very telling exception for Asia–Pacific and the rise of China. This reality and prospect, Stephen Walt (2019, p. 33) writes, likely requires of the United States that it works to "maintain and deepen its current security ties with Australia, Japan, the Philippines, and South Korea and continue to nurture its strategic partnerships with India, Singapore, and Vietnam."

Troops to Balance China

Thus, most *Leavers* become *Stayers* with respect to Asia–Pacific, and balancing China is the preferred policy option. What is presumably perceived as "encirclement" in Beijing, and "deterrence" in Washington, is also pretty close to the strategy that the United States is, and has for some time been, pursuing. Despite Donald Trump's occasional gripes about various base hosts (more about which below), the main ones in Asia–Pacific retained—or even increased—their importance under his tenure.

At the same time, new agreements with lesser states were consolidated or sought by Washington. Australia is a case in point. The decision to ramp up U.S. presence in its Northern Territory, as part of the "Asia pivot," was taken by the Obama administration as early as in 2011 (Shanahan, 2016). What was envisioned, and what has later in large part been realized, was a gradual increase of rotational troops, clearly to "hedge against the growing clout of China" (*The Economist*, 2018). Plans are ongoing for substantial further investment by the United States in military infrastructures, such as airfields, ports, accommodation facilities, and training ranges (Reuters, 2019d). Meanwhile, just north of Darwin, in Papua New Guinea, the United States is apparently working with Australia to develop a naval base on Manus Island, a response to China's expanding diplomatic and military reach in the Pacific (BBC, 2018). The same rationale undergirds efforts by Washington to renew the Compact of Free Association, a deal with three Pacific island states—the Marshall Islands, Palau, and Micronesia—that gives the Pentagon exclusive access to their airspace and maritime territories (Packham & Barrett, 2019). This comes in addition to the considerable military investments long since conducted by the United States on a string of Pacific islands that are formally parts of U.S. territory, such as Wake Island, Johnston Atoll, Northern Mariana Islands, and, of course, Guam (U.S. Department of Defense, 2018). In December 2020, moreover, the U.S. Congress authorized a $2.2 billion funding of the Pacific Deterrence Initiative, with the yearly amount soon expected to double (*The Economist*, 2021). Further to the west, we find the tiny city-state of Singapore. Its enduring strategic location is almost unprecedented, abutting one of the world's most important maritime bottlenecks. Although the huge, purpose-built Changi Naval Base stood ready to host U.S. aircraft carriers and submarines over two decades ago (Kaplan, 2014, p. 96), Singapore has recently gained even added importance to Washington as a staging hub for a range of military vessels and spy planes used to check China's growing control over the vast South China Sea (Bowie, 2017).

The Philippines is also central to the U.S. projection of military power in the South China Sea. In early 2019, it was announced that new military facilities for hosting visiting U.S. troops were completed on the main island of Luzon, while others would soon be built on the island of Palawan (Robson, 2019). The plans were part of an Obama-era deal that would allow for a U.S. military presence, and U.S. operational control of, five bases adjacent to the South China Sea (Tilghman, 2016). The military-base issue is a thorny one in the Philippines, though, and for decades there has been much back and forth with respect to the forward deployment of U.S. troops. A relative rarity for a base host, the Philippines formally told Washington to vacate its military facilities in the early 1990s, only to reverse that decision in November 2015 (French, 2017, pp. 75–76, 178). Then, in October 2016, the newly installed President of the Philippines, Rodrigo Duterte, expressed his desire to revoke this deal with the Americans and for all U.S. troops to leave the country within a couple of years (Rich, 2016). This happened even as claims were that the president had really no intention to follow through on this considering in particular the importance of the U.S. troops in Manila's fight against *jihadist* groups (Tan & Ward, 2017). The issue then lay more or less dormant for a few years until Duterte, in January 2020, amid an unrelated diplomatic spat with Washington, even more forcefully stated that the U.S. bases in the country would be closed (Reuters, 2020b). Although President Trump stated that he was "fine" with the decision as it would allow the United States to "save a lot of money" (Olson, 2020), it seemingly failed to please U.S. Defense Secretary Mark Esper, who lamented that the decision was particularly unfortunate in a situation where the United States and its partners were attempting to force China to follow "international rules of order" in the region (Lema et al., 2020).

Withdrawal Threats and Bargaining Over Cost-Sharing

It rarely happens that military-base agreements between the United States and host states are abrogated. And, after much back and forth in the last year of the Trump presidency and the first months of the Biden administration, it looked like the U.S.-Philippines defense pact would survive (Lema, 2020). On the other hand, it is not uncommon to witness opposition to these deals among segments of the host-country population—although it is far more questionable whether, in most states, public opinion *as a whole* would prefer the American military to leave (Allen et al., 2020). Protests against U.S. forces, and against military-base expansion, have been particularly rife in Okinawa, Japan, as evidenced by several (non-binding) local referendum results (Ono & Kelly, 2019; Sieg, 2019). Still, the government in Tokyo is quite unperturbed by the localized base opposition. Narratives of deterrence, and of the absolute indispensability of the U.S. military bases and troops to Japan's security, continue to override all other concerns (O'Shea, 2019).

More problematic for the Japanese government was the election as U.S. President of Donald Trump, a man who had for decades complained that "[t]he Japanese cajole us, they bow to us, they tell us how great we are and then they pick our pockets" (Plaskin, 1990). Similar charges were also brought by President Trump to the U.S.-Japanese negotiations over military-base costs. Trump was reportedly seeking a quadrupling of Japan's contributions; and he also *reportedly* threatened that a failure of Tokyo to acquiesce could lead to an abrogation of the U.S.-Japan defense treaty and the withdrawal of U.S. troops from Japan (Jacobs, 2019; Johnson, 2020). Nothing came of this, however; and a few weeks after Trump left office, a deal on cost-sharing was reached between the Biden administration and Japan (Johnson & Sugiyama, 2021). U.S. troop numbers in Japan were now, as we know, considerably higher than they were when Donald Trump had assumed the presidency.

Trump had clearly *indicated* that a troop withdrawal was indeed a real prospect. But in the specific case of Japan, there is little to suggest that this threat was particularly credible, at least not within the confines of a single presidential term. Things were a bit different with respect to South Korea, which was also forced to endure what one analyst calls the "amoral transactionalism" of a U.S. president whose basic inclination was "to play off linkages between the economic sphere and security commitments to obtain short-term gains" (Dian, 2018). In November 2019, it was reported that President Trump had demanded that South Korea take on a much greater share of the costs, insisting that the U.S. presence was "$5 billion worth of protection"—which would amount to a five-fold increase in expenses for South Korea (Shin, 2019). This was seen in Seoul as an egregious demand, and the huge gap between the two sides' viewpoints soon led to the United States breaking off negotiations (Kim, 2019). Although talks soon restarted, the deal was only reached when Trump's successor had firmly entered the White House (Shin & Stewart, 2021).

The base-cost negotiation crisis may first and foremost have revolved around pecuniary issues. At the same time, the bargaining also involved what was regarded as implicit threats by Donald Trump that U.S. troops could be withdrawn depending on how forthcoming Seoul ultimately proved to be. In November 2019, news reports claimed that the United States was ready to pull back an Army brigade—around 4000 troops—if their demands were not met (Al Jazeera, 2019). While these claims were refuted by Washington, President Trump soon added to the uncertainty by ambiguously stating that it was debatable whether having troops in South Korea served U.S. national interests: "I can go either way," he said somewhat cryptically (Reuters, 2019f). Furthermore, a July 2020 article in *The Wall Street Journal* claimed that the Pentagon had presented Trump with a plan to cut the U.S. presence in South Korea, as preparations for the eventuality that negotiations proved unsuccessful (Gordon & Lubold, 2020b). Such implied threats had been issued previously as well. In March 2018, angry about the U.S. trade deficit with Seoul, Trump had allegedly linked this issue with the matter of bases in Korea (Croucher, 2018). Earlier that year, *NBC News* reported that, according to several former and current U.S. officials, the president was ready to order the withdrawal of troops from South Korea. He only decided not to go through with this after having been dissuaded by then-White House chief of staff John Kelly, in what was described as a "heated exchange" (Lee et al., 2018).

Whether President Trump ever really intended to execute a withdrawal, we will probably never know for certain. What we do know, though, is that there existed—as there still exist—a plenitude of barriers against any pullout. One such barrier stems from legislation enacted by the U.S. Congress, which explicitly forbids any significant reduction in the number of U.S. soldiers in South Korea unless the secretary of defense certifies that a withdrawal is in the national-security interest of the United States (Zengerle, 2019b). It is not at all straightforward to make such a certification when one considers the existence of a nuclear-armed, authoritarian North Korea—and, not least, a rapidly rising China. Yet another constraint, common to most cases of forward-deployed troops, is the sheer "stickiness" of bases, or their "legacy costs"—including the geopolitical costs that might well occur sometime down the line—, which to a considerable degree increases the attractiveness of the status quo (Kelly, 2019).

The Leavers Are Stayers

One specific U.S. base in South Korea exemplifies the contrast between, on the one hand, the relatively frequent hints that the United States was about to withdraw, and, on the other hand, the sense that life at base, so to speak, continues unperturbed. Camp Humphreys, which was already among the largest—if not already *the* largest (Persio, 2017)—overseas U.S. bases when Trump entered office has now definitely assumed this title; a significant expansion of these facilities was conducted even as ties between Washington and Seoul were deeply strained (Cloud, 2019). More broadly, the threats to withdraw from Asia–Pacific hosts, from Japan and from South Korea, never materialized, and they might well have been inherently incredible threats from the start. The *acts* one observes—not only in South Korea, but also in Japan, in Australia, in Guam, in the Pacific Ocean, and elsewhere in the region—is generally one of expansion, of further investments in military facilities, of new security deals being reached with regional states, of long-term plans of a continuation or even a bolstering of U.S. military presence, and of quite significant U.S. troop *increases*.

The fact of the matter is that, with respect to U.S. military-force projection in Asia–Pacific, sundry forces pull in the same direction, which is the direction that the *Stayers* prefer and promote. Indeed, as suggested earlier, most *Leavers* aren't really *Leavers* at all when it comes to this part of the world. The United States constructed an elaborate network of bilateral alliances and military-base agreements during, and in the years following, the Second World War and the Korean War. This network represents the status quo to which the vast majority of the U.S. foreign-policy community are deeply committed. Most host countries are similarly, in fact often even deeper, committed to the prevailing state of affairs; their biggest fear is that the United States would one day abandon them (Cha, 2000). This goes a long way to explain why a major power such as Japan continues to act as Washington's servile "client state" intent on "performing the tasks expected of a subordinate state, steering the country into a new and deeper phase of Clientilism" (McCormack, 2018, p. 5).

The U.S. base network used to be directed against other great-power adversaries. Times have certainly changed, but a new rival—China—has risen in quite dramatic fashion, and the base network continues to be of immense perceived value to the United States. First and foremost, it works to suppress numerous *possible* adverse geopolitical chain reactions, substituting the known (if uneasy) status quo for the *potentially* catastrophic unknown. Alternatives to the existing order are almost inevitably judged to represent something worse. An August 2020 editorial in Bloomberg (2020) is representative in that it implicitly reasons that a withdrawal of U.S. troops from South Korea *might* lead to Seoul choosing to bandwagon with China; it *might* spawn serious conflict between Japan and South Korea and it *might* incite South Korea into developing a nuclear bomb. More broadly and weightier, U.S. retrenchment from Asia–Pacific *might* pave the way for Chinese regional hegemony, which in turn *might* lead China to project its power globally with far less friction, which *might* spell the end of U.S. global predominance—and of *Pax Americana*. To hinder these fundamentally hypothetical mights and ifs ever to manifest was, is, and will continue to be the main purpose and rationale for the U.S. military bases and forward-deployed troops—as even otherwise steadfast *Leavers* would acknowledge.

References

Al Ansari, S. (2017, October 9). The case for moving Al-Udeid airbase from Qatar. *The Hill*. https://thehill.com/opinion/international/354555-the-case-for-moving-al-udeid-airbase-from-qatar

Ali, I. (2018, November 15). Pentagon to cut troops in Africa as focus shifts to China, Russia. *Reuters*. https://www.reuters.com/article/us-usa-africa-military/pentagon-to-cut-troops-in-africa-as-focus-shifts-to-china-russia-idUSKCN1NK2N5

Ali, I. (2019a, October 20). All U.S. troops withdrawing from Syria expected to go to western Iraq: Pentagon chief. *Reuters*. https://www.reuters.com/article/us-syria-security-usa/all-u-s-troops-withdrawing-from-syria-expected-to-go-to-western-iraq-pentagon-chief-idUSKBN1WZ01S

Ali, I. (2019b, December 17). Pentagon chief: Need to speak with Turkish counterpart to understand Erdogan base comments. *Reuters*. https://www.reuters.com/article/us-usa-turkey-pentagon/pentagon-chief-need-to-speak-with-turkish-counterpart-to-understand-erdogan-base-comments-idUSKBN1YL02W

Ali, I., & Landay, J. (2018, December 21). Planned drawdown in Afghanistan imperils U.S. push for peace. *Reuters*. https://www.reuters.com/article/us-usa-afghanistan-military-analysis/planned-drawdown-in-afghanistan-imperils-u-s-push-for-peace-idUSKCN1OK2EQ

Ali, I., & Stewart, P. (2018, December 20). U.S. plans for more than 5,000 troops to be withdrawn from Afghanistan. *Reuters*. https://www.reuters.com/article/us-usa-afghanistan-trump/u-s-plans-for-more-than-5000-troops-to-be-withdrawn-from-afghanistan-idUSKCN1OJ2VN

Ali, I., & Stewart, P. (2020a, April 22). Trump tells Navy to destroy Iranian gunboats if they 'harass' American ships. *Reuters*. https://www.reuters.com/article/us-usa-iran-military/trump-tells-navy-to-destroy-iranian-gunboats-if-they-harass-american-ships-idUSKCN2241UK

Ali, I., & Stewart, P. (2020b, October 8). Trump unchained? Afghan troop surprise shows pre-election impulse to upend policy. *Reuters*. https://www.reuters.com/article/uk-usa-afghanistan-trump-military-idINKBN26T3DG

Al Jazeera. (2004, May 20). *The Greater Middle East initiative*. https://www.aljazeera.com/news/2004/5/20/the-greater-middle-east-initiative

Al Jazeera. (2019, November 21). *US denies claims it is pulling up to 4,000 troops from S Korea*. https://www.aljazeera.com/news/2019/11/considers-pulling-4000-troops-south-korea-report-191121033711289.html

Al Jazeera. (2020, May 27). *Trump says he wants full Afghanistan pullout but sets no timeline*. https://www.aljazeera.com/news/2020/5/27/trump-says-he-wants-full-afghanistan-pullout-but-sets-no-timeline

Al Jazeera (2021, January 15). *US downsizes its troops in Afghanistan to 2,500*. https://www.aljazeera.com/news/2021/1/15/us-meets-its-goal-of-downsizing-in-afghanistan-to-2500-troops

Allen, M. A., Flynn, M. E., Martinez Machain, C., & Stravers, A. (2020): Outside the wire: U.S. military deployments and public opinion in host states. *American Political Science Review*, *114*(2), 326–341.

Allison, G. (2017). *Destined for war: Can America and China escape Thucydides's trap?* Houghton Mifflin Harcourt.

Ames, P. (2016, September 29). China's Atlantic stopover worries Washington. *Politico*. https://www.politico.eu/article/chinas-atlantic-stopover-terceira-worries-washington-li-keqiang-united-states/

Anna, C. (2020, November 26). 'Why now?' Dismay as US considers troop pullout from Somalia. *Associated Press*. https://apnews.com/article/islamic-state-group-elections-africa-somalia-kenya-6fad3fe2b14858274daf34a29a78dbe7

Anna, C. (2021, January 17). US military says its troop removal from Somalia is complete. *Associated Press*. https://apnews.com/article/donald-trump-troop-withdrawals-national-elections-elections-africa-a5c53d5c7cb0fa3473e7df98ee66ece1

Associated Press. (2018, September 24). *National Security adviser Bolton: US troops not leaving Syria 'as long as Iranian troops are outside Iranian borders.'* https://apnews.com/article/95b9dbdd449a4513b9abe858bd50c53a

Baker, P., & Jakes, L. (2019, October 7). Trump throws Middle East policy into turmoil over Syria. *New York Times*. https://www.nytimes.com/2019/10/07/us/politics/turkey-syria-trump.html

Baldor, L. C. (2021, February 5). Biden halts Trump-ordered US troops cuts in Germany. *Associated Press*. https://apnews.com/article/joe-biden-donald-trump-military-facilities-europe-lloyd-austin-ff57f288a1bb3e5a38e3253ea0b94d80

Barkey, H. J. (2019). The Kurdish awakening: Unity, betrayal, and the future of the Middle East. *Foreign Affairs*, *98*(2), 107–118.

Barnes, J. E. (2019, December 30). U.S. Launches Airstrikes on Iranian-Backed Forces in Iraq and Syria. *New York Times*. https://www.nytimes.com/2019/12/29/world/middleeast/us-airstrikes-iran-iraq-syria.html

Barnes, J. E., & Jervell, E. E. (2017, January 16). Amid Trump's NATO blasts, U.S. Marines' Norway mission goes on. *Wall Street Journal*. https://www.wsj.com/articles/amid-trumps-nato-blasts-u-s-marines-norway-mission-goes-on-1484591359

BBC. (2017a, August 22). *Trump rules out Afghan troops withdrawal*. https://www.bbc.com/news/world-us-canada-41008116

BBC. (2017b, October 5). *US soldiers killed in ambush in Niger*. https://www.bbc.com/news/world-africa-41507337

BBC. (2018, November 17). *US to join Australia in Papua New Guinea naval base plan*. https://www.bbc.com/news/world-asia-46247446
BBC. (2019, September 3). *Afghanistan war: US-Taliban deal would see 5,400 troops withdraw*. https://www.bbc.com/news/world-asia-49559493
BBC. (2020, March 9). *Afghanistan conflict: US begins withdrawing troops*. https://www.bbc.com/news/world-asia-51810498
Birnbaum, M., & Rucker, P. (2018, 12 July). Trump says NATO nations make major new defense spending commitments after he upends summit. *Washington Post*. https://www.washingtonpost.com/world/europe/trump-upends-nato-summit-demandingimmediate-spending-increases-or-he-willdo-his-own-thing/2018/07/12/a3818cc6-7f0a-11e8-a63f-7b5d2aba7ac5_story.html?utm_term=.d83f5a4e60bf
Black, J. (2019, April 1). *A ruling from the Hague threatens a U.S. and British military base in Asia*. RAND Corporation. Retrieved 29 August, 2021, from https://www.rand.org/blog/2019/04/a-ruling-from-the-hague-threatens-a-us-and-british.html
Bloomberg. (2020, August 7). *Don't use U.S. troops as bargaining chips*. https://www.bloomberg.com/opinion/articles/2020-08-07/u-s-shouldn-t-withdraw-troops-from-south-korea-to-suit-trump
Borger, J. (2020, July 29). US to pull 12,000 troops out of Germany as Trump blasts 'delinquent' Berlin. *The Guardian*. https://www.theguardian.com/us-news/2020/jul/29/us-germany-troop-withdrawal-donald-trump
Bowie, N. (2017, October 25). Lee-Trump visit underscores Singapore's balancing act. *Asia Times*. https://asiatimes.com/2017/10/lee-trump-visit-underscores-singapores-balancing-act/
Brands, H. (2018, June 3). How many US troops are needed in Korea? *Korea Herald*. Retrieved January 11, 2021, from http://www.koreaherald.com/view.php?ud=20180603000063&ACE_SEARCH=1
Brennan, D. (2020, December 22). Iran presses U.S. to exit Iraq and Afghanistan, maneuvers to influence both. *Newsweek*. https://www.newsweek.com/iran-presses-us-exit-iraq-afghanistan-maneuvers-influence-both-1556651
Browne, R. (2020a, February 2). Top US general in Africa warns of growing threats as Trump administration weighs cutting troop numbers. *CNN*. https://edition.cnn.com/2020/02/02/politics/africa-threats-trump-troops/index.html
Browne, R. (2020b, August 7). US to end deployment of US Marines in Norway after boosting it in 2018 amid Russian tensions. *CNN*. https://edition.cnn.com/2020/08/07/politics/us-marines-norway/index.html
Brumberg, D. (2021, July 15). *US Middle East influence in Afghanistan's shadow*. Arab Center. Retrieved September 3, 2021, from https://arabcenterdc.org/resource/us-middle-east-influence-in-afghanistans-shadow/

Buckton, M. (2020, November 10). Japan ready to increase costs for US military presence. *Taiwan Times*. https://thetaiwantimes.com/japan-ready-to-increase-costs-for-us-military-presence/

Burns, R. (2020, November 19). US general says IS in Iraq and Syria still long-term threat. *Associated Press*. https://apnews.com/article/politics-islamic-state-group-syria-iraq-c15ed37959f39e7d40f7c613fc8cb5e8

Burrows, I. (2017, August 10). Guam: America's military base in the western Pacific. *ABC News*. https://www.abc.net.au/news/2017-08-09/where-is-guam-and-how-many-us-troops-are-there/8788566

Cha, V. D. (2000). Abandonment, entrapment, and neoclassical realism in Asia: The United States, Japan, and Korea. *International Studies Quarterly, 44*(2), 261–291.

Cha, V. D. (2009–2010). Powerplay: Origins of the U.S. alliance system in Asia. *International Security*, *34*(3), 158–196.

Cha, V. D. (2019). The unintended consequences of success: U.S. retrenchment from Korea? *Korean Journal of Defense Analysis*, *31*(2), 165–191.

Christensen, T. J. (2016). *The China challenge: Shaping the choices of a rising power*. W. W. Norton.

Cloud, D. S. (2019, March 22). South Korea pays for vast U.S. Army base expansion, but Trump wants it to pay more. *Los Angeles Times*. https://www.latimes.com/politics/la-na-pol-korea-trump-allies-payments-20190322-story.html

Cloud, D. S. (2021, March 12). Inside U.S. troops' stronghold in Syria, a question of how long Biden will keep them there. *Los Angeles Times*. https://www.latimes.com/politics/story/2021-03-12/us-troops-syria-civil-war-biden

Cloud, D. S., & Wilkinson, T. (2020, July 2). White House tentatively agrees to leave some troops in Afghanistan past U.S. election. *Los Angeles Times*. https://www.latimes.com/politics/story/2020-07-02/afghanistan-trump-withdrawal-on-track

Coffey, L. (2019, April 5). *Oman: America's unique Arab ally in the Middle East*. Heritage Foundation, Issue Brief, p. 4, Retrieved August 17, 2021, from https://www.heritage.org/sites/default/files/2019-04/IB4950.pdf

Cohen, H. J. (2020, May 13). Pulling troops out of Africa could mean another endless war. *War on the Rocks*. https://warontherocks.com/2020/05/pulling-troops-out-of-africa-could-mean-another-endless-war/

Coker, C. (2015). *The improbable war: China, the United States and the continuing logic of great power conflict*. Oxford University Press.

Connable, B., Dobbins, J., Shatz, H. J., Cohen, R. S., & Wasser, B. (2020, May). *Weighing U.S. troop withdrawal from Iraq: Strategic risks and recommendations*. RAND Corporation, Retrieved September 1, 2021, from https://www.rand.org/content/dam/rand/pubs/perspectives/PE300/PE362/RAND_PE362.pdf

Cookson, J. R. (2020, November 15). Fewer US troops are needed in Europe no matter the president. *The Hill*. https://thehill.com/opinion/international/526051-fewer-us-troops-are-needed-in-europe-no-matter-the-president

Cooper, H. (2020, January 14). Plan to cut U.S. troops in West Africa draws criticism from Europe. *New York Times*. https://www.nytimes.com/2020/01/14/world/africa/milley-troops-withdraw.html

Cooper, H., Gibbons-Neff, T., Savage, C., & Schmitt, E. (2019, December 24). Pentagon eyes Africa drawdown as first step in global troop shift. *New York Times*. https://www.nytimes.com/2019/12/24/world/africa/esper-troops-africa-china.html

Cornish, C., & al-Omar, A. (2020, June 9). Isis operations increase in Iraq as coalition withdraws. *Financial Times*. https://www.ft.com/content/abbdcd29-fe66-4be2-b35e-efcfca536ce1

Cotovio, V., & Pleitgen, F. (2018, November 1). NATO back on the hunt for Russian submarines in the Arctic. *CNN*. https://lite.cnn.com/en/article/h_811a3cb43409832a2d26c1f45173b8ff

Croucher, S. (2018, March 15). Did Trump just threaten to withdraw U.S. troops from South Korea? *Newsweek*. https://www.newsweek.com/did-trump-just-threaten-withdraw-us-troops-south-korea-845943

Cunningham, E. (2020, April 23). Top Iranian general warns U.S. against 'dangerous behavior' in Persian Gulf after Trump threat. *Washington Post*. https://www.washingtonpost.com/world/top-iranian-general-warns-us-against-dangerous-behavior-in-persian-gulf-after-trump-threat/2020/04/23/659b44c0-8551-11ea-81a3-9690c9881111_story.html

Dahir, A. L., & Schmitt, E. (2020, November 18). Somalia worries that a U.S. withdrawal will be disastrous. *New York Times*. https://www.nytimes.com/2020/11/18/world/africa/somalia-us-troops.html

Davison, J. (2020, September 28). Threat to evacuate U.S. diplomats from Iraq raises fear of war. *Reuters*. https://www.reuters.com/article/usa-iraq-iran-int-idUSKBN26J1Z4

deGrandpre, A., & Snow, S. (2017, May 14). In Turkey, new demands to evict US forces from Incirlik Air Base. *Military Times*. https://www.militarytimes.com/news/pentagon-congress/2017/05/14/in-turkey-new-demands-to-evict-us-forces-from-incirlik-air-base/

Detsch, J. (2020, March 19). Congress pushes for Syria funding boost. *Al-Monitor*. https://www.google.com/search?q=al+monitor&rlz=1C1CHBF_noNO757NO757&oq=al+monitor&aqs=chrome..69i57j0i512l5j69i60l2.2017j0j7&sourceid=chrome&ie=UTF-8

DeYoung, K. (2018, December 21). With Syria decision, Trump's foreign policy doctrine emerges. *Washington Post*. https://www.washingtonpost.com/world/national-security/with-syria-decision-trumps-foreign-policy-doctrine-emerges/2018/12/20/ad055810-0487-11e9-9122-82e98f91ee6f_story.html

DeYoung, K. (2020, March 1). Peace deal with Taliban sets timetable for U.S. exit but no long-term solutions for Afghans. *Washington Post*. https://www.washingtonpost.com/national-security/afghanistan-peace-deal-taliban/2020/02/29/c2d2ae14-5af2-11ea-9000-f3cffee23036_story.html

DeYoung, K., Lamothe, D., Ryan, M., & Birnbaum, M. (2019, October 26). Trump decided to leave troops in Syria after conversations about oil, officials say. *Washington Post*. https://www.washingtonpost.com/world/us-defense-secretary-mark-esper-says-us-will-leave-forces-in-syria-to-defend-oil-fields-from-islamic-state/2019/10/25/fd131f1a-f723-11e9-829d-87b12c2f85dd_story.html

Dian, M. (2018). Trump's mixed signals toward North Korea and US-led alliances in East Asia. *International Spectator, 53*(4), 112–128.

Dobbins, J., Campbell, J. H., Mann, S., & Miller, L. E. (2019, January). *Consequences of a precipitous U.S. withdrawal from Afghanistan*. RAND Corporation. Retrieved September 20, 2021, from https://www.rand.org/pubs/perspectives/PE326.html

Ekathimerini. (2018, August 21). *AHI hails US investment at Souda Bay naval base*. http://www.ekathimerini.com/231904/article/ekathimerini/news/ahi-hails-us-investment-at-souda-bay-naval-base

Emmott, R., & Chalmers, J. (2020, June 8). Trump troop pullout would still leave hefty U.S. footprint in Europe. *Reuters*. https://www.reuters.com/article/us-usa-germany-military-analysis-idUSKBN23F29P

Euronews. (2017, October 17). Romania takes key role as NATO prepares to face off with Russia. https://www.euronews.com/2017/10/17/romania-takes-key-role-as-nato-prepares-to-face-off-with-russia

Fabian, B., Gunzinger, M., van Tol, J., Cohn, J., & Evans, G. (2019). *Strengthening the defense of NATO's eastern frontier*. Center for Strategic and Budgetary Assessments. Retrieved February 3, 2020, from https://csbaonline.org/uploads/documents/Stengthening_the_Defense_of_NATOs_Eastern_Frontier_WEB_1.pdf

Falk, T. O. (2020, November, 22). What the US troop withdrawal means for Iraq. *Al Jazeera*. https://www.aljazeera.com/news/2020/11/22/what-the-us-troop-withdrawal-means-for-iraq

France24. (2019, October 9). *Trump: US going into Middle East was 'worst decision ever.'* https://www.france24.com/en/20191009-trump-us-going-into-middle-east-was-worst-decision-ever

French, H. W. (2017). *Everything under the heavens: How the past helps shape China's push for global power*. Vintage.

Gaffey, C. (2017, April 19). Why is Trump sending more U.S. troops to Somalia? *Newsweek*. https://www.newsweek.com/us-troops-somalia-donald-trump-al-shabab-586004

Gaouette, N., & Browne, R. (2017, July 17). Trump changes tune, flexes US muscle overseas. *CNN*. https://edition.cnn.com/2017/07/17/politics/trump-world-policeman/index.html

Garamone, J. (2017, August 21). *President unveils new Afghanistan, South Asia strategy.* U.S. Department of Defense, DOD News. Retrieved March 12, 2020, from https://www.defense.gov/News/News-Stories/Article/Article/1284964/president-unveils-new-afghanistan-south-asia-strategy/

Garamone, J. (2021a, January 15). *U.S. completes troop-level drawdown in Afghanistan, Iraq*. U.S. Department of Defense, DOD News. Retrieved September 15, 2021, from https://www.defense.gov/Explore/News/Article/Article/2473884/us-completes-troop-level-drawdown-in-afghanistan-iraq/

Garamone, J. (2021b, February 18). *NATO defense leaders agree to increase Iraqi mission, defer decision on Afghanistan*. U.S. Department of Defense, DoD News. Retrieved October 1, 2021, from https://www.defense.gov/News/News-Stories/Article/Article/2507564/nato-defense-leaders-agree-to-increase-iraqi-mission-defers-decision-on-afghani/

Gause, F. G., III. (2019). Should we stay or should we go? The United States and the Middle East. *Survival, 61*(5), 7–24.

Gearan, A, & Sonne, P. (2019, June 13). Trump elevates Poland at the expense of Germany with troop decision. *Washington Post*. https://www.washingtonpost.com/politics/trump-elevates-poland-at-the-expense-of-germany-with-troop-decision/2019/06/12/335cf8a0-8d36-11e9-adf3-f70f78c156e8_story.html

George, S., & Tassal, A. (2020, October 10). With U.S. troops gone, Taliban expands influence in one Afghan province. *Washington Post*. https://www.washingtonpost.com/world/2020/10/10/afghanistan-us-troop-withdrawal/

Gibbons-Neff, T. (2020, November 12). U.S. troops are still leaving, but Afghans hope Biden will help. *New York Times*. https://www.nytimes.com/2020/11/12/world/asia/biden-afghanistan-withdrawal.html

Gibbons-Neff, T., Cooper, H., & Schmitt, E. (2021, March 14). U.S. has 1,000 more troops in Afghanistan than it disclosed. *New York Times*. https://www.nytimes.com/2021/03/14/world/asia/us-troops-afghanistan.html

Gibbons-Neff, T., & Mashal, M. (2019, October 21). U.S. is quietly reducing its troop force in Afghanistan. *New York Times*. https://www.nytimes.com/2019/10/21/world/asia/afghanistan-troop-reduction.html

Golby, J. (2020, November 30). Trump makes a bad situation worse in Afghanistan. *The Atlantic*. https://www.theatlantic.com/ideas/archive/2020/11/trump-leaving-biden-mess-afghanistan/617229/

Gordon, M. R., & Lubold, G. (2020a, June 5). Trump to pull thousands of U.S. troops from Germany. *Wall Street Journal*. Retrieved from https://www.wsj.com/articles/trump-directs-u-s-troop-reduction-in-germany-11591375651

Gordon, M. R., & Lubold, G. (2020b, July 17). Trump administration weighs troop cut in South Korea. *Wall Street Journal*. Retrieved from https://www.wsj.com/articles/trump-administration-weighs-troop-cut-in-south-korea-11595005050

Gordon, M. R., & Wise, L. (2020, December 3). Congress moves to block Trump's troop cuts in Germany. *Wall Street Journal*. https://www.wsj.com/articles/congress-moves-to-block-trumps-troop-cuts-in-germany-11607052172

Gordon, M. R., & Youssef, N. A. (2020, July 29). Pentagon to move nearly 12,000 U.S. troops from Germany. *Wall Street Journal*. https://www.wsj.com/articles/pentagon-to-detail-plans-for-u-s-troop-cuts-in-germany-11596026413

Gould, J. (2020, June 30). Congress moves to block Trump's Germany troop withdrawal plans. *Defense News*. https://www.defensenews.com/congress/2020/06/30/congress-moves-to-block-trumps-germany-troop-withdrawal-plans/

Gould, J., & Insinna, V. (2018, June 25). Poking the bear: US Air Force builds in Russia's backyard. *Defense News*. https://www.defensenews.com/smr/nato-priorities/2018/06/25/poking-the-bear-us-air-force-builds-in-russias-backyard/

Goure, D. (2016, March). *Souda Bay: NATO's military gem in the Eastern Mediterranean*. Lexington Institute. Retrieved September 14, 2021, from https://www.lexingtoninstitute.org/wp-content/uploads/2016/03/Souda-Bay-NATOs-Military-Gem.pdf

Gross, J. A (2017, September 18). In first, US establishes permanent military base in Israel. *Times of Israel*. https://www.timesofisrael.com/in-first-us-establishes-permanent-military-base-in-israel/

Grygiel, J. J. (2006). *Great powers and geopolitical change*. Johns Hopkins University Press.

Haley, N. (2019, October 7). *Twitter message*. Retrieved October 11, 2021, from https://twitter.com/nikkihaley/status/1181191973367160834

Haltiwanger, J. (2017, November 21). Trump's secret war? U.S. military's presence in Middle East has grown 33 percent in past four months. *Newsweek*. https://www.newsweek.com/trumps-secret-war-us-militarys-presence-middle-east-has-grown-33-percent-past-718089

Holland, S. (2020, June 5). Trump orders big U.S. troop cut in Germany, official says. *Reuters*. https://www.reuters.com/article/us-usa-germany-military-idUSKBN23C2PZ

Holmes, O. (2017, August 9). What is the US military's presence near North Korea? *The Guardian*. https://www.theguardian.com/us-news/2017/aug/09/what-is-the-us-militarys-presence-in-south-east-asia

Hussein, M., & Haddad, M. (2021, September 10). Infographic: US military presence around the world. *Al Jazeera*. https://www.aljazeera.com/news/2021/9/10/infographic-us-military-presence-around-the-world-interactive

Ioannou, L. (2017, October 17). Trump's message to Tsipras: US may boost investments in Greece, a key NATO ally. *CNBC*. https://www.cnbc.com/2017/10/17/trumps-message-to-tsipras-us-may-up-investment-in-nato-ally-greece.html

Jacobs, J. (2019, June 24). Trump muses privately about ending postwar Japan defense pact. *Bloomberg*. https://www.bloomberg.com/news/articles/2019-06-25/trump-muses-privately-about-ending-postwar-japan-defense-pact

Johnson, J. (2020, June 22). Trump demanded Japan cough up $8 billion for U.S. troops—Or risk pullout, Bolton says. *Japan Times*. https://www.japantimes.co.jp/news/2020/06/22/national/us-donald-trump-japan-troops-john-bolton/

Johnson, J., & Sugiyama, S. (2021, February 17). Japan and U.S. reach agreement on cost of hosting American troops. *Japan Times*. https://www.japantimes.co.jp/news/2021/02/17/national/japan-us-troop-cost/

Jones, S. (2018, March 30). Trump: 'We'll be coming out of Syria, like, very soon.' *CNS News*. https://www.cnsnews.com/news/article/susan-jones/trump-well-be-coming-out-syria-very-soon

Kafura, C. (2019, September 18). *Americans divided over US military forces in Germany*. Chicago Council on Global Affairs. Retrieved September 1, 2021, from https://www.thechicagocouncil.org/research/public-opinion-survey/americans-divided-over-us-military-forces-germany

Kaplan, R. D. (2012). *The revenge of geography: What the map tells us about coming conflicts and the battle against fate*. Random House.

Kaplan, R. D. (2014). *Asia's cauldron: The South China Sea and the end of a stable Pacific*. Random House.

Karam, J. (2017, July 14). Trump says US has alternatives to Qatar's Al Udeid airbase. *The National*. https://www.thenational.ae/world/mena/trump-says-us-has-alternatives-to-qatar-s-al-udeid-airbase-1.608878

Karlin, M., & Dalton, M. (2017, August 4). *It's long past time to rethink U.S. military posture in the Gulf*. Brookings. Retrieved April 25, 2020, from https://www.brookings.edu/blog/markaz/2017/08/04/its-long-past-time-to-rethink-u-s-military-posture-in-the-gulf/

Kates, G. (2017, September 20). Defense Dept. decision gives China chance for foothold between U.S. and Europe. *CBS News*. https://www.cbsnews.com/news/azores-withdrawal-gives-china-foothold-between-us-europe/

Kelly, R. E. (2019, June 10). Adapting to a nuclear North Korea is better than swapping away U.S. regional assets. *National Interest*. https://nationalinterest.org/blog/korea-watch/adapting-nuclear-north-korea-better-swapping-away-us-regional-assets-62082

Khan, I. (2020, September 26). Opinion: Imran Khan: Peace is within reach in Afghanistan. A hasty international withdrawal would be unwise. *Washington Post*. https://www.washingtonpost.com/opinions/2020/09/26/imran-khan-peace-is-within-reach-afghanistan-hasty-international-withdrawal-would-be-unwise/

Kheel, R. (2020a, July 1). House panel votes to constrain Afghan drawdown, ask for assessment on 'incentives' to attack US troops. *The Hill*. https://thehill.com/policy/defense/505568-house-panel-votes-to-constrain-afghan-drawdown-ask-for-assessment-on

Kheel, R. (2020b, August 6). Poll: About three quarters support bringing troops home from Iraq, Afghanistan. *The Hill*. https://thehill.com/policy/defense/510851-poll-about-three-quarters-support-bringing-troops-home-from-iraq-afghanistan

Kim, M. J. (2019, November 19). U.S. breaks off talks with South Korea over costs of military alliance. *Washington Post*. https://www.washingtonpost.com/world/asia-pacific/us-breaks-off-talks-with-south-korea-over-costs-of-military-alliance/2019/11/19/949773d8-0a9b-11ea-8054-289aef6e38a3_story.html

Knight, B. (2020, June 16). US military in Germany: What you need to know. *Deutsche Welle*. https://www.dw.com/en/us-military-in-germany-what-you-need-to-know/a-49998340

Kohnavard, N. (2020, March 16). Iraq military bases: US pulling out of three key sites. *BBC*. https://www.bbc.com/news/world-middle-east-51914600

Kramer, M. (2016). Israel and the post-American Middle East: Why the status quo is sustainable. *Foreign Affairs, 95*(4), 51–56.

Kristian, B. (2020, September 16). Donald Trump should withdraw all U.S. troops from Afghanistan now. *National Interest*. https://nationalinterest.org/blog/skeptics/donald-trump-should-withdraw-all-us-troops-afghanistan-now-169023

Lamothe, D. (2021, January 10). These U.S. troops survived one of the greatest crises of the Trump era. A year later, they're still coping. *Washington Post*. https://www.washingtonpost.com/national-security/us-military-iran-missile-attack/2021/01/10/651c3930-4fb0-11eb-b2e8-3339e73d9da2_story.html

Lamothe, D., Dawsey, J., Ryan, M., & Sonne, P. (2018, December 21). Trump orders major military withdrawal from Afghanistan as Mattis departs. *Washington Post*. https://www.washingtonpost.com/world/national-security/trump-agitating-for-major military-withdrawal-from-afghanistan-advisers-say/2018/12/20/0c35f874-04a3-11e9-b5df-5d3874f1ac36_story.html

Lamothe, D., Ryan, M., DeYoung, K., & George, S. (2020, November 14). Defense secretary sent classified memo to White House about Afghanistan before Trump fired him. *Washington Post*. https://www.washingtonpost.com/national-security/trump-pentagon-afghan-war/2020/11/13/5ac54c7e-25cb-11eb-8599-406466ad1b8e_story.html

Landay, J. (2019, September 3). Full U.S. pullout from Afghanistan could ignite 'total civil war': ex-U.S. envoys. *Reuters*. https://www.reuters.com/article/us-usa-afghanistan-formerenvoys/full-us-pullout-from-afghanistan-could-ignite-total-civil-war-ex-us-envoys-idUSKCN1VO21Y
Landler, M., Cooper, H., & Schmitt, E. (2018, December 19). Trump to withdraw U.S. forces from Syria, declaring 'we have won against ISIS.' *New York Times*. https://www.nytimes.com/2018/12/19/us/politics/trump-syria-turkey-troop-withdrawal.html
Landler, M., & Schmitt, E. (2017, May 23). Trump administration is split on adding troops in Afghanistan. *New York Times*. https://www.nytimes.com/2017/05/23/world/europe/saudi-arabia-arms-deal-nato.html
Lee, C. E., Kube, C., Welker, K., & Ruhle, S. (2018, April 30). Kelly thinks he's saving U.S. from disaster, calls Trump 'idiot,' say White House staffers. *NBC News*. https://www.nbcnews.com/politics/white-house/kelly-thinks-he-s-saving-u-s-disaster-calls-trump-n868961
Lema, K. (2020, February 28). Door not shut for saving U.S.-Philippines defense pact—Envoy. *Reuters*. https://www.reuters.com/article/us-philippines-usa-defence/door-not-shut-for-saving-u-s-philippines-defense-pact-envoy-idUSKCN20M0ZJ
Lema, K., Petty, M., & Stewart, P. (2020, February 11). Duterte terminates Philippines troop pact, U.S. calls move 'unfortunate.' *Reuters*. https://www.reuters.com/article/us-philippines-usa-defence/duterte-terminates-philippines-troop-pact-u-s-calls-move-unfortunate-idUSKBN2050E9
Lendon, B. (2020, August 22). Trump's planned troop pullout from Germany has many military holes. *CNN*. https://edition.cnn.com/2020/08/22/europe/europe-us-troop-movement-analysis-intl-hnk/index.html
Loveluck, L. (2020, January 4). Iran vows revenge for U.S. killing of top commander as Trump sends more troops. *Washington Post*. https://www.washingtonpost.com/world/middle_east/iran-vows-revenge-for-us-killing-of-top-commander-as-trump-sends-more-troops/2020/01/03/35224048-2e47-11ea-bffe-020c88b3f120_story.html
Loveluck, L., & Ryan, M. (2020, March 28). Militia attacks on Americans in Iraq are becoming more audacious. The U.S. is wrestling with how to respond. *Washington Post*. https://www.washingtonpost.com/world/middle_east/militia-attacks-on-americans-in-iraq-becoming-more-audacious-us-wrestling-with-how-to-respond/2020/03/27/7b31d76c-6d38-11ea-a156-0048b62cdb51_story.html
Luca, A. M. (2018, August 30). US Air Force set to upgrade base in Romania. *Balkan Insight*. https://balkaninsight.com/2018/08/30/us-air-force-to-spend-40-million-on-base-in-transylvania-08-27-2018/
Lynch, M. (2016). Belligerent minimalism: The Trump administration and the Middle East. *Washington Quarterly, 39*(4), 127–144.

MacDonald, P. K. (2018). America first? Explaining continuity and change in Trump's foreign policy. *Political Science Quarterly, 133*(3), 401–435.

Mackinder, H. J. (1904). The geographical pivot of history. *The Geographical Journal, 170*(4), 421–444.

Mackinder, H. J. (1943). The round world and the winning of the peace. *Foreign Affairs, 21*(4), 595–605.

Mahan, A. T. (1890). *The influence of sea power upon history, 1660–1783*. Little, Brown.

Mance, H. (2016, November 16). Extended US lease blocks Chagossians' return home. *Financial Times*. https://www.ft.com/content/abbc879a-ac1d-11e6-ba7d-76378e4fef24

Mason, J., & Mohammed, A. (2020, June 15). Trump says U.S. to pull some troops from Germany over NATO spending feud. *Reuters*. https://www.reuters.com/article/us-usa-germany-military-trump/trump-says-u-s-to-pull-some-troops-from-germany-over-nato-spending-feud-idUSKBN23M2VG

Maza, C. (2018, September 5). U.S. to expand military presence in Greece as tensions rise with Russia and Turkey. *Newsweek*. https://www.newsweek.com/us-expand-military-presence-greece-tensions-rise-russia-and-turkey-1104915

McCarthy, N. (2017, June 7). Where U.S. troops are in the Middle East. *Forbes*. https://www.forbes.com/sites/niallmccarthy/2017/06/07/qatar-hosts-largest-u-s-base-in-the-middle-east-despite-allegedly-funding-extremism-infographic/#437608b13dc7

McCormack, G. (2018). Grappling with clientelism: The Japanese state and Okinawa under Abe Shinzo. *Asia-Pacific Journal, 16*(23), 1–16.

McDaniel, R. (2013, June 24). *No "Plan B": U.S. strategic access in the Middle East and the question of Bahrain*. Brookings. Retrieved April 13, 2020, from https://www.brookings.edu/research/no-plan-b-u-s-strategic-access-in-the-middle-east-and-the-question-of-bahrain/

McLeary, P. (2020, July 29). Trump contradicts SecDef on Germany withdrawal. *Breaking Defense*. https://breakingdefense.com/2020/07/trump-contradicts-secdef-on-germany-withdrawal-money-trumps-strategy/

Mearsheimer, J. J. (2014, January 2). America unhinged. *National Interest*. https://nationalinterest.org/article/america-unhinged-9639

Medcalf, R. (2020). *Indo-Pacific empire: China, America and the contest for the world's pivotal region*. Manchester University Press.

Mehta, A., & Copp, T. (2017, August 21). Trump Afghanistan strategy calls for more troops, regional pressure. *Military Times*. https://www.militarytimes.com/news/2017/08/22/trump-afghanistan-strategy-calls-for-more-troops-regional-pressure/

Middle East Monitor. (2017, August 30). *US admits using military base in UAE to strike Syria, Iraq*. https://www.middleeastmonitor.com/20170830-us-admits-using-military-base-in-uae-to-strike-syria-iraq/

Middle East Monitor. (2020a, February 5). *US troops intercept Russian patrols near oil-rich Syria region*. https://www.middleeastmonitor.com/20200205-us-troops-intercept-russian-patrols-near-oil-rich-syria-region/

Middle East Monitor. (2020b, May 15). *Sources: US troop reinforcements arrive to secure oil region in Syria*. https://www.middleeastmonitor.com/20200515-sources-us-troop-reinforcements-arrive-to-secure-oil-region-in-syria/

Militarybases.com. (2021). *Spangdahlem Air Force Base in Trier, Germany*. Retrieved September 15, 2021, from https://militarybases.com/overseas/germany/spangdahlem/

Mitchell, E. (2020, January 19). Lawmakers push back at Pentagon's possible Africa drawdown. *The Hill*. https://thehill.com/policy/defense/478988-lawmakers-push-back-at-pentagons-possible-africa-drawdown

Moore, J. (2017, November 17). Trump's Afghanistan troop surge is complete, raising total number of U.S. servicement to 14,000. *Newsweek*. https://www.newsweek.com/trumps-afghanistan-troop-surge-complete-raising-total-number-us-servicemen-714588

Morgan, W. (2017, November 19). U.S. military builds up in land of 'Black Hawk Down' disaster. *Politico*. https://www.politico.com/story/2017/11/19/troops-somalia-military-buildup-247668

Nagasawa, T., & Miyasaka, S. (2020, July 5). Thousands of US troops will shift to Asia-Pacific to guard against China. *Nikkei*. https://asia.nikkei.com/Politics/International-relations/Thousands-of-US-troops-will-shift-to-Asia-Pacific-to-guard-against-China

Nazeh, M., & Al-Sudani, T. (2020, August 23). U.S.-led troops withdraw from Iraq's Taji base. *Reuters*. https://www.reuters.com/article/us-iraq-security-usa-military/u-s-led-troops-withdraw-from-iraqs-taji-base-idUSKBN25J088

New York Times. (2016, April 27). *Transcript: Donald Trump's foreign policy speech*. https://www.nytimes.com/2016/04/28/us/politics/transcript-trump-foreign-policy.html

New York Times. (2017, May 25). *In Brussels, Trump scolds allies on cost-sharing, and stays vague on Article 5*. https://www.nytimes.com/2017/05/25/world/europe/trump-brussels-nato.html

O'Brien, C. (2020, June 9). House Republicans oppose withdrawing US troops from Germany. *Politico*. https://www.politico.eu/article/house-republicans-oppose-withdrawing-us-troops-from-germany/

O'Connor, T. (2017, September 5). North Korea blasts U.S. military influence over 'puppet' South, laughs off war threats. *Newsweek*. https://www.newsweek.com/north-korea-blast-us-influence-puppet-south-laugh-war-threats-659729

O'Connor, T. (2020, March 19). New Iraq group claiming attacks on U.S. troops says it can strike Israel. *Newsweek*. https://www.newsweek.com/new-iraq-group-claiming-attacks-us-troops-says-it-can-strike-israel-1493335

O'Connor, T., & Jamali, N. (2020, September 3). U.S. troops in Syria stuck fighting 'forgotten war' for oil as Russia advances around them. *Newsweek*. https://www.newsweek.com/us-troops-syria-stuck-forgotten-war-oil-russia-advances-1529436

O'Hanlon, M. E., & Allawi, S. (2020, May 18). *Order from chaos: Iraq and the US still need each other*. Brookings. Retrieved May 29, 2021, from https://www.brookings.edu/blog/order-from-chaos/2020/05/18/iraq-and-the-us-still-need-each-other/

Olsen, H. (2020, June 17). Opinion: Trump is right to reduce troops in Germany. *Washington Post*. https://www.washingtonpost.com/opinions/2020/06/17/trump-is-right-pull-troops-out-germany/

Olson, W. (2020, February 12). Trump 'fine' with Philippines' plan to end pact allowing US troops to train there. *Stars and Stripes*. https://www.stripes.com/news/pacific/trump-fine-with-philippines-plan-to-end-pact-allowing-us-troops-to-train-there-1.618505

Ono, M., & Kelly, T. (2019, April 3). Outnumbered and elderly, Okinawa protesters oppose U.S. military runway. *Reuters*. https://www.reuters.com/article/us-japan-okinawa-henoko/outnumbered-and-elderly-okinawa-protesters-oppose-u-s-military-runway-idUSKCN1RF0KC

O'Shea, P. (2019). Strategic narratives and US military bases in Japan: How 'deterrence' makes the Marine base on Okinawa 'indispensable.' *Media, War & Conflict, 12*(4), 450–467.

Ottens, N. (2018, July 27). *When America first meets Italy first*. Atlantic Council. Retrieved February 18, 2020, from https://www.atlanticcouncil.org/blogs/new-atlanticist/when-america-first-meets-italy-first/

Packham, C., & Barrett, J. (2019, August 5). U.S. seeks to renew Pacific islands security pact to foil China. *Reuters*. https://www.reuters.com/article/us-micronesia-usa-pompeo/u-s-seeks-to-renew-pacific-islands-security-pact-to-foil-china-idUSKCN1UV0UV

Parker, M., & Wingrove, J. (2020, June 21). Trump says he's cutting troops in Germany over pipeline and NATO. *Bloomberg*. https://www.bloomberg.com/news/articles/2020-06-21/trump-says-he-s-cutting-troops-in-germany-over-pipeline-and-nato

Pawlyk, O. (2017, August 28). Air Force acknowledges clandestine base in UAE. *Military.com*. https://www.military.com/dodbuzz/2017/08/28/air-force-acknowledges-clandestine-base-in-uae

Persio, S. L. (2017, November 7). Trump visits largest U.S. military overseas base and boasts he could have built it for less. *Newsweek*. https://www.newsweek.com/trump-visits-largest-us-military-overseas-base-boasts-703678

Phillips, M. M. (2020, March 15). U.S. weighs troop cuts in Africa, leaving allies to confront growing militant threat. *Wall Street Journal*. https://www.wsj.com/articles/u-s-weighs-troop-cuts-in-africa-leaving-allies-to-confront-growing-militant-threat-11584291493

Plaskin, G. (1990, March 1). The Playboy interview with Donald Trump. *Playboy Magazine*. https://www.playboy.com/read/playboy-interview-donald-trump-1990

Politi, J. (2016, February 22). US drones to strike Isis targets in Libya from base in Italy. *Financial Times*. https://www.ft.com/content/da388300-d9ad-11e5-a72f-1e7744c66818

Posen, B. R. (2018). The rise of illiberal hegemony: Trump's surprising grand strategy. *Foreign Affairs, 97*(2), 20–27.

Plucinska, J., & Ali, I. (2020, June 10). U.S.-Polish Fort Trump project crumbles. *Reuters*. https://www.reuters.com/article/us-poland-usa-defense-forttrump-idUSKBN23H36P

Pramuk, J. (2017a, August 8). Trump warns North Korea threats 'will be met with fire and fury.' *CNBC*. https://www.cnbc.com/2017/08/08/trump-warns-north-korea-threats-will-be-met-with-fire-and-fury.html

Pramuk, J. (2017b, August 21). What Trump said about Afghanistan before he became president. *CNBC*. https://www.cnbc.com/2017/08/21/what-trump-said-about-afghanistan-before-he-became-president.html

Psaropoulos, J. (2019, October 5). US, Greece hail strategic relationship after signing defence deal. *Al Jazeera*. https://www.aljazeera.com/news/2019/10/greece-hail-strategic-relationship-signing-defence-deal-191005175854999.html

Rapp-Hooper, M. (2020, July 29). Withdrawing thousands of troops from Germany will be costly for U.S. *Washington Post*. https://www.washingtonpost.com/politics/2020/06/09/trump-will-withdraw-9500-us-troops-germany-that-will-be-costly-us/

Rasheed, A., Aboulenein, A., & Mason, J. (2020, January 5). Iraq wants foreign troops out after air strike; Trump threatens sanctions. *Reuters*. https://www.reuters.com/article/us-iraq-security/iraq-wants-foreign-troops-out-after-air-strike-trump-threatens-sanctions-idUSKBN1Z409A

Rasheed, A., & Ali, I. (2019, December 31). Protesters burn security post at U.S. Embassy in Iraq; Pentagon sending more troops to region. *Reuters*. https://www.reuters.com/article/us-iraq-security-usa/protesters-burn-security-post-at-u-s-embassy-in-iraq-pentagon-sending-more-troops-to-region-idUSKBN1YZ1D7

Rasheed, A., & Ismail, A. (2020, November 17). Rockets fall inside Baghdad's Green Zone, child killed: Military. *Reuters*. https://www.reuters.com/article/us-iraq-security-rockets/rockets-fall-inside-baghdads-green-zone-child-killed-military-idUSKBN27X2RZ

Reuters. (2018, August 27). *Qatar to expand air base hosting major U.S. military facility*. https://www.reuters.com/article/us-qatar-airbase/qatar-to-expand-air-base-hosting-major-u-s-military-facility-idUSKCN1LC1TJ

Reuters. (2019a, May 8). *U.S. will need forces in Afghanistan until no insurgency left: Dunford*. https://www.reuters.com/article/us-usa-afghanistan-military/us-will-need-forces-in-afghanistan-until-no-insurgency-left-dunford-idUSKCN1SE2ER

Reuters. (2019b, June 20). *Iran's Zarif tweets that downed U.S. drone took off from UAE*. https://www.reuters.com/article/us-mideast-iran-drone-emirates/irans-zarif-tweets-that-downed-us-drone-took-off-from-uae-idUSKCN1TL2ID

Reuters. (2019c, July 2). *Trump says he is worried about terrorist attacks if U.S. troops leave Afghanistan*. https://www.reuters.com/article/us-usa-trump-afghanistan/trump-says-he-is-worried-about-terrorist-attacks-if-u-s-troops-leave-afghanistan-idUSKCN1TX042

Reuters. (2019d, July 30). *Australia says U.S. plans to build military infrastructure*. https://www.reuters.com/article/us-australia-usa/australia-says-u-s-plans-to-build-military-infrastructure-idUSKCN1UP0GY

Reuters. (2019e, August 29). *Trump: U.S. will maintain presence in Afghanistan even if deal reached with Taliban*. https://www.reuters.com/article/us-usa-afghanistan-trump/trump-u-s-will-maintain-presence-in-afghanistan-even-if-deal-reached-with-taliban-idUSKCN1VJ1R8

Reuters. (2019f, December 3). *Trump says North Korea's Kim 'likes sending rockets up.'* https://www.reuters.com/article/us-usa-northkorea-trump/trump-says-north-koreas-kim-sure-likes-sending-rockets-up-idUSKBN1Y71CT

Reuters. (2020a, January 3). *Pentagon to deploy thousands of additional troops to Middle East*. https://www.reuters.com/article/us-iraq-security-usa-middleeast/pentagon-to-deploy-thousands-of-additional-troops-to-middle-east-idUSKBN1Z21U4

Reuters. (2020b, January 23). *Philippines' Duterte threatens to end military deal with the United States*. https://www.reuters.com/article/us-philippines-usa-duterte/philippines-duterte-threatens-to-end-military-deal-with-the-united-states-idUSKBN1ZM2IZ

Reuters. (2020c, March 26). *U.S.-led forces depart Iraqi military base near Mosul in drawdown*. https://www.reuters.com/article/us-iraq-security-usa/u-s-led-forces-depart-iraqi-military-base-near-mosul-in-drawdown-idUSKBN21D1UR

Reuters. (2020d, July 24). *Rockets hit military base south of Baghdad, no casualties: Military*. https://www.reuters.com/article/us-iraq-security-idUSKCN24P2CQ
Rich, M. (2016, October 26). Rodrigo Duterte, in Japan, calls for U.S. troops to exit Philippines in 2 years. *New York Times*. https://www.nytimes.com/2016/10/27/world/asia/philippines-president-rodrigo-duterte-japan.html
Robson, S. (2019, January 31). Facility for US forces opens on Philippines' main island; another slated for Palawan. *Stars and Stripes*. https://www.stripes.com/news/pacific/facility-for-us-forces-opens-on-philippines-main-island-another-slated-for-palawan-1.566695
Rosen, K. R. (2021, May 31). The inside story of how Trump 'Kept the Oil' in Syria and lost. *Daily Beast*. https://www.thedailybeast.com/how-delta-crescent-fell-short-in-trumps-keep-the-oil-plan-in-syria?ref=scroll
Rucker, P., & Costa, R. (2016, March 21). Trump questions need for NATO, outlines noninterventionist foreign policy. *Washington Post*. https://www.washingtonpost.com/news/post-politics/wp/2016/03/21/donald-trump-reveals-foreign-policy-team-in-meeting-with-the-washington-post/
Ryan, M. (2020, May 18). For most of Afghanistan war, U.S. 'never really fought to win,' Trump declares. *Washington Post*. https://www.washingtonpost.com/national-security/for-most-of-afghanistan-war-america-never-really-fought-to-win-trump-declares/2020/05/18/9ab85afa-992c-11ea-b60c-3be060a4f8e1_story.html
Ryan, M., DeYoung, K., & Morris, L. (2020, July 29). Pentagon plan will move troops from Germany to Italy, Belgium and back to U.S. *Washington Post*. https://www.washingtonpost.com/national-security/us-troop-withdrawal-germany/2020/07/29/f5d23982-d19f-11ea-af07-1d058ca137ae_story.html
Saab, B. Y. (2018, August 22). Relocating the fifth fleet? *American Interest*. https://www.the-american-interest.com/2018/08/22/relocating-the-fifth-fleet/
Safi, M., Holmes, O., & Abdul-Ahad, G. (2020, January 8). Iran launches missiles at Iraq airbases hosting US and coalition troops. *The Guardian*. https://www.theguardian.com/world/2020/jan/08/suleimani-assassination-two-us-airbases-in-iraq-hit-by-missiles-in-retaliation
Salim, M., Sly, L., & Morello, C. (2019, December 30). U.S. airstrikes on Iranian-backed militia draw condemnation, retaliation threats in Iraq. *Washington Post*. https://www.washingtonpost.com/world/middle_east/us-airstrikes-on-iran-backed-militia-draw-condemnation-retaliation-threats-in-iraq/2019/12/30/d13a10be-2af0-11ea-bffe-020c88b3f120_story.html
Schmitt, E. (2018, April 22). A shadowy war's newest front: A drone base rising from Saharan dust. *New York Times*. https://www.nytimes.com/2018/04/22/us/politics/drone-base-niger.html

Schmitt, E. (2019, March 1). Where terrorism is rising in Africa and the U.S. is leaving. *New York Times*. https://www.nytimes.com/2019/03/01/world/africa/africa-terror-attacks.html

Schmitt, E. (2020a, October 15). Trump's tweets on troop withdrawals unnerve Pentagon. *New York Times*. https://www.nytimes.com/2020/10/15/us/politics/trump-troop-withdrawals.html

Schmitt, E. (2020b, December 10). To deter Iranian attacks on U.S. troops, Pentagon orders B-52 flights to Middle East. *New York Times*. https://www.nytimes.com/2020/12/10/world/middleeast/bombers-iran-deterrence.html

Schmitt, E., & Haberman, M. (2019, October 20). Trump said to favor leaving a few hundred troops in eastern Syria. *New York Times*. https://www.nytimes.com/2019/10/20/world/middleeast/trump-troops-syria-turkey.html

Schmitt, E., Haberman, M., & Wong, E. (2019, October, 7). President endorses Turkish military operation in Syria, shifting U.S. policy. *New York Times*. https://www.nytimes.com/2019/10/07/us/politics/trump-turkey-syria.html?action=click&module=Top%20Stories&pgtype=Homepage

Schreer, B. (2019). Trump, NATO and the future of Europe's defence. *RUSI Journal, 164*(1), 10–17.

Sediqi, A. Q., & Cornwell, A. (2020, February 29). U.S. and Taliban sign troop withdrawal deal; now comes the hard part. *Reuters*. https://www.reuters.com/article/us-usa-afghanistan-taliban/u-s-taliban-sign-historic-troop-withdrawal-deal-in-doha-idUSKBN20N06R

Seldin, J. (2020, July 31). US pulling Africa Command from Germany. *VOA News*. https://www.voanews.com/a/usa_us-pulling-africa-command-germany/6193739.html

Seligman, L., & Lefebre, B. (2020, August 3). Little-known U.S. firm secures deal for Syrian oil. *Politico*. https://www.politico.com/news/2020/08/03/delta-crescent-energy-syrian-oil-391033

Shalizi, H., Sediqi, A. Q., & Jain, R. (2020, May 1). Taliban step up attacks on Afghan forces since signing U.S. deal: data. *Reuters*. https://www.reuters.com/article/us-health-coronavirus-afghanistan-taliba-idUSKBN22D5S7

Shambaugh, D. (2021). *Where great powers meet: America and China in Southeast Asia*. Oxford University Press.

Shanahan, D. (2016, October 6). Obama's Northern Territory Asia pivot locked in. *The Australian*. https://www.theaustralian.com.au/national-affairs/defence/obamas-northern-territory-asia-pivot-locked-in/news-story/78adc1115af0056b8c736ad6f761cf28

Shin, H. (2019, November 13). South Korean workers protest as military cost-sharing dispute with U.S. threatens jobs. *Reuters*. https://www.reuters.com/article/us-southkorea-usa-military/south-korean-workers-protest-as-military-cost-sharing-dispute-with-u-s-threatens-jobs-idUSKBN2171EU

Shin, H., & Stewart, P. (2021, March 10). South Korea agrees to biggest increase in its share of cost for U.S. troops in years. *Reuters*. https://www.reuters.com/article/us-southkorea-usa-defence-idUSKBN2B20LC

Shinkman, P. D. (2017, May 25). Trump's war budget has $500M for foreign countries' bases. *U.S.News*. https://www.usnews.com/news/world/articles/2017-05-25/trumps-war-budget-calls-for-half-a-billion-dollars-for-foreign-countries-bases

Shlapak, D. A., & Johnson, M. W. (2016). *Reinforcing deterrence on NATO's eastern flank: Wargaming the defense of the Baltics*. RAND Corporation. Retrieved August 12, 2021, from https://apps.dtic.mil/sti/pdfs/AD1002535.pdf

Sieg, L. (2019, February 25). Japan to push ahead with U.S. base relocation despite Okinawa referendum result. *Reuters*. https://www.reuters.com/article/us-japan-okinawa/japan-to-push-ahead-with-u-s-base-relocation-despite-okinawa-referendum-result-idUSKCN1QE06Q

Silove, N. (2016). The pivot before the pivot: U.S. strategy to preserve the power balance in Asia. *International Security*, *40*(4), 45–88.

Snow, S. (2020, May 14). Tensions with Iran and its proxies led to accelerated transfer of bases housing coalition troops to Iraqi forces. *Military Times*. https://www.militarytimes.com/flashpoints/2020/05/14/tensions-with-iran-and-its-proxies-led-to-accelerated-transfer-of-bases-housing-coalition-troops-to-iraqi-forces/

Sonne, P., Dawsey, J., & Ryan, M. (2018, December, 21). Mattis resigns after clash with Trump over troop withdrawal from Syria and Afghanistan. *Washington Post*. https://www.washingtonpost.com/world/national-security/trump-announces-mattis-will-leave-as-defense-secretary-at-the-end-of-february/2018/12/20/e1a846ee-e147-11e8-ab2c-b31dcd53ca6b_story.html

South, T. (2020, June 19). New Iran rocket attacks won't force US hand in troop drawdown, CENTCOM commander says. *Military Times*. https://www.militarytimes.com/news/mtr/2020/06/18/new-iran-rocket-attacks-wont-force-us-hand-in-troop-drawdown-general-says/

Spykman, N. J. (2017). *America's strategy in world politics: The United States and the balance of power*. Routledge. (Original work published 1942.)

Stavridis, J. (2020, September 10). Trump's 'ending' the Iraq War is a gift to Iran. *Bloomberg*. https://www.bloombergquint.com/opinion/trump-s-iraq-troop-withdrawal-is-good-for-iran-russia-syria

Stewart, P. (2017, May 16). U.S. signs new defense accord with Gulf ally UAE. *Reuters*. https://www.reuters.com/article/us-usa-emirates-military-idUSKCN18C1TN

Stewart, P. (2019a, March 24). With an eye on Iran, U.S. clinches strategic port deal with Oman. *Reuters*. https://www.reuters.com/article/us-usa-oman-military/with-an-eye-on-iran-u-s-clinches-strategic-port-deal-with-oman-idUSKCN1R50JD

Stewart, P. (2019b, December 5). Exclusive: U.S. military completes pullback from northeast Syria, Esper says. *Reuters*. https://www.reuters.com/article/us-mideast-crisis-usa-syria-exclusive/exclusive-u-s-military-completes-pullback-from-northeast-syria-esper-says-idUSKBN1Y90CU

Stewart, P. (2020, January 24). Pentagon could start global troop rebalance this year, Esper says. *Reuters*. https://www.reuters.com/article/us-usa-pentagon-troops/pentagon-could-start-global-troop-rebalance-this-year-esper-says-idUSKBN1ZN037

Stewart, P., & Ali, I. (2019, February 5). As U.S. withdraws, top general warns on Islamic State threat in Syria. *Reuters*. https://www.reuters.com/article/us-mideast-crisis-syria-usa/as-u-s-withdraws-top-general-warns-on-islamic-state-threat-in-syria-idUSKCN1PU1P5

Stewart, P., & Ali, I. (2020, July 29). U.S. to withdraw about 12,000 troops from Germany but nearly half to stay in Europe. *Reuters*. https://www.reuters.com/article/uk-usa-trump-germany-military-idUKKCN24U20B

Stravers, A., & El Kurd, D. (2020). Strategic autocracy: American military forces and regime type. *Journal of Global Security Studies, 5*(3), 427–442.

Tan, R., & Ward, A. (2017, June 27). Rodrigo Duterte loves bashing the US—But he also loves US troops fighting terrorists for him. *Vox*. https://www.vox.com/world/2017/6/27/15833572/duterte-philippines-mindanao-marawi-abu-sayyaf-us-troops

Tawfeeq, M., & Starr, B. (2020, September 30). Rockets strike Iraqi base where US troops are located, US defense official says. *CNN*. https://edition.cnn.com/2020/09/30/politics/rockets-us-base-iraq/index.html

Taylor, A. (2019a, July 16). Secret locations of U.S. nuclear weapons in Europe accidentally included in report from NATO parliament. *Washington Post*. https://www.washingtonpost.com/world/2019/07/16/location-us-nuclear-weapons-europe-accidentally-revealed-report-nato-body/

Taylor, A. (2019b, August 21). As Trump tries to end 'endless wars,' America's biggest Mideast base is getting bigger. *Washington Post*. https://www.washingtonpost.com/world/as-trump-tries-to-end-endless-wars-americas-biggest-mideast-base-is-getting-bigger/2019/08/20/47ac5854-bab4-11e9-8e83-4e6687e99814_story.html

Taylor, A., & Birnbaum, M. (2020, November 17). In rare rebuke of Trump, NATO chief warns against troop cuts in Afghanistan. *Washington Post*. https://www.washingtonpost.com/world/2020/11/17/jens-stoltenberg-afghanistan-trump/

The Economist. (2018, February 6). *Why US Marines are deployed to Australia's far north*. https://www.economist.com/democracy-in-america/2018/02/06/why-us-marines-are-deployed-to-australias-far-north

The Economist. (2019, October 10). *America pours a thousand more troops into Poland*. https://www.economist.com/europe/2019/10/10/america-pours-a-thousand-more-troops-into-poland

The Economist. (2020, June 25). *Going home: Donald Trump's baffling proposal to withdraw troops from Germany*. https://www.economist.com/europe/2020/06/25/donald-trumps-baffling-proposal-to-withdraw-troops-from-germany

The Economist. (2021, March 13). *Dispersal orders: America's top brass responds to the threat of China in the Pacific*. https://www.economist.com/asia/2021/03/11/americas-top-brass-responds-to-the-threat-of-china-in-the-pacific

The Guardian. (2017, June 16). *US sending almost 4,000 extra forces to Afghanistan, Trump official says*. https://www.theguardian.com/us-news/2017/jun/15/us-troops-afghanistan-trump-administration

The White House. (2015, February). *National Security strategy*. Retrieved March 30, 2020, from https://obamawhitehouse.archives.gov/sites/default/files/docs/2015_national_security_strategy_2.pdf

The White House. (2017, December). *National security strategy of the United States of America*. Retrieved April 23, 2020, from https://trumpwhitehouse.archives.gov/wp-content/uploads/2017/12/NSS-Final-12-18-2017-0905.pdf

The White House. (2021, February 4). *Remarks by President Biden on America's place in the world*. Retrieved April 27, 2021, from https://www.whitehouse.gov/briefing-room/speeches-remarks/2021/02/04/remarks-by-president-biden-on-americas-place-in-the-world/

Tilghman, A. (2016, March 21). The U.S. military is moving into these 5 bases in the Philippines. *Military Times*. https://www.militarytimes.com/news/your-military/2016/03/21/the-u-s-military-is-moving-into-these-5-bases-in-the-philippines/

Trevithick, J. (2019, January 14). Docs show US to massively expand footprint at Jordanian Air Base amid spats with Turkey, Iraq. *The Drive*. https://www.thedrive.com/the-war-zone/25955/docs-show-us-to-massively-expand-footprint-at-jordanian-air-base-amid-spats-with-turkey-iraq

U.S. Department of Defense. (2018). *Base structure report—Fiscal year 2018 baseline*. Retrieved March 15, 2021, from https://www.acq.osd.mil/eie/Downloads/BSI/Base%20Structure%20Report%20FY18.pdf

Vandiver, J. (2017, February 2). Germans irked by demands for financial support for US bases. *Stars and Stripes*. https://www.stripes.com/news/germans-irked-by-demands-for-financial-support-for-us-bases-1.452175

Vandiver, J. (2018a, February 1). Qatar upgrading facilities for US families at Al Udeid. *Stars and Stripes*. https://www.stripes.com/news/qatar-upgrading-facilities-for-us-families-at-al-udeid-1.509474

Vandiver, J. (2018b, 11 June). NATO's new southern hub in Italy to be fully operational in July. *Stars and Stripes*. https://www.stripes.com/news/nato-s-new-southern-hub-in-italy-to-be-fully-operational-in-july-1.532131

VanOpdorp, D. (2020, August 5). Donald Trump accuses Germany of 'making a fortune' off US soldiers. *Deutsche Welle*. https://www.dw.com/en/trump-nato-germany/a-54451443

Votel, J., & Dent, E. (2019, October 8). The danger of abandoning our partners. *The Atlantic*. https://www.theatlantic.com/politics/archive/2019/10/danger-abandoning-our-partners/599632/

Wall Street Journal. (2020, May 17). *Hopeful news in Afghanistan*. https://www.wsj.com/articles/hopeful-news-in-afghanistan-11589746435

Walt, S. M. (2019). The end of hubris and the new age of American restraint. *Foreign Affairs, 98*(3), 26–35.

Washington Post. (2020, March 1). *This is exactly the wrong moment to pull U.S. troops out of Africa*. https://www.washingtonpost.com/opinions/global-opinions/this-is-exactly-the-wrong-moment-to-pull-us-troops-out-of-africa/2020/03/01/01f203b2-5a5b-11ea-9b35-def5a027d470_story.html

Westbrook, T. (2017, April 18). U.S. Marines arrive in Darwin for Australia, China exercises. *Reuters*. https://www.reuters.com/article/us-australia-usa-defence-idUSKBN17K09N

Williams, K. B. (2020, November 12). Outgoing Syria envoy admits hiding US troop numbers; praises Trump's Mideast record. *Defense One*. https://www.defenseone.com/threats/2020/11/outgoing-syria-envoy-admits-hiding-us-troop-numbers-praises-trumps-mideast-record/170012/

Witte, G., & Stanley-Becker, I. (2018, December 21). 'A morning of alarm': Mattis departure sends shock waves abroad. *Washington Post*. https://www.washingtonpost.com/nation/2018/12/21/mattis-departure-sends-shock-waves-abroad-allies-question-us-approach-global-crises/

Wong, E., & Specia, M. (2020, January 10). U.S. says it won't discuss withdrawing troops from Iraq, defying Baghdad's request. *New York Times*. https://www.nytimes.com/2020/01/10/world/middleeast/us-troops.html

Wroughton, L., & Jain, R. (2019, July 29). Trump wants forces reduced in Afghanistan by next U.S. election: Pompeo. *Reuters*. https://www.reuters.com/article/us-usa-afghanistan-pompeo/trump-wants-forces-reduced-in-afghanistan-by-next-us-election-pompeo-idUSKCN1UO22R
Yost, D. S. (1995). The future of U.S. overseas presence. *Joint Forces Quarterly* (Summer), 70–82.
Youssef, N. A. (2018, September 11). U.S. eyes military expansion in Greece amid strains with Turkey. *Wall Street Journal*. https://www.wsj.com/articles/u-s-military-looks-toward-greece-amid-strains-with-turkey-1536696318
Youssef, N. A. (2020, August 4). U.S. Sending 200 Troops to Poland, Part of a Gradual Increase. *Wall Street Journal*. https://www.wsj.com/articles/u-s-sending-200-troops-to-poland-part-of-a-gradual-increase-11596579781
Youssef, N. A., & Phillips, M. M. (2020, November, 20). U.S. considers moving forces out of Somalia. *Wall Street Journal*. https://www.wsj.com/articles/u-s-considers-moving-forces-out-of-somalia-11605911141
Yusuf, A. (2021, January 13). Trump's withdrawal from Somalia is a security threat. Biden should reverse it. *Foreign Policy*. https://foreignpolicy.com/2021/01/13/trump-somalia-troop-withdrawal-security-threat-biden-reverse/
Żaryn, S. (2020, December 15). The US presence in Poland is a pain in Russia's neck. *Military.com*. https://www.military.com/daily-news/opinions/2020/12/15/us-presence-poland-pain-russias-neck.html
Zengerle, P. (2019a, February 5). Senate breaks from Trump with Syria troop vote. *Reuters*. https://www.reuters.com/article/us-mideast-crisis-syria-congress/senate-breaks-from-trump-with-syria-troop-vote-idUSKCN1PT2DH
Zengerle, P. (2019b, December 10). U.S. lawmakers reach deal on massive defense bill, eye Russia, Turkey, China. *Reuters*. https://www.reuters.com/article/us-usa-defense-congress/u-s-lawmakers-reach-deal-on-massive-defense-bill-eye-russia-turkey-china-idUSKBN1YE07K
Zengerle, P. (2020a, June 18). U.S. lawmakers offer bill to block Trump from German troop withdrawal. *Reuters*. https://www.reuters.com/article/us-usa-germany-military-congress-idUSKBN23P2LA
Zengerle, P. (2020b, June 23). Republican lawmakers urge Trump to reconsider Germany troop reduction plan. *Reuters*. https://www.reuters.com/article/us-usa-germany-military-lawmakers-idUSKBN23U1BI
Zengerle, P. (2020c, December 4). U.S. Congress defense bill defies Trump on Germany withdrawal, base names. *Reuters*. https://www.reuters.com/article/us-usa-defense-congress-idUSKBN28E0Y5
Zenko, M. (2018, October 18). *US military policy in the Middle East: An appraisal*. Chatham House. Retrieved February 14, 2020, from https://reader.chathamhouse.org/us-military-policy-middle-east-appraisal

CHAPTER 8

Conclusion: Whither U.S. Overseas Troops?

*

Chapters 6 and 7 have shown that, when Donald Trump left the presidential office, U.S. overseas troops generally were where they were when he entered it. They were also there in roughly comparable numbers, save for Asia–Pacific, which saw a considerable *increase* in U.S. troop deployments in the period. The president did propose withdrawals from several host countries, but few of these plans came to fruition. Donald Trump was and is a *Leaver*, but he didn't actually make the United States leave. The main exceptions to this basic pattern are Afghanistan and, to a lesser degree, Iraq. Notably, along with Syria, these were the same countries in which the United States had been, and in some ways still was, at war—wars that, in the cases of Afghanistan and Iraq, started almost two decades ago. In that sense, Trump did in the end manage to start a process that could conceivably, sometime in the near or far future, bring the United States' Middle Eastern wars to an end. It is still so that the deployment rationales for troops engaged in actual fighting differ from the rationales of troops not normally engaged in combat (Jakobsen & Jakobsen, 2019, p. 156); and it is not overly surprising that the troops are eventually pulled out from the former. The war-deployment rationales also tend to change markedly over time. The United States had thousands of forces stationed in Panama following the 1989 invasion; virtually all of these

J. Jakobsen, *The Geopolitics of U.S. Overseas Troops and Withdrawal*, Palgrave Studies in International Relations,
https://doi.org/10.1007/978-3-030-94488-9_8

were withdrawn a decade later. The same was the case in Croatia, Macedonia, Serbia, and Bosnia and Herzegovina: the thousands of U.S. troops that were present there during and following the Bosnian and Kosovo Wars in the 1990s were, for the most part, withdrawn as realities on the ground changed. But the rest of the U.S. overseas military-base network remained intact then, and it also remained intact when Donald Trump's presidency was over.

The Biden Presidency and the Afghan Debacle

In other words, the new U.S. President, Joe Biden, who assumed office on January 20, 2021, inherited what was pretty much the same set of military bases, and the same number of U.S. troops stationed at the same locations, as Trump had inherited four years previously. There was surely little or nothing to indicate that Biden would follow the same retrenchment—or *attempted* retrenchment—track as his predecessor. "America is back," was the refrain of his election campaign (DeYoung, 2020). The country was ready yet again to lead and to organize the world, and to make sure that international vacuums were not filled by Washington's enemies (Sanger, 2020). Once inaugurated, the new president reassured allies, most of which were exasperated after four highly tumultuous years, that their alliance ties to the United States were and would remain solid (Madhani, 2021). Neither would the issue of U.S. forward-deployed troops be "politicized" on his watch (Williams, 2021).

Joe Biden and U.S. Overseas Troops—Early Indications

One of the first moves by President Biden was to instruct his national-security team to start working on a Global Posture Review, which would be led by his Defense Secretary Lloyd Austin. The aim of the review, whose conclusions were to be finalized in late 2021, was to provide an overarching assessment of the global role of the United States and its military forces and bases to ensure, in Biden's words, that "our military footprint is appropriately aligned with our foreign policy and national security priorities" (The White House, 2021). Until that assessment was ready, little change was to be expected in troop levels or locations. Some officially announced clues appeared in March, when Biden (2021) released the Interim National Security Strategic Guidance. That document was meant to reflect the president's, and hence the United States', geostrategic vision. As expected, it stressed that America must reinvigorate its leadership abroad, engage deeply and consistently with its many allies, and incessantly work to counter the challenge posed by China, Russia, Iran, North Korea, and transnational terrorist groups.

On the specific issue of U.S. troop deployments, though, the Guidance essentially indicated that it would follow up at least one of the processes that Trump had started, namely, to disengage the country from the "'forever wars' that have cost thousands of lives and trillions of dollars." The Biden administration would endeavor to "right-size" U.S. presence in the Middle East, "to the level required to disrupt international terrorist networks, deter Iranian aggression, and protect other vital U.S. interests," while U.S. force projection would "be most robust in the Indo-Pacific and Europe" (Biden, 2021, p. 15). "Most robust" could reasonably be taken to mean either that the status quo would be preserved in the latter two regions, or that a strengthening of the U.S. force posture was in the cards. To "right-size" force projection in the Middle East, for its part, would probably entail ambitions to scale back U.S. military presence somewhat. This, however, was an ambition that Biden shared with the three previous winners of the U.S. presidential elections, and none of his predecessors had really succeeded in executing any significant retrenchment.

More explicitly, the Biden administration declared that it would work to end the United States' two-decade war in Afghanistan, a matter to which I will return shortly. Afghanistan apart, the first months of the new administration saw little change in troop deployments, and the policy pursued was certainly not one of retrenchment. According to the Pentagon's official data of "permanently assigned, active-duty troops" for September 30, 2021, the first eight months of Biden's presidency had seen a small *increase*—by some 4700—in the overall tally of U.S. overseas troops (again, apart from Afghanistan, for which troop numbers were still not officially published) (U.S. Department of Defense, 2021). Numbers for a couple of countries had risen markedly since the end of 2020: Australia now had 1768 U.S. troops (mostly Marines) stationed on its soil, while Saudi Arabia's numbers had risen by nearly one thousand, to 1396. All other changes were minuscule, including among the biggest hosts: Japan saw a very small reduction, South Korea and Guam a small increase. In Europe, all of the biggest hosts—Germany, Italy, and the United Kingdom—saw a small increase.

Germany, which the year before had been castigated by Donald Trump, who set in motion withdrawal plans that were never actually realized, saw U.S. troop numbers rise by 993 (or 2.9%) in the first three-quarters of 2021. Trump's withdrawal plans were almost immediately shelved for good by the new administration, which instead promised it would soon ship additional military personnel to Germany (Reuters, 2021a). Moreover, Greece signed an expanded defense agreement with Washington in October 2021 (France24, 2021), while Romania was set to receive considerable funding from the United States aimed at creating a new hub for NATO aircraft patrolling the Black Sea (Barberá, 2021). Further north, the United States and Norway signed an agreement that would let Washington construct facilities at one naval base and several airports (Reuters et al., 2021). In Asia, Japan and South Korea soon reached deals with the new U.S. administration on cost sharing for military bases (Al Jazeera, 2021a; Nikkei Asia, 2021). In September, it looked certain that the relationship between the United States and its long-term ally Australia would be strengthened even further, through the ramp-up of the contingent of U.S. Marines stationed in Darwin (Al Jazeera, 2021b) and through the creation of a controversial new three-way defense alliance, which also included the United Kingdom (Megerian, 2021). Furthermore, the Philippines moved to stabilize its quite wobbly military relationship to Washington, finally restoring the defense agreement that governs U.S. troop access to the country (Ali & Lema, 2021). Even Taiwan, it was reported in late October, now had a few U.S. troops temporarily and controversially present on the island, where they conducted training with local forces (Ripley et al., 2021).

In the Middle East, though, early signals were that the Biden administration would terminate its support for Saudi Arabia's six-year long war in Yemen (BBC, 2021). The United States also moved to withdraw the antimissile batteries (along with some attendant U.S. troops) that Biden's predecessor had deployed to Saudi Arabia and other states in the Gulf region amid increased tensions with Iran (Lubold & Strobel, 2021). In Iraq, the United States still found itself enmeshed in a "shadow war" against Iran and its proxies (Sommerville, 2021). Even so, President Biden announced in late July that U.S. combat missions in the country would cease by the year's end, although the 2500 U.S. troops present there looked set to stay in order to provide training and assistance to local forces (Burns et al., 2021). Any further withdrawal would likely also complicate U.S. operations in neighboring Syria, where indications were that all U.S. troops would remain to assist their local Kurdish allies in their fight against the Islamic State (Seligman, 2021). Overall, it seemed clear, Washington's regional military-base network would remain intact. According to *The Economist* (2021d), by midsummer 2021, the United States had some 60,000 troops stationed in the Middle East.

Withdrawal from Afghanistan

That tally presumably included a few troops remaining in Afghanistan (even if that country is formally and geographically located in South Asia). But it was in Afghanistan that real change would occur, and it was here that the domino principle would effectively be put to the empirical test. Under President Biden, the United States finally withdrew its last remaining soldiers. After two decades, and after having spent over $2 trillion, the U.S. war in Afghanistan had come to an end—a very dramatic end—on August 30, 2021 (Youssef & Lubold, 2021).

The backdrop to this finale was the decision by Biden's predecessor to finalize a February 2020 peace deal with the Taliban, the Islamist, predominantly Pashtun group which had ruled Afghanistan in a rather draconical way from 1996 until the United States removed it from power in fall 2001. Under the terms of the deal, the United States would immediately start reducing its troop numbers in the country, which it did, and completely withdraw its military by May 1, 2001 (Lee & Tucker, 2021). This deadline was not met by Trump's successor. However, Joe Biden might have been a *Stayer* in most respects, but on the issue of Afghanistan, he had actually been a reasonably credible *Leaver* for over a decade (Baker, 2009). Moreover, his retrenchment inclinations, which were evidently not wholly unambiguous, were greatly augmented by the momentum created by the previous administration's peace deal and troop reductions. President Biden would later argue that he had in reality faced a binary choice between two glaring evils (Baker, 2021). He could either pull out all remaining U.S. troops, which would risk setting in motion a series of adverse geopolitical chain reactions; or he could ignore the deal with the Taliban and let the troops stay, which would, by his own assessment, almost certainly lead to the war escalating, which would soon necessitate sending even more troops to Afghanistan.

The specter of Afghanistan turning out a real "forever war" for the United States weighed heavily enough to make Biden choose the former option. In mid-April, the administration announced that all U.S. troops would leave Afghanistan by September 11, the twentieth anniversary of the Al-Qaeda terrorist attacks that had spurred the U.S. military engagement there in the first place (Ryan & DeYoung, 2021). Following the standard playbook of cases of proposed withdrawal, the broader foreign-policy community—as well as U.S. allies—reacted strongly and overwhelmingly negatively to the decision. A full-blown civil war would result, many foresaw, and one in which the Taliban would certainly emerge as winners, effectively setting the clock back twenty years (Barnes et al., 2021). Meanwhile, the governance vacuum that would inevitably be created would give transnational terrorist groups—notably, Al-Qaeda and the Islamic State's South Asian branch—ample opportunities to organize and to carry out attacks, even against the U.S. homeland (Reuters, 2021b). Many saw a troop pullout as a victory for other major powers—such as China and Russia (*The Economist*, 2021a)—while neighboring India feared that it would invite regional "disruptors," by which it presumably meant Pakistan, to fill the void (Reuters & Miglani, 2021). Others pointed out that troops on the ground in Afghanistan were, for intelligence and counterterrorism purposes, an absolute necessity given the lack of military-basing rights in the vicinity (*The Economist*, 2021b). Still others highlighted the severe damage that a withdrawal—or "capitulation"—would inflict on U.S. credibility (Dobbins, 2021).

An Afghan Domino and the Manifestation of Geopolitical Chain Reactions

Such warnings about the *possibility* of adverse geopolitical chain reactions being set in motion by a troop withdrawal failed to make the president reverse his decision. By early July, the withdrawal was reportedly quite close to being finalized (Reuters, 2021c); President Biden now announced that all remaining U.S. troops would be pulled out by August 31 (Miller & Madhani, 2021). The endgame was near, and it would be more dramatic than virtually anyone—even the most ardent *Stayers*—had foreseen. In the first half of August, a string of major Afghan cities had fallen to the Taliban; these were developments the speed of which astonished both analysts and policymakers (*The Economist*, 2021e). U.S. intelligence agencies had earlier, in late March, warned the president that, if U.S. troops fully withdrew, the Islamist group would likely gain control over Afghanistan "within two or three years" (Barnes et al., 2021). As things turned out, it only took four months from the 9/11 deadline was set and until Afghanistan had once again fallen under the sway of the Taliban.

President Biden's decision, announced in mid-April, had changed the on-the-ground dynamics, signaling that the United States would now for real withdraw almost no matter what. This, in turn, augmented the Taliban's confidence and dealt a hammer blow to the morale of the Afghan security forces, whose dependence on U.S. military support became glaringly obvious as events unfolded. The government forces simply disintegrated as the Taliban's momentum accelerated. The United States scrambled fervently to evacuate its many remaining citizens—military and non-military alike—as quickly as possible, as did other countries (Reuters, 2021d). Numerous ordinary Afghans also attempted to flee their country, leading to tragic scenes of people clinging on to U.S. military planes, some falling to their death (Harding & Doherty, 2021). On August 26, the Islamic State exploited the commotion around Kabul Airport, conducting a suicide attack that killed close to two hundred people, including thirteen U.S. military personnel (Burke, 2021; Hubbard et al., 2021). The United States launched a drone attack the day after; it believed it had hit Islamic State terrorists, but it later acknowledged that it had instead mistakenly killed ten Afghan civilians, including seven children (Schmitt & Cooper, 2021). On August 30, the last of the U.S. troops left Afghanistan.

Despite the unprecedented chaos that had ensued, President Biden and his administration steadfastly maintained that withdrawal had been the right decision. They largely blamed the calamity on the demoralized and fleeing Afghan leaders and security forces, arguing that it was not the United States' responsibility to fight and to witness its soldiers die in a war when the willingness to fight among Washington's local allies was nearly non-existent (Pannett et al., 2021). To withdraw was necessary, the president held, even if he added that there is "never a good time to withdraw U.S. forces" (Suliman, 2021).

It is safe to say, though, that not everyone was impressed by these retrenchment justifications. Again, gloomy predictions abounded, largely centering on the same *potential* falling dominoes as had so long been feared. The difference was that, by early September, *some* of the troubling effects of the withdrawal had become patently evident, even if most still belonged in the realm of hypotheticals and possibilities. The U.S.-supported government in Kabul had surely and emphatically collapsed, with the Taliban rushing in to fill the vacuum. This was a portent of a future Afghanistan once more governed by Islamist fundamentalists disregarding basic human rights and, most ominously for Americans, *perhaps* providing yet again a secure base from which terrorist groups could organize attacks. The "insurance policy against another 9/11" would evaporate if the troops were pulled out, Republican Senator Lindsay Graham (R-SC) had warned when the withdrawal decision was originally made (Holland & Stewart, 2021). And even if Al-Qaeda seemed to remain the main long-term concern with respect to the security of the U.S. homeland, the immediate post-withdrawal phase saw a sharp rise in terrorist attacks within Afghanistan conducted by the Islamic State (Human Rights Watch, 2021). The troop withdrawal, largely perceived as a U.S. defeat, was even said to have given considerable impetus to *jihadists* as far away as in West Africa (Paquette & Noack, 2021).

The broader implications of the Afghan debacle for the international prestige and reputation of the United States were also a huge worry. This was the United States' "Saigon moment," many claimed, referring to the equally chaotic 1975 exit from a lost war in Vietnam (Schnell, 2021). *The Economist* (2021f), also likening events to Saigon, called the calamitous days in August 2021 "a defining geopolitical moment" as they made it egregiously plain that the superpower had been defeated in a war against minnows. That those U.S. allies which had long had a presence in Afghanistan were indignant that Washington had so hastily decided to withdraw without consulting them seemed to matter less than the question marks that arose among virtually *all* allies concerning U.S. resolve and trustworthiness (Gallo, 2021; Reuters, 2021e; Sly, 2021).

At the end of the day, the assessment of the Afghan withdrawal rests on a counterfactual: What would have happened if the Biden administration had decided instead to ignore the peace deal and keep the troops in Afghanistan? The president seemed adamant that a "middle ground" did not in fact exist: either all troops had to leave, or a major troop surge had to be ordered (Baker, 2021). Others, including high-ranking Pentagon officials, disagreed. In testimony to the U.S. Senate Armed Services Committee, chairman of the Joint Chiefs of Staff General Mark A. Milley, and CENTCOM chief General Frank McKenzie, revealed that they had recommended that the president keep a minimum of 2500 troops in Afghanistan (Demirjian et al., 2021). The reasoning behind the recommendation, which was backed by several other U.S. military leaders and officials, was that even a relatively small military contingent was *vastly* better than none at all, as a few thousand U.S. troops would likely have been sufficient to help Afghan government forces stave off a Taliban takeover and to secure the admittedly uneasy status quo for many years (Baker, 2021; Shane III, 2021). In other words, attaining "victory" was not the real, or at least not a very realistic, objective of the troops that were eventually pulled out; *avoiding very detrimental outcomes*—even if most of these were, and still are, eminently hypothetical ones—was.

Ten Conclusions from the Book

Even if the Afghan case in the end resulted in a complete pullout of U.S. forces, it does not refute the overall arguments and findings presented in this book. To the contrary, each of the many phases of the withdrawal process—from Donald Trump's early months as president and until August 30, 2021—brimmed with warnings about all the things that *could* go wrong if the United States abandoned Afghanistan. These warnings were, for a while, a major factor contributing to halting the withdrawal. Indeed, Donald Trump initially increased troop numbers in Afghanistan due to classical domino concerns conveyed by his military advisors (BBC, 2017; MacDonald, 2018, pp. 427–428), and the actual drawdown only began in spring 2020. Eventually, though, the warnings and the pressure from the broader foreign-policy community, and from U.S. allies, proved insufficient: the withdrawal process seemed to have gained irreversible momentum. Furthermore, Afghanistan is a case—in many ways the *only* case in the period covered herein (with the partial exceptions of Syria and Iraq)—where we have gotten a clear glimpse of the *manifestation* of falling dominoes or adverse geopolitical chain reactions. In that sense, the Afghan withdrawal provides us with a natural experiment of the consequences of retrenchment, even if most of those consequences are as yet uncertain.

A Leaver's Clash with the Stayers

Afghanistan is thus a case that nicely (though also very tragically) illustrates some of the key points presented in this book. In fact, if I summarize my arguments and empirical findings, all ten conclusions that emerge can be quite usefully illuminated by the drawn-out withdrawal story from Afghanistan. The first conclusion, which is entirely empirically based, is that the United States, in the Trump era, did not retrench in any really meaningful way. The exceptions largely prove the rule: the Trump administration must surely take much of the credit (or blame) for the Afghan withdrawal, which had already come a long way when Joe Biden assumed office. A few thousand troops were eventually also pulled out of Iraq. There, the retrenchment story pretty much stops. Granted, the United States also left Somalia near the end of 2020. However, it was Trump himself who had initially decided to send U.S. troops there, to fight the al-Shabaab; and the Somali-based troops were not withdrawn from the African continent, but were instead relocated to neighboring countries. As for Syria, the president had repeatedly signaled that U.S. troops would leave; and while some of the troops did actually leave in late 2019, most remained, and they remained in larger numbers than when Trump had entered the Oval Office. With respect to the U.S. troops stationed at bases in countries that were not ridden by recent or ongoing war, they virtually all remained where they were. In early 2021, the United States still had 60,000 or so military-service personnel in the Middle East; the quite considerable tally of troops in Europe stayed almost exactly the same; and the Asia–Pacific region saw U.S. troop numbers rise by a third over the period.

Second, despite that Trump, when judged solely by outcomes, turned out to resemble a *Stayer*, the overall evidence does suggest that he was very much the *Leaver* that many feared (and some hoped) he would be. Trump clearly *wanted* to bring U.S. soldiers out of Afghanistan, Iraq, Syria, Somalia, and Germany; and he did, with a reasonable amount of credibility, threaten to withdraw from other hosts, such as South Korea and Saudi Arabia. More broadly, Trump appeared to pose a severe challenge to numerous U.S. alliances and security partnerships: he intermittently threatened to leave NATO, to abandon South Korea, and to rethink the U.S.-Japan alliance and the many security partnerships with allies around the Persian Gulf. As we now know, all these alliances survived to live another day, but the allies themselves were certainly not entirely certain that they would. Numerous scholars also highlighted the "Fortress America" dimension of the president's ambitions, which could very well lead to a major drawdown of U.S. overseas troops unless the president's inclinations were tempered by a sufficient number of countervailing forces (Brands, 2017; Clarke & Ricketts, 2017). Donald Trump represents, so to speak, one half of U.S. citizens and voters; viz., that half whose position on the issue of the global role of the United States is imbued by isolationist desires. In this, I may again draw on Nicholas Spykman's book from the early 1940s; it revolved in no small part around the tension and debate between two halves of geostrategic thinking among Americans. One half championed isolation, the other intervention, in what, according to Spykman (2017 [1942], p. 5) "is the oldest issue in American foreign policy."

The third conclusion arising from my study is that every single case of proposed withdrawal was met by a flurry of warnings from the broader U.S. foreign-policy community—or the Blob—which exerted a considerable amount of pressure on the administration to abstain from retrenching. This group of voices—the "carriers" of foreign-policy ideas (Meibauer, 2020)—includes, among others, foreign-policy analysts and scholars, the mainstream media, members of the U.S. Congress, members of the U.S. government and bureaucracy, intelligence agencies, and think tanks and lobby groups. They constitute a large and loud and powerful, if also uncoordinated, crowd. When U.S. troops are seemingly on their way out of Syria, or Afghanistan, or Germany, or South Korea, or some other place—*any* place really—members of the Blob write scathing journal articles and newspaper op-eds, they depict doomsday-like scenarios in television interviews, they hold speeches and engage in debates where they point to all the horrible consequences that a troop withdrawal *might* have, and they enact legislation seeking to put a final stop to any retrenchment aspirations. And, for the most part, they succeed. They managed to stop the withdrawal from Germany; they made the president dither continuously over the long-promised pullout from Syria, a country which the United States never actually left; they likely contributed to pre-empt any move that would start a process of retrenchment from South Korea, from Japan, from West Africa, and from U.S. allies in the Gulf; and, for a few years, they helped delay a withdrawal from Afghanistan, though they couldn't in the end prevent it.

System, Psychology, and the Specter of Falling Dominoes

The fourth conclusion is that these warnings are conveyed by a diverse group of *Stayers* who first and foremost draw on the system level of analysis to construct their narratives and rationales explaining why a withdrawal must be avoided at virtually all costs. Other constraints on retrenchment, such as sheer institutional inertia and the physical and pecuniary realities of military bases and sunk costs, do play their part; but they usually do so in combination with system-level properties and developments—or *hypothetical* developments—and these system-level factors play a dominating role whenever a troop withdrawal really seems to be in the cards. As the neoclassical-realist school argues, the complex and murky signals that emanate from the international system are filtered through the minds of these carriers of ideas, who interpret them likely as best as they can, ultimately deriving from them recommendations about U.S. foreign policy. Whether explicitly or implicitly, members of the foreign-policy community describe an anarchical world wherein power is the key currency—along with a considerable dose of history and geography. Absent any central power to resolve international conflicts, the highly interconnected world is generally interpreted as a dangerous one where one single misstep—such as a pullout of some U.S. troops from somewhere—*may* set in motion a process, or an avalanche, whose end result *might* be very detrimental to U.S. national security.

Fifth, power, history, and geography may be the underlying factors from which the *Stayers* of the foreign-policy community take their cue when they interpret international politics. Still, psychological mechanisms undoubtedly play their own vital part in terms of giving the interpretations and policy recommendations a very specific direction. This is the direction of negativity, of caution, of resistance to change. Many international-security scholars grumble about the threat-inflation and alarmist behavior that permeate the Blob (e.g., Layne, 2017; Mueller & Stewart, 2012; Walt, 2018). On the other hand, the field of behavioral psychology has long since presented convincing arguments and empirical evidence to suggest that human beings—and, by implication, states—are commonly inflicted by a "negativity bias." This means that negative phenomena and developments (including hypothetical ones) in one's external environment tend to be weighted heavier than positive ones (Johnson & Tierney, 2018–2019; Rozin & Royzman, 2001). In all likelihood a trait rooted in our evolutionary past (McDermott et al., 2008), people tend to be loss averse around some reference point—usually the status quo—to which they are psychologically committed (Kahneman & Tversky, 1979). Such assertions, arguments, and findings mesh well with political-realist theory, which generally paints a very bleak overall picture of international politics. In particular, realism's "defensive" branch argues that states are overwhelmingly positionalists; that is, they value what they already possess more than what they don't possess but could potentially acquire (Grieco, 1990, p. 40; Schweller, 1996; Waltz, 1979, p. 126). Indeed, political realism, taken as a whole, tend to exhibit, if sometimes only tacitly, a profound understanding of why states are apt to institute "safety-first" policies based on worst-case-scenario assumptions. Prospect theory and political realism, *especially in conjunction*, offer a very convincing explanation of why members of the U.S. foreign-policy community are so profoundly reluctant to go along with any retrenchment suggestion.

Sixth, more often than not, the process whereby the murky systemic impulses are filtered through the minds of recipients who are heavily influenced by a negativity bias produces warnings about the very real *possibility* that an array of adverse geopolitical chain reactions *might* ensue if the United States pulls out its troops from somewhere or anywhere. That is to say, the *Stayers* regularly, and sometimes seemingly inevitably, stress the specter of falling dominoes. The existing U.S. military-base network is the known entity, the entity that represents normality and stability in the eyes of the Blob. The unknown is exactly that: unknown; and it is this unknown entity that will manifest once U.S. troops have withdrawn. The upshot is ostensibly *potentially* calamitous: "One thing can lead to another," as Michael Mandelbaum (1988, p. 137) laconically states in his analysis of why the United States has been so prone to seek to extend its influence abroad. To wit, the basic reasoning—again whether it is only implicitly expressed—is that one harmful thing *may* lead to another *unless the United States stays where it is.*

Power, Credibility, and Geography

Seventh, it follows that most members of the foreign-policy community want the United States to shape and to control its external environment—i.e., to maximize its influence abroad—as much as it can, not least to steady wobbling dominoes. These dominoes fall into various categories, depending on case-specific context. Almost always a prominent worry, though, is the fear that major-power adversaries *might* exploit a U.S. troop withdrawal in order to extend their own influence in what is perceived to be a zero-sum game of international power and security (Nieman et al., 2021). Since power tends to beget power, the adversary's initial progress *might* lead to it progressing even further thereafter. Also of prominence is the fear that smaller states, such as recalcitrant regional powers harboring nuclear-weapons ambitions, *might* become emboldened by U.S. retrenchment, which *might*, in turn, make even more states, including U.S. allies, consider going nuclear. Lastly, in large parts of the world, the "vacuums" created by a U.S. withdrawal *might*, it is commonly feared, make the threat posed by transnational terrorist groups multiply and metastasize. The U.S. foreign-policy community fully accept the view, famously expressed in 1998 by then U.S. Secretary of State Madeleine Albright, that the United States is "the indispensable nation" (U.S. Department of State, 2021). U.S. leadership and military-force projection, they hold, are absolute necessities for purposes of securing a stable world order through deterring bad actors and reassuring those states that are as inclined as the United States to prefer that the status quo prevails. In a world devoid of U.S. foreign military bases, Russia *might* stamp its authoritarian authority on much of Europe; the Middle East *might* descend into an inimitable chaos of interstate rivalry, intrastate conflict, and terrorism; and China *might* soon attain regional hegemony, which *might* eventually lead to Beijing dominating world affairs.

Eighth, what we have also learned from recent years' withdrawal cases and debates is that nearly all *Stayers* are enormously concerned about the reputation and credibility, or perceptions of the resolve and power of, the United States. Indeed, this runs as a common thread through the last seven decades of U.S. history. The *Stayers* at the time of the Vietnam War did not primarily worry about the consequences of a U.S. troop withdrawal on Vietnam per se; they worried more about the losses that would be inflicted on the prestige of the United States. The superpower rivalry, many feared, *could* easily be lost to the Soviet Union, or to China, or to both; if, that is, the United States failed, first, to scale up their war efforts in Vietnam and, second, to show sufficient staying power. This was what President Eisenhower meant by the "'falling domino' principle": Vietnam was the first, shaky domino whose *possible* fall *could* lead to "a disintegration that would have the most profound influences" and cause losses that "are just incalculable to the free world" (U.S. Department of State, Office of the Historian, 2021). The same reasonings and narratives saturate debates about U.S. troop withdrawals from Afghanistan, from Syria, from Iraq, from Somalia, from West Africa, from Germany, and from South Korea. If we presume that retrenchment entails a variant of "loss" for the United States, and if we follow the assumption of Robert Jervis (2017, p. 87), that "[l]osses may not scale in their international impact" so "it is possible that in terms of reputation and credibility a small loss would cost the country not much less than a significantly larger one," the implication that follows is this: Even in cases where the actual number of troops that might be pulled out is quite modest, such as in Syria and Africa, the presumed effects on *perceptions* of U.S. credibility are judged to be ***potentially*** disastrous, which is exactly the view of the *Stayers*.

Ninth, modest troop withdrawals or base closures are not only believed to hamper U.S. credibility in a non-material sense; they also ***risk*** damaging the effectiveness of the entire U.S. military-base network. This network is still rooted in a fundamentally geographic logic superbly reflected decades ago in the writings of classical geopoliticians such as Halford Mackinder, Nicholas Spykman, and Alfred Thayer Mahan. The main mission of the troops that the United States has deployed overseas has always been to forestall any expansionary moves by great Eurasian powers. It still is. Now, Russia and China—and smaller regional powers such as Iran—are to be balanced by way of a network of military bases and tens of thousands of U.S. troops primarily stationed along the rimland of Eurasia. These bases come in addition to an equally impressive network of small nodes along the main oceanic "highways." Even if mostly only unspoken, the value of this entire system of points on the map is given a prominent place in the narratives of the *Stayers*, who are loath to see any single node or bridgehead or base or battalion disappear. The military-base network, or so is the message one may extract from their arguments, forms a coherent whole in a world system that makes up a complex, interconnected strategic whole where just one missing peace ***might*** set in motion events that ***could*** ultimately make the very favorable status quo disintegrate.

The tenth and last of the conclusions from my study is that the United States' command of material power is still overwhelming. Granted, America finds itself in relative decline, as it has done for many years now. But anything else would have been wholly unnatural. During the height of unipolarity a couple of decades ago, the United States ruled absolutely supreme on nearly all parameters of power, including non-material ones. As the centuries-long history of the rise and fall of great powers has taught us, such one-sidedness is fundamentally unsustainable. The decline of America, and the concomitant relative rise of China and a few others, is a slow process, it is gradual, and it originates in a previous period, not long ago, where U.S. relative power was truly remarkable. One may surely debate what metrics are most suitable to get a grasp of the true power of a nation—if there is such thing as true power—and one may also emphasize the importance of perceptions. Still, there is no hiding the fact the United States continues to command nearly one quarter of world GDP (measured at official exchange rates). Neither should one disregard that the United States accounts for nearly 40% of world military spending.

This is a share that is unprecedented in modern world history—if, that is, one does not compare it with previous versions of the United States. But even if one did, one would find that the number is the same as that for the U.S. anno 1999, arguably around the height of unipolarity. *In addition* comes the abundance of U.S. alliances and security partnership and, of course, its vast network of military bases and attendant forward-deployed troops. The latter are part of the power resources of the United States; but they also represent power-as-influence. And, as held by political realism, in general, and neoclassical realist theory in particular, a nation commanding huge power resources would normally want to exert influence commensurate with its potential; it would want, as the *Stayers* clearly expect and even take for granted that it would and should, maximize influence in order to shape and control a *potentially* unruly external environment as much as possible.

What Could Make America Retrench?

Prospects of U.S. retrenchment should not be dismissed out of hand, however. But exactly what developments or events, then, could possibly make the United States scale back its military-force projection abroad? One obvious candidate is a second round of Donald Trump as president, or of someone of his ilk. For it is clear that the outcome of the 2016 presidential election was in part impacted by the way that Donald Trump's isolationist impulses and rhetoric resonated with a substantial share of U.S. voters. On the other hand, as has become patently clear, the first round of Trump failed to bring about any meaningful retrenchment. Any eventual second round is also certain to involve the very same counterforces that so vigorously appeared in the first. Moreover, the American public has for generations been ambivalent, divided squarely down the middle, on the issue of the global role of the United States.

What one might by now suspect is something very close to an innate trait of the U.S. public may actually work to increase the likelihood that the United States remains deeply engaged in global affairs. It is fully conceivable that Robert Kagan (2012, p. 18) is right when he states that "it was precisely Americans' limitations and hesitations that made them such an attractive leader of the transatlantic 'empire'." This combines with the United States also being, from the viewpoint of (most of) the rest of the world, a faraway sea power that is less likely to pose a mortal threat and more likely to represent a very useful ally against more proximate perils (Levy & Thompson, 2010). This points to forces that counteract a second possible cause of U.S. retrenchment, namely, that Washington will be "pushed" to leave. This also seems a highly unlikely prospect for the foreseeable future. The United States is very rarely forced to pack up its military forces and leave a host country. During Trump's tenure, apart from the countries in which the U.S. troops had originally been part of an invasion force (notably, Iraq, Syria, and Afghanistan), it is really only the Philippines—or, rather, the president of the Philippines—that has been openly critical of a U.S. military presence, which in any case has been meager since the early 1990s. The Philippines has apparently chosen to hedge its security bets by engaging in a sort of quasi-bandwagoning with China. But recent indications are that Manila has become somewhat disillusioned with China's behavior, which suggests that the relationship between the Philippines and the United States is about to strengthen again (*The Economist*, 2021c).

Third and relatedly, most *Leavers* and most offshore balancers would advise U.S. policymakers to "pass the buck" to regional powers, which would then themselves, in the first instance, be tasked with, and bear the costs of, countering any potential menace in their neighborhood (Mearsheimer & Walt, 2016, p. 71). Insofar as there exist such alternative balancers, or even one or more "powerful and closely allied 'successor states'" to America (Haynes, 2015, p. 490), Washington could withdraw the bulk of its forces even from vital regions, instead keeping them "over the horizon." However, while sounding plausible in theory, in practice it is highly doubtful that any potent buck recipient actually exists. As for Asia, even steadfast *Leavers* do not unequivocally recommend any major U.S. retrenchment considering the rapid rise of China, which is a plausible prospective regional hegemon (Walt, 2018, p. 269). In the Middle East, the United States did for a long time pursue something akin to an offshore-balancing strategy, only to be sucked into the region for real following Iraq's invasion of Kuwait in 1990. Thereafter, several U.S. presidents have expressed a desire to retrench, only to discover that extricating the country from the Middle East is incredibly difficult, in no small part because there is a dearth of reliable allies to which the buck can be passed. Europe, for its part, harbors enduring ambitions to achieve "strategic autonomy" from the United States. However, it still seems a very long way from acquiring the necessary military wherewithal and political coherence to allow, or to compel, NATO and the United States to assume a backseat role in European security (Meijer & Brooks, 2021).

A fourth, more fundamental reason why great powers sooner or later are forced to scale back their global activities are profound changes in the balance of power. Insofar as one's own command of economic and military resources is declining relative to the power of other states, the status quo may be unsustainable. Empires rise and empires fall depending not least on changes in relative power. As for falling ones, "overstretching" compounds the problem: a tendency "for the economic costs of maintaining the status quo to rise faster than the economic capacity to support the status quo" (Gilpin, 1981, p. 156) has befallen an array of empires throughout history, including the British and Soviet ones. The United States, though, is not like the Great Britain of yesteryear, whose quite enormous possessions around the globe rested on a relatively limited economic and demographic home base. Already at the end of the nineteenth century, it was clear that other great powers, notably Germany and the United States, had caught up with Britain on many parameters signifying economic potential, such as energy consumption, steel production, and manufacturing output (Kennedy, 1988, pp. 257–259). Even so, the Empire lived on, and it even expanded after the calamity that was the First World War, before it finally entered a process of fairly rapid unraveling following World War II. In contrast, the United States still finds itself in a completely different and more prominent league with respect to most power metrics. And it is certainly not even close to being plagued by the same economic woes that the Soviet Union suffered from the 1970s until the final, overdue collapse of the superpower in late 1991.

Moreover, compared to the British and Soviet empires, the U.S. "empire" is not really a true empire at all in the traditional understanding of the word; it is generally not geared toward territorial and political control, and it has, for the most part, avoided falling into what George Modelski (1978, p. 230) calls the trap of "territoriality." It is in many respects an empire "on the cheap," an empire whose relative cost effectiveness has long, and still is, aided by it also being largely "an empire by invitation," which reduces much of the friction costs that other hegemons have had to pay (Lundestad, 1986). And neither has the United States to contend with massive security threats near its own doorstep, and little or nothing indicates that it will in the foreseeable future. Such threats represented a severe drain on resources and attention for both the United Kingdom and the Soviet Union, and they were also what helped force a drawdown of the Dutch maritime empire in the late seventeenth century. While it is clear that the United States really *is* in relative decline, this is a very slow and gradual process, and one that seems far more manageable than the decline experienced by most former empires and hegemons. The ongoing "shift in economic gravity" toward Asia, writes the international relations scholar Christopher Coker (2015, p. 50), "is not the same as a historical power shift, and that is the main difference between Britain's decline after 1870 and America's today."

A fifth and final possible cause of real U.S. retrenchment is one whose likelihood of occurring is considerably more difficult to assess, even if it is probably the most plausible candidate. This is the possibility that some dramatic event, a "crushing international setback" or "catastrophic foreign policy disaster," in the words of Stephen Walt (2018, pp. 19, 282)—a "black swan" or something that constitute a similar unforeseen and unforeseeable surprise situated "outside the realm of regular expectations" (Taleb, 2010 [2007], p. xxii)—will either prompt the *Stayers* to rethink U.S. grand strategy or by itself leave Washington with no choice but to abandon its informal empire. The outbreak of a major war, especially a multi-front one, could produce such an effect. The British Empire, though ultimately successful on the battlefield, found itself increasingly overstretched from the late nineteenth century and through the two world wars of the twentieth. Even severe international tensions short of any shooting war can be energy-sapping enough to affect adversaries' and allies', and one's own, *perceptions* of one's capabilities and resolve. Once more, we are then in the realm of credibility, which is a tenuous phenomenon—or, at least, it is believed by many or most to be so. Members of the United States' foreign-policy establishment were obsessed with U.S. credibility throughout the four decades of the Cold War history; indeed, they were likely more obsessed with credibility, with avoiding humiliation and retreat even on relatively minor issues, than with the threat posed by communism itself (Gaddis, 2005 [1982], pp. 211–212, 239). Appearing to be weak in and over Berlin, Korea, Formosa, Cuba, or Vietnam could easily, it was widely believed, mark the beginning of the end of the U.S. Empire.

*

Withdrawing forward-deployed U.S. troops can also be interpreted as weakness. It is surely so that most members of the U.S. foreign-policy community reckon that *others*—allies and adversaries alike—will perceive withdrawal as such. This means that in the times ahead we can expect Washington's foes to test U.S. resolve whenever they are given the opportunity, with the same hope that the Soviet Union once held of starting a chain-reaction process that would ultimately lead to major U.S. retrenchment. However, it is, among other things, for this exact reason that we can expect the *Stayers* of the Blob to continue to argue and to work forcefully for a continuation of the current state of affairs—the status quo—and for a perpetuation of U.S. global engagement, U.S. hegemony, and the far-flung network of U.S. overseas military bases and troops.

References

Ali, I., & Lema, K. (2021, July 30). Philippines' Duterte fully restores key U.S. troop pact. *Reuters*. https://www.reuters.com/world/asia-pacific/us-aims-shore-up-philippine-ties-troop-pact-future-lingers-2021-07-29/

Al Jazeera. (2021a, February 17). *Japan, US extend agreement on military base payments*. https://www.aljazeera.com/news/2021/2/17/japan-us-extend-agreement-on-military-base-payments

Al Jazeera. (2021b, September 17). *Australia says more US troops to come, plans missile project*. https://www.aljazeera.com/news/2021/9/17/australia-says-more-us-troops-to-come-eyes-missile-work

Baker, P. (2009, October 13). Biden no longer a lone voice on Afghanistan. *New York Times*. https://www.nytimes.com/2009/10/14/world/14biden.html

Baker, P. (2021, September 1). All in or all out? Biden saw no middle ground in Afghanistan. *New York Times*. https://www.nytimes.com/2021/08/28/us/politics/trump-taliban-biden-afghanistan.html

Barberá, M. G. (2021, May 18). US to turn Romanian airbase into NATO Black Sea hub. *Balkan Insight*. https://balkaninsight.com/2021/05/18/us-to-turn-romanian-airbase-into-nato-black-sea-hub/

Barnes, J. E., Gibbons-Neff, T., & Schmitt, E. (2021, March 26). Officials try to sway Biden using intelligence on potential for Taliban takeover of Afghanistan. *New York Times*. https://www.nytimes.com/2021/03/26/us/politics/biden-afghanistan-intelligence.html

BBC. (2017, August 22). *Trump rules out Afghan troops withdrawal*. https://www.bbc.com/news/world-us-canada-41008116

BBC. (2021, February 5). *Yemen war: Joe Biden ends support for operations in foreign policy reset*. https://www.bbc.com/news/world-middle-east-55941588

Biden, J. R., Jr. (2021, March). *Interim national security strategic guidance*. The White House. Retrieved September 9, 2021, from https://www.whitehouse.gov/wp-content/uploads/2021/03/NSC-1v2.pdf

Brands, H. (2017). U.S. grand strategy in an age of nationalism: Fortress America and its alternatives. *Washington Quarterly*, *40*(1), 73–94.

Burke, J. (2021, August 26). Islamic State claims responsibility for Kabul airport blasts. *The Guardian*. https://www.theguardian.com/world/2021/aug/26/isis-affiliate-iskp-is-prime-suspect-for-kabul-airport-suicide-bomb

Burns, R., Madhani, A., & Abdul-Zahra, Q. (2021, July 27). Biden says US combat mission in Iraq to conclude by year end. *Associated Press*. https://apnews.com/article/joe-biden-government-and-politics-middle-east-iraq-islamic-state-group-9397d9996703d7416f857165072a0a05

Clarke, M., & Ricketts, A. (2017). Donald Trump and American foreign policy: The return of the Jacksonian tradition. *Comparative Strategy, 36*(4), 366–379.

Coker, C. (2015). *The improbable war: China, the United States and the continuing logic of great power conflict*. Oxford University Press.

Demirjian, K., Horton, A., Wagner, J., & Sonmez, F. (2021, September 28). Military leaders, refusing to fault Biden, say troop withdrawal ensured Afghanistan's collapse. *Washington Post*. https://www.washingtonpost.com/national-security/afghanistan-hearing-milley-austin-mckenzie/2021/09/28/75d1557e-2086-11ec-9309-b743b79abc59_story.html

DeYoung, K. (2020, October 21). Biden foreign policy begins with telling the world: 'America's back'. *Washington Post*. https://www.washingtonpost.com/national-security/biden-foreign-policy-begins-with-telling-the-world-americas-back/2020/10/21/2fc0e528-1348-11eb-bc10-40b25382f1be_story.html

Dobbins, J. (2021, April 26). *To lose a war*. RAND Corporation. Retrieved August 12, 2021, from https://www.rand.org/blog/2021/04/to-lose-a-war.html

France24. (2021, October, 14). *US, Greece expand defense ties amid Mediterranean tensions*. https://www.france24.com/en/live-news/20211014-us-greece-expand-defense-ties-amid-mediterranean-tensions

Gaddis, J. L. (2005). *Strategies of containment: A critical appraisal of American national security policy during the Cold War*. Oxford University Press (Original work published 1982).

Gallo, W. (2021, August 20). How the Afghanistan withdrawal looks from South Korea, America's other 'forever war.' *VOA News*. https://www.voanews.com/a/usa_how-afghanistan-withdrawal-looks-south-korea-americas-other-forever-war/6209777.html

Gilpin, R. (1981). *War and change in world politics*. Cambridge University Press.

Grieco, J. M. (1990). *Cooperation among nations: Europe, America, and non-tariff barriers to trade*. Cornell University Press.

Harding, L., & Doherty, B. (2021, August 16). Kabul airport: Footage appears to show Afghans falling from plane after takeoff. *The Guardian*. https://www.theguardian.com/world/2021/aug/16/kabul-airport-chaos-and-panic-as-afghans-and-foreigners-attempt-to-flee-the-capital

Haynes, K. (2015). Decline and devolution: The sources of strategic military retrenchment. *International Studies Quarterly, 59*(3), 490–502.
Holland, S., & Stewart, P. (2021, April 14). 'Time to end the forever war': Biden to start U.S. Afghanistan pullout on May 1. *Reuters*. https://www.reuters.com/world/asia-pacific/biden-ready-announce-us-withdrawal-even-peace-eludes-afghanistan-2021-04-14/
Hubbard, B., Schmitt, E., & Rosenberg, M. (2021, August 26). After decades of war, ISIS and Al Qaeda can still wreak havoc. *New York Times*. https://www.nytimes.com/2021/08/26/world/asia/Afghanistan-isis-qaeda.html
Human Rights Watch. (2021, October 25). *Afghanistan: Surge in Islamic State attacks on Shia*. Retrieved November 2, 2021, from https://www.hrw.org/news/2021/10/25/afghanistan-surge-islamic-state-attacks-shia
Jakobsen, J., & Jakobsen, T. G. (2019). Tripwires and free-riders: Do forward-deployed U.S. troops reduce the willingness of host-country citizens to fight for their country? *Contemporary Security Policy, 40*(2), 135–164.
Jervis, R. (2017). *How statesmen think: The psychology of international politics*. Princeton University Press.
Johnson, D. D. P., & Tierney, D. (2018–2019). Bad world: The negativity bias in international politics. *International Security, 43*(3), 96–140.
Kagan, R. (2012). *The world America made*. Vintage.
Kahneman, D., & Tversky, A. (1979). Prospect theory: An analysis of decision under risk. *Econometrica, 47*(2), 263–291.
Kennedy, P. (1988). *The rise and fall of the great powers*. Random House.
Layne, C. (2017). The US foreign policy establishment and grand strategy: How American elites obstruct strategic adjustment. *International Politics, 54*(3), 260–275.
Lee, M., & Tucker, E. (2021, August 19). Was Biden handcuffed by Trump's Taliban deal in Doha? *Associated Press*. https://apnews.com/article/joe-biden-middle-east-taliban-doha-e6f48507848aef2ee849154604aa11be
Levy, J. S., & Thompson, W. R. (2010). Balancing on land and at sea: Do states ally against the leading global power? *International Security, 35*(1), 7–43.
Lubold, G., & Strobel, W. P. (2021, April 1). Biden trimming forces sent to Mideast to help Saudi Arabia. *Wall Street Journal*. https://www.wsj.com/articles/biden-trimming-forces-sent-to-mideast-to-help-saudi-arabia-11617279687
Lundestad, G. (1986). Empire by invitation? The United States and Western Europe, 1945–1952. *Journal of Peace Research, 23*(3), 263–277.
MacDonald, P. K. (2018). America first? Explaining continuity and change in Trump's foreign policy. *Political Science Quarterly, 133*(3), 401–435.
Madhani, A. (2021, February 20). Biden declares 'America is back' in welcome words to allies. *Associated Press*. https://apnews.com/article/biden-foreign-policy-g7-summit-munich-cc10859afd0f542fd268c0a7ddcd9bb6

Mandelbaum, M. (1988). *The fate of nations: The search for national security in the nineteenth and twentieth centuries*. Cambridge University Press.

McDermott, R., Fowler, J. H., & Smirnov, O. (2008). On the evolutionary origin of prospect theory preferences. *Journal of Politics, 70*(2), 335–350.

Mearsheimer, J. J., & Walt, S. M. (2016). The case for offshore balancing: A superior U.S. grand strategy. *Foreign Affairs, 95*(4), 70–83.

Megerian, C. (2021, September 15). U.S. will share nuclear submarine technology with Australia in new defense partnership. *Los Angeles Times*. https://www.latimes.com/politics/story/2021-09-15/biden-announce-us-will-share-nuclear-submarine-technology-with-australia

Meibauer, G. (2020). Interests, ideas, and the study of state behaviour in neoclassical realism. *Review of International Studies, 46*(1), 20–36.

Meijer, H., & Brooks, S. G. (2021). Illusions of autonomy: Why Europe cannot provide for its security if the United States pulls back. *International Security, 45*(4), 7–43.

Miller, Z., & Madhani, A. (2021, July 8). 'Overdue': Biden sets Aug. 31 for US exit from Afghanistan. *Associated Press*. https://apnews.com/article/joe-biden-afghanistan-government-and-politics-86f939c746c7bc56bb9f11f095a95366

Modelski, G. (1978). The long cycle of global politics and the nation-state. *Comparative Studies in Society and History, 20*(2), 214–235.

Mueller, J., & Stewart, M. G. (2012). The terrorism delusion: America's overwrought response to September 11. *International Security, 37*(1), 81–110.

Nieman, M. D., Martinez Machain, C., Chyzh, O., & Bell, S. R. (2021). An international game of risk: Troop placement and major power competition. *Journal of Politics, 83*(4), 1307–1321.

Nikkei Asia. (2021, March 8). *South Korea agrees to boost funding for US troops*. https://asia.nikkei.com/Politics/International-relations/South-Korea-agrees-to-boost-funding-for-US-troops

Pannett, R., Hassan, J., Ang, K., Cunningham, E., Ryan, M., Parker, C., & Lamothe, D. (2021, August 16). Biden defends decision to withdraw from Afghanistan after Taliban's rapid return to power. *Washington Post*. https://www.washingtonpost.com/world/2021/08/16/afghanistan-kabul-taliban-live-updates/.

Paquette, D., & Noack, R. (2021, August 27). France's drawdown in West Africa fuels local extremists' hope for a Taliban-style victory. *Washington Post*. https://www.washingtonpost.com/world/2021/08/27/west-africa-militants-afghanistan/

Reuters. (2021a, April 13). *U.S. to base additional troops in Germany in reversal of Trump policy*. https://www.reuters.com/world/europe/us-base-additional-troops-germany-reversal-trump-policy-2021-04-13/

Reuters. (2021b, June 17). *Militant groups could pose threat to U.S. in two years from Afghanistan-Pentagon leaders*. https://www.reuters.com/world/us/terrorist-groups-could-pose-threat-us-two-years-afghanistan-pentagon-leaders-2021-06-17/

Reuters. (2021c, July 6). *U.S. military withdrawal from Afghanistan more than 90 % complete*. https://www.reuters.com/world/asia-pacific/us-military-withdrawal-afghanistan-more-than-90-complete-2021-07-06/

Reuters. (2021d, August 30). *Factbox: Evacuations from Afghanistan by country*. https://www.reuters.com/world/evacuations-afghanistan-by-country-2021-08-26/

Reuters. (2021e, September 13). *Afghan withdrawal raises questions about United States, Gulf Arab official says*. https://www.reuters.com/world/asia-pacific/afghan-withdrawal-raises-questions-about-united-states-gulf-arab-official-says-2021-09-13/

Reuters & Miglani, S. (2021, April 15). *India's military chief concerned about US pullout from Afghanistan*. https://www.reuters.com/world/asia-pacific/indias-military-chief-concerned-about-us-military-pullout-afghanistan-2021-04-15/

Reuters, Solsvik, T., & Adomaitis, N. (2021, April 16). *Norway to allow U.S. military to build on its soil in new accord*. https://www.reuters.com/world/europe/norway-may-allow-us-military-build-its-soil-revised-cooperation-deal-2021-04-16/

Ripley, W., Cheung, E., & Westcott, B. (2021, October 28). Taiwan's President says the threat from China is increasing 'every day' and confirms presence of US military trainers on the island. *CNN*. https://edition.cnn.com/2021/10/27/asia/tsai-ingwen-taiwan-china-interview-intl-hnk/index.html

Rozin, P., & Royzman, E. B. (2001). Negativity bias, negativity dominance, and contagion. *Personality and Social Psychology Review*, *5*(4), 296–320.

Ryan, M., & DeYoung, K. (2021, April 13). Biden will withdraw all U.S. forces from Afghanistan by Sept. 11, 2021. *Washington Post*. https://www.washingtonpost.com/national-security/biden-us-troop-withdrawal-afghanistan/2021/04/13/918c3cae-9beb-11eb-8a83-3bc1fa69c2e8_story.html

Sanger, D. E. (2020, November 9). The end of 'America First': How Biden says he will re-engage with the world. *New York Times*. https://www.nytimes.com/2020/11/09/us/politics/biden-foreign-policy.html

Schmitt, E., & Cooper, H. (2021, September 17). Pentagon acknowledges Aug. 29 drone strike in Afghanistan was a tragic mistake that killed 10 civilians. *New York Times*. https://www.nytimes.com/2021/09/17/us/politics/pentagon-drone-strike-afghanistan.html

Schnell, M. (2021, August 29). McConnell rips Biden's Afghanistan withdrawal decision. *The Hill*. https://thehill.com/homenews/sunday-talk-shows/569896-mcconnell-afghanistan-one-of-the-worst-foreign-policy-decisions-in?rl=1

Schweller, R. L. (1996). Neorealism's status-quo bias: What security dilemma? *Security Studies, 5*(3), 90–121.

Seligman, L. (2021, July 27). Troops to stay put in Syria even as Biden seeks to end America's 'forever wars.' *Politico.* https://www.politico.com/news/2021/07/27/troops-to-stay-in-syria-biden-500848

Shane III, L. (2021, October 5). Former administration officials argue over whether keeping US troops in Afghanistan would have helped. *Military Times.* https://www.militarytimes.com/flashpoints/afghanistan/2021/10/05/former-administration-officials-argue-over-whether-keeping-us-troops-in-afghanistan-would-have-helped/

Sly, L. (2021, August 15). Afghanistan's collapse leaves allies questioning U.S. resolve on other fronts. *Washington Post.* https://www.washingtonpost.com/world/asia_pacific/afghanistan-chaos-blame-us/2021/08/14/0d4e5ab2-fd3e-11eb-911c-524bc8b68f17_story.html

Sommerville, Q. (2021, June 8). Iraqis suffer as US–Iran shadow war shifts gear. *BBC.* https://www.bbc.com/news/world-middle-east-57399895

Spykman, N. J. (2017). *America's strategy in world politics: The United States and the balance of power.* Routledge (Original work published 1942).

Suliman, A. (2021, August 20). Nearly 20 years of war, 10 days to fall: Afghanistan, by the numbers. *Washington Post.* https://www.washingtonpost.com/world/2021/08/20/afghanistan-war-key-numbers/

Taleb, N. N. (2010). *The Black Swan: The impact of the highly improbable.* Random House (Original work published 2007).

The Economist. (2021a, April 17). Retreat from Kabul: Joe Biden offers unconvincing reasons for ending America's longest war. https://www.economist.com/united-states/2021/04/17/joe-biden-offers-unconvincing-reasons-for-ending-americas-longest-war

The Economist. (2021b, April 17). Ending the forever war: Joe Biden calls time on America's longest-ever war. https://www.economist.com/asia/2021/04/13/joe-biden-calls-time-on-americas-longest-ever-war

The Economist. (2021c, May 27). The Philippines is repairing its relationship with America. https://www.economist.com/asia/2021/05/27/the-philippines-is-repairing-its-relationship-with-america

The Economist. (2021d, July 31). Eyeing the exits: Why Joe Biden is ending America's combat mission in Iraq. https://www.economist.com/united-states/2021/07/27/why-joe-biden-is-ending-americas-combat-mission-in-iraq

The Economist. (2021e, August 13). The endgame: Afghanistan is disintegrating as the Taliban gain momentum. https://www.economist.com/asia/afghanistan-is-disintegrating-as-the-taliban-gain-momentum/21803598

The Economist. (2021f, August 21). The Taliban are back in town: From Saigon to Kabul: What America's Afghan fiasco means for the world. https://www.economist.com/briefing/2021/08/21/from-saigon-to-kabul-what-americas-afghan-fiasco-means-for-the-world

The White House. (2021, February 4). *Remarks by President Biden on America's place in the world*. Retrieved April 27, 2021, from https://www.whitehouse.gov/briefing-room/speeches-remarks/2021/02/04/remarks-by-president-biden-on-americas-place-in-the-world/

U.S. Department of Defense. (2021). Military and civilian personnel by service/agency by state/country. *U.S. Defense Manpower Data Center*. Retrieved November 23, 2021, from https://dwp.dmdc.osd.mil/dwp/app/dod-data-reports/workforce-reports

U.S. Department of State. (2021). *Secretary of State Madeleine K. Albright: Interview on NBC-TV "The Today Show" with Matt Lauer*. Retrieved October 15, 2021, from https://1997-2001.state.gov/statements/1998/980219a.html

U.S. Department of State, Office of the Historian. (2021). *Foreign relations of the United States, 1952–1954, Indochina, Volume XIII, Part 1*. Retrieved March 13, 2020, from https://history.state.gov/historicaldocuments/frus1952-54v13p1/d716

Walt, S. M. (2018). *The hell of good intentions: America's foreign policy elite and the decline of U.S. primacy*. Farrar, Straus and Giroux.

Waltz, K. N. (1979). *Theory of international politics*. McGraw-Hill.

Williams, K. B. (2021, February 10). Biden: 'I will never politicize' US troops. *Defense One*. https://www.defenseone.com/policy/2021/02/biden-i-will-never-politicize-us-troops/171996/

Youssef, N. A., & Lubold, G. (2021, August 30). Last U.S. troops leave Afghanistan after nearly 20 years. *Wall Street Journal*. https://www.wsj.com/articles/last-u-s-troops-leave-afghanistan-after-nearly-20-years-11630355853

Index

J. Jakobsen, *The Geopolitics of U.S. Overseas Troops and Withdrawal*, Palgrave Studies in International Relations, https://doi.org/10.1007/978-3-030-94488-9

The manufacturer's authorised representative in the EU is Springer
Nature Customer Service Centre GmbH, Europaplatz 3, 69115 Heidelberg,
Germany. If you have any concerns regarding our products, please
contact ProductSafety@springernature.com

Printed and bound by CPI Group (UK) Ltd, Croydon, CR0 4YY
30/09/2026
02225704-0005